Myelin Repair and Neuroprotection in Multiple Sclerosis

Ian D. Duncan · Robin J.M. Franklin
Editors

Myelin Repair and Neuroprotection in Multiple Sclerosis

Editors
Ian D. Duncan
Department of Medical Sciences
University of Wisconsin-Madison
School of Veterinary Medicine
2015 Linden Drive West
Madison, Wisconsin, USA

Robin J.M. Franklin
Wellcome Trust and MRC
Cambridge Stem Cell Institute and
Department of Veterinary Medicine
University of Cambridge
Madingley Road
Cambridge, CB3 0ES, UK

ISBN 978-1-4614-2217-4   ISBN 978-1-4614-2218-1 (eBook)
DOI 10.1007/978-1-4614-2218-1
Springer New York Heidelberg Dordrecht London

Library of Congress Control Number: 2012944174

© Springer Science+Business Media New York 2013
This work is subject to copyright. All rights are reserved by the Publisher, whether the whole or part of the material is concerned, specifically the rights of translation, reprinting, reuse of illustrations, recitation, broadcasting, reproduction on microfilms or in any other physical way, and transmission or information storage and retrieval, electronic adaptation, computer software, or by similar or dissimilar methodology now known or hereafter developed. Exempted from this legal reservation are brief excerpts in connection with reviews or scholarly analysis or material supplied specifically for the purpose of being entered and executed on a computer system, for exclusive use by the purchaser of the work. Duplication of this publication or parts thereof is permitted only under the provisions of the Copyright Law of the Publisher's location, in its current version, and permission for use must always be obtained from Springer. Permissions for use may be obtained through RightsLink at the Copyright Clearance Center. Violations are liable to prosecution under the respective Copyright Law.
The use of general descriptive names, registered names, trademarks, service marks, etc. in this publication does not imply, even in the absence of a specific statement, that such names are exempt from the relevant protective laws and regulations and therefore free for general use.
While the advice and information in this book are believed to be true and accurate at the date of publication, neither the authors nor the editors nor the publisher can accept any legal responsibility for any errors or omissions that may be made. The publisher makes no warranty, express or implied, with respect to the material contained herein.

Printed on acid-free paper

Springer is part of Springer Science+Business Media (www.springer.com)

Foreword

Each year thousands of individuals are diagnosed with multiple sclerosis (MS). These individuals and the several million already living with the disease face the many uncertainties associated with MS. Fortunately, therapeutic options are available for some individuals with relapsing–remitting forms of the disease. Many patients, however, struggle with loss of function either as a direct or as a secondary result of nervous system damage associated with MS. These patients and their loved ones are asking—"When will we see myelin repair treatments?"

The notion that we might see development of myelin repair treatments is not as far-fetched as it might seem. Since the publication of John Prineas' seminal work in 1992, we have known that the MS lesion is capable of undergoing natural repair. As elegantly illustrated in this volume, the scientific community has built on this and many other discoveries to expand our knowledge and lay the foundation for a plethora of translational studies.

Today we are poised to see development and testing of the first generation of myelin repair treatments in people with MS. Indeed, we may soon see a world where physicians have the ability to prescribe a combination of treatments that halt inflammation, protect the nervous system from damage, and repair the damage found in MS. Thanks to the global collaborative efforts of the MS research community, this vision is not just blind hope but fast becoming a reality.

New York, NY, USA   Timothy Coetzee, Ph.D.

Introduction

The last decade has witnessed two significant advances in the way we think and deal with multiple sclerosis. The first, which one may describe as "immunological", has been the dramatic advances in the development and use of disease modifying drugs that target the disease causing immune response. The second, which can be described as being "neurobiological", has been the emerging story of the intricate interdependency between the axon and its myelin sheath. While the former has provided evermore effective means of treating the initial acute inflammatory stages of the disease, the second opens up the enticing possibility of developing new therapies that repair and promote remyelination damage and prevent the progressive axonal loss that typifies the chronic progressive stages of the disease. Together with immunomodulatory therapies, these will provide a more complete set of therapeutic weapons with which to treat the disease.

We may be close to achieving these goals based on recent major advances in identifying molecules that may promote endogenous oligodendrocyte progenitors and in the generation of these cells in vitro from either embryonic stem cells or by reprogramming adult somatic cells to produce induced pluripotent cells. With these new, exciting opportunities, what do we need to consider before clinical translation can occur? In this book, experts on glial cell development and regeneration, MS pathology and remyelination, neuroprotection, imaging, and immune regulation discuss the background knowledge of how and when such therapies might be applied, and ultimately what outcome measures might be used to evaluate their efficacy. These are exciting times in science and MS may be the disease where breakthroughs in regenerative therapies will be first "past the post".

Madison, WI, USA Ian D. Duncan

Cambridge, UK Robin J.M. Franklin

Contents

1. **Development of Oligodendrocytes in the Vertebrate CNS** 1
   Robert H. Miller

2. **Demyelination and Remyelination in Multiple Sclerosis** 23
   Lars Bø, Margaret Esiri, Nikos Evangelou, and Tanja Kuhlmann

3. **Microglial Function in MS Pathology** 47
   Trevor J. Kilpatrick and Vilija G. Jokubaitis

4. **Endogenous Remyelination in the CNS** 71
   Robin J.M. Franklin, Chao Zhao, Catherine Lubetzki,
   and Charles ffrench-Constant

5. **Exogenous Cell Myelin Repair and Neuroprotection
   in Multiple Sclerosis** 93
   Ian D. Duncan and Yoichi Kondo

6. **A Peripheral Alternative to Central Nervous
   System Myelin Repair** 129
   V. Zujovic and A. Baron Van Evercooren

7. **Immune Modulation and Repair Following
   Neural Stem Cell Transplantation** 153
   Tamir Ben-Hur, Stefano Pluchino, and Gianvito Martino

8. **Axonal Protection with Sodium Channel Blocking
   Agents in Models of Multiple Sclerosis** 179
   Joel A. Black, Kenneth J. Smith, and Stephen G. Waxman

9. **Effects of Current Medical Therapies on Reparative
   and Neuroprotective Functions in Multiple Sclerosis** 203
   Jack P. Antel and Veronique E. Miron

| 10 | **Imaging of Demyelination and Remyelination in Multiple Sclerosis** | 233 |

Douglas L. Arnold, Catherine M. Dalton, Klaus Schmierer, G. Bruce Pike, and David H. Miller

| 11 | **Designing Clinical Trials to Test Neuroprotective Therapies in Multiple Sclerosis** | 255 |

P. Connick, M. Kolappan, A. Compston, and S. Chandran

Index 277

Contributors

Jack P. Antel Neuroimmunology Unit, Montreal Neurological Institute, McGill University, Montreal, QC, Canada

Douglas L. Arnold McConnell Brain Imaging Centre, WB323, Montreal Neurological Institute, Montreal, QC, Canada

A. Baron Van Evercooren ICM, Centre de Recherche de l'Institut du Cerveau et de la Moelle Epinière, UPMC-Paris6, UMR_S 975, Inserm U 975, CNRS UMR 7225, 47 bd de l'Hôpital, Paris, France

Tamir Ben-Hur Department of Neurology, The Agnes Ginges Center for Human Neurogenetics, Hadassah-Hebrew University Medical Center, Jerusalem, Israel

Joel A. Black Department of Neurology, Yale University School of Medicine, New Haven, CT, USA

Paralyzed Veterans of America/United Spinal Association, Center for Neuroscience and Regeneration Research, Yale University School of Medicine, West Haven, CT, USA

Rehabilitation Research Center, VA Connecticut Healthcare System, West Haven, CT, USA

Lars Bø National Competence Center for Multiple Sclerosis, Department of Neurology, Haukeland University Hospital, Bergen, Norway

Department of UK Clinical Medicine, University of Bergen, Bergen, Norway

G. Bruce Pike McConnell Brain Imaging Centre, WB-315, Montreal Neurological Institute, Montreal, QC, Canada

S. Chandran Centre for Neuroregeneration, University of Edinburgh, Edinburgh, UK

A. Compston Department of Clinical Neurosciences, School of Clinical Medicine, Addenbrookes Hospital, Cambridge, UK

P. Connick Department of Clinical Neurosciences, School of Clinical Medicine, Addenbrookes Hospital, Cambridge, UK

Catherine M. Dalton Department of Neuroinflammation, The Institute of Neurology (Queen Square), University College London, London, UK

Ian D. Duncan Department of Medical Sciences, School of Veterinary Medicine, University of Wisconsin-Madison, Madison, WI, USA

Margaret Esiri Neuropathology Department, West Wing, John Radcliffe Hospital, Oxford, UK

Department of Clinical Neurology, Oxford University, Oxford, UK

Nikos Evangelou Division of Clinical Neurology, Nottingham University Hospital, University of Nottingham, Nottingham, UK

Charles ffrench-Constant MRC Centre for Regenerative Medicine/Scottish Centre for Regenerative Medicine, The University of Edinburgh, Edinburgh, UK

University of Edinburgh Centre for Translational Research, Centre for Inflammation Research, Queen's Medical Research Institute, The University of Edinburgh, Edinburgh, UK

Robin J.M. Franklin Wellcome Trust and MRC, Cambridge Stem Cell Institute, University of Cambridge, Cambridge, UK

Vilija G. Jokubaitis Centre for Neuroscience, The University of Melbourne, Parkville, VIC, Australia

Melbourne Brain Centre, Florey Neuroscience Institutes, The University of Melbourne, Parkville, VIC, Australia

Trevor J. Kilpatrick Melbourne Neuroscience Institute, The University of Melbourne, Parkville, VIC, Australia

Multiple Sclerosis Division, Florey Neuroscience Institutes, The University of Melbourne, Parkville, VIC, Australia

M. Kolappan NMR Research Unit, Department of Neuroinflammation, The Institute of Neurology (Queen Square), University College London, London, UK

Yoichi Kondo Department of Medical Sciences, School of Veterinary Medicine, University of Wisconsin-Madison, Madison, WI, USA

Tanja Kuhlmann Institute of Neuropathology, University Hospital Münster, Münster, Germany

Catherine Lubetzki Départment de Neurologie et CR-ICM, AP-HP, Inserm 975, UPMC, Hôpital de la Salpêtrière, Paris, France

Gianvito Martino Neuroimmunology Unit, Division of Neuroscience, San Raffaele Scientific Institute, Milan, Italy

Institute of Experimental Neurology, Division of Neuroscience, San Raffaele Scientific Institute, Milan, Italy

David H. Miller Department of Neuroinflammation, The Institute of Neurology (Queen Square), University College London, London, UK

Robert H. Miller Center for Translational Neuroscience, Department of Neurosciences, Case Western Reserve University, Cleveland, OH, USA

Veronique E. Miron Scottish Centre for Regenerative Medicine, The University of Edinburgh, Edinburgh, UK

Stefano Pluchino CNS Repair Unit, Division of Neuroscience, San Raffaele Scientific Institute, Milan, Italy

Institute of Experimental Neurology, Division of Neuroscience, San Raffaele Scientific Institute, Milan, Italy

Department of Clinical Neurosciences, Cambridge Centre for Brain Repair and Cambridge Stem Cell Initiative, Cambridge, UK

Klaus Schmierer Centre for Neuroscience & Trauma, Blizard Institute, Barts and The London School of Medicine & Dentistry, London, UK

Centre for Neuroscience & Trauma, Blizard Institute, Barts and The London School of Medicine & Dentistry, London, UK

Kenneth J. Smith Department of Neuroinflammation, The Institute of Neurology (Queen Square), University College London, London, UK

Stephen G. Waxman Department of Neurology, Yale University School of Medicine, New Haven, CT, USA

Paralyzed Veterans of America/United Spinal Association, Center for Neuroscience and Regeneration Research, Yale University School of Medicine, West Haven, CT, USA

Rehabilitation Research Center, VA Connecticut Healthcare System, West Haven, CT, USA

Chao Zhao Wellcome Trust and MRC Cambridge Stem Cell Institute, University of Cambridge, Cambridge, UK

V. Zujovic ICM, Centre de Recherche de l'Institut du Cerveau et de la Moelle Epinière, UPMC-Paris6, UMR_S 975, Inserm U 975, CNRS UMR 7225, 47 bd de l'Hôpital, Paris, France

Chapter 1
Development of Oligodendrocytes in the Vertebrate CNS

Robert H. Miller

1.1 Introduction

One characteristic of the vertebrate CNS is the presence of myelin, a fatty insulation that surrounds individual axons and facilitates the rapid conduction and enhances the efficiency of electrical impulses along the length of the axon. Oligodendrocytes are the primary cellular source of myelin in the CNS, and their development has been extensively studied over the last 2 decades. Compact myelin appears as concentric rings of modified plasma membrane that ensheaths a segment or internode along an axon. The gap between two adjacent internodes on the same axon is known as the Node of Ranvier and is a complex structure that maintains the integrity of the axon–glial unit. The discontinuous nature of myelin and the localized concentration of ion channels at Nodes of Ranvier result in accelerated conduction of electrical signals along axons and a lowering of the threshold for propagating such signals. Not all axons are myelinated, and some axons have regions that are both unmyelinated and myelinated. The functioning of the adult nervous system is however dependent on the appropriate level of myelination, and diseases or injuries that reduce the extent of local myelination result in functional deficits.

The development of molecular markers that allowed for the specific identification of oligodendrocytes and their precursors facilitated the characterization of environmental and cell intrinsic cues regulating their development. Initial in vitro studies identified oligodendrocyte precursors and provided a framework to guide in vivo analyses of oligodendrocyte precursor appearance, migration, differentiation, and myelination. Currently, the oligodendrocyte lineage is perhaps the best understood of any cell lineage in the vertebrate CNS, and a major future challenge is to harness

R.H. Miller (✉)
Center for Translational Neuroscience, Department of Neurosciences,
Case Western Reserve University, Cleveland, OH 44106, USA
e-mail: Rhm3@case.edu

this information to develop new approaches to enhance myelin repair in the setting of CNS disease. While multiple factors have been implicated in the regulation of oligodendrocyte development and myelination, it is not clear whether there is redundancy in the molecular pathways or whether there are multiple ways to generate a myelinating oligodendrocyte. Furthermore, the relative effects of different signaling pathways appear to be context dependent even within the same cell population, and while stimulation of a specific pathway can promote or inhibit the appearance of early OPCs, later in the lineage, stimulation of the same pathway may regulate cell differentiation or the onset of myelination. Such observations suggest that the behavior of cells of the oligodendrocyte lineage is likely regulated by networks of signals that are integrated by the cell to evoke the appropriate response.

1.2 Regulation of Early Oligodendrocyte Precursor Appearance

1.2.1 Lineage Relationships of OPCs: Insights from In Vitro Studies

The application of cell culture approaches to neural development allowed the identification and characterization of oligodendrocytes and their precursors (Raff 1989; Raff et al. 1978). Early studies in the optic nerve of the rat defined a population of oligodendrocyte precursors and revealed general insights into the control of their proliferation, migration, and differentiation (Raff 1989; Raff et al. 1984). Subsequent studies in other regions of the CNS have added increasing levels of complexity to our understanding of OPC development but have largely supported the fundamental observations (Miller 2002). In culture, oligodendrocyte precursors (OPCs) have a characteristic bipolar morphology, are highly migratory, and express a number of specific cell surface characteristics. For example, OPCs express a range of gangliosides that bind the monoclonal antibody A2B5 (Raff et al. 1983) as well as the proteoglycan NG2 (Nishiyama et al. 1996) and a distinct class of platelet-derived growth factor receptors (PDGF alpha receptors) (Richardson et al. 1988). While there is extensive overlap in the expression of these different markers, it is not universal. It may be that selective expression of individual markers identifies different steps in the emergence of OPCs (Baracskay et al. 2007) or that there are distinct spatial and temporal subsets of OPCs. As OPCs mature, they begin to express cell surface molecules including galactosulfatide that bind the monoclonal antibody O4 (Sommer and Schachner 1981) and develop a more complex morphology with multiple process. Such cells have been termed pro-oligodendrocytes (Bansal et al. 1992), and they subsequently cease division and migration before differentiating into immature oligodendrocytes that express the major myelin lipid galactocerebroside (Raff et al. 1978). Antibodies to galactocerebroside are not the only useful indicator of differentiated oligodendrocytes. Other markers such as CC1 have been

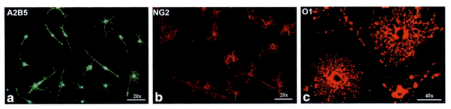

Immunopanned A2B5+ cell cultures stained with antibodies of A2B5, NG2 and O1.

Fig. 1.1 In cell cultures, oligodendrocyte precursors (OPCs) develop through a series of stages that can be identified by cell-type-specific markers. (**a**) Immature oligodendrocyte precursors are largely bipolar cells that are highly migratory and label on their surface with the monoclonal antibody A2B5 that recognizes a series of gangliosides. (**b**) A largely overlapping but somewhat distinct population of OPCs can be identified through the expression of the NG2 epitope. NG2 cells are similar to A2B5+ OPCs; however, NG2 is also expressed on a population of pericytes and neural stem cells. Immunopanning with mAb A2B5 or NG2 antibodies generates a relatively pure population of OPCs that can be utilized for multiple experiments. (**c**) When OPCs cease division, they differentiate into oligodendrocytes, which express galactocerebroside on their surface. Galactocerebroside can be identified by labeling with monoclonal antibody O1. Upon differentiation, oligodendrocytes develop multiple processes, cease to be highly migratory, and ultimately generate cells capable of myelination. $Bar=20$ mm in **a** and **b** and 10 mm in **c**

used extensively to identify and characterize these cells both in vivo and in vitro. With further maturation, oligodendrocytes express high levels of all the major components of myelin, including myelin basic protein (MBP), proteolipid protein (PLP), and myelin oligodendrocyte glycoprotein (MOG). One remarkable characteristic of OPCs in vitro is that in response to specific environmental cues, they have the potential to generate a subpopulation of astrocytes (Raff 1989; Raff et al. 1983) (termed type 2 astrocytes) that can be identified by the expression of glial fibrillary acidic protein (GFAP), a characteristic of all astrocytes. Unexpectedly, OPCs in vitro can even be induced to revert to neural stem-like cells and generate neurons under appropriate culture conditions (Kondo and Raff 2000). Recent in vivo lineage-tracing studies suggest that NG2+ or PDGFαR+ cells rarely generate cell types other than oligodendrocytes or their precursors during normal development, but OPCs may be induced to give rise to neurons or oligodendrocytes in some pathological conditions (Zhu et al. 2008).

The ability to purify OPCs through panning with antibodies directed against cell-type-specific antigens (Barres et al. 1994) and subsequently grow them in culture provided invaluable insights into the control of their proliferation and differentiation (Fig. 1.1). Several growth factors have been identified that promote the proliferation of OPCs. These include platelet-derived growth factor (PDGF) that binds to the PDGFαR on the surface of OPCs (Richardson et al. 1988) and basic fibroblast growth factor (FGF2). Not only does PDGF promote the proliferation of OPCs but it also supports their survival and allows the cells to differentiate into oligodendrocytes. Addition of both PDGF and FGF to cultures of OPCs not only stimulates their proliferation but also maintains the cells in a proliferative state and inhibits their differentiation into oligodendrocytes (Bogler et al. 1990). Such culture studies continue to provide a framework and critical insights that have been leveraged in developing a more detailed understanding of the pattern of development of OPCs and myelination in the intact CNS.

1.2.2 OPCs Arise in Distinct Locations in the Developing CNS

Oligodendrocytes are relatively ubiquitously distributed in the adult CNS. While the highest density of oligodendrocytes and myelin is in white matter axon tracts, there is substantial myelination in gray matter and a large population of morphologically identified and phenotypically characterized oligodendrocytes. Classical studies suggested that in any specific region of the CNS, the three major cell types of the CNS—neurons, astrocyte, and oligodendrocytes—were derived from cells of neural tube as a result of periods of coordinated cell induction, proliferation, and differentiation. In general, birth-dating studies indicate that neuronal populations arise first followed by astrocytes and then oligodendrocytes (Jacobson 1978). Remarkably, the timing of the relative appearance of different CNS cell types is largely recapitulated in cultures of neonatal neural cells (Abney et al. 1981) and even neural stem cells. Using an approach pioneered by Reynolds and Weiss (1992), culturing dissociated neural cells on a nonadherent substrate results in the death of the vast majority of cells. Some of the residual cells, however, form neurospheres that can be passaged repeatedly and when replated give rise to neurons, astrocytes, and oligodendrocytes. Such preparations have been extensively studied, and in general, neurospheres derived from virtually all regions of the developing and adult CNS have the capability to give rise to cells of the oligodendrocyte lineage, although the number of OPCs is relatively small compared to the numbers of neurons or astrocytes and they are often the last cell type to arise. It seems likely that this reproducible timing of cell development is a reflection of the interactions occurring between cells in the neurosphere culture rather than a cell intrinsic property of neural stem cells. While useful for defining the properties of neural stem cells and potential molecular regulators of early oligodendrocyte development, neurosphere studies provide few insights into the localization of initial appearance of OPCs during normal development.

The application of specific markers and growth factor receptors expressed by OPCs to in vivo preparation combined with appropriate in vitro assays allowed a detailed description of the timing and localization of OPC's initial appearance in different regions of the CNS. In the spinal cord, for example, separation of dorsal and ventral regions along the sulcus limitans early in embryogenesis suggested that OPCs originated selectively from ventral tissues (Warf et al. 1991), an unexpected result given the wide spread distribution of oligodendrocytes in the adult CNS. Subsequent studies identified an initial source of OPCs at the ventral midline directly above the floor plate. In mouse, the localized expression of PDGF alpha receptor (PDGFαR)-expressing cells revealed the earliest OPCs around embryonic day 12 (Pringle and Richardson 1993), while in chick embryos, OPCs first appear in the ventral ventricular zone around stage 25 (Ono et al. 1995; Orentas and Miller 1996). In both species, there is a close correlation between the appearance of OPCs with the development of motor neurons from the same embryonic domain that reflects the potential shared ancestry of the two cell types (Rowitch 2004). This localized ventral origin of OPCs is a consequence of signaling between neighboring tissues. For example, in chick embryos, transplantation of an additional notochord adjacent to lateral or dorsal regions of the neural tube results in the local induction of an additional floor plate, motor neuron, and OPCs (Jessell 2000; Ono et al. 1995; Orentas and Miller 1996). The accurate spatial and temporal characterization of the initial appearance of OPCs in the spinal cord facilitated the identification of major regulators of OPC induction, including environmental cues such as Sonic hedgehog (Shh) and intracellular cues such as the expression of distinct cohorts of transcription factors including olig1 and olig2 that orchestrate oligodendrocyte development (Rowitch 2004). The ventral midline is not the only source of OPCs in the developing spinal cord. A smaller source of OPCs arises from a discrete location in the dorsal spinal cord later in development (Cai et al. 2005), while intermediate regions of the spinal cord germinal zone do not appear to give rise to oligodendrocytes although they generate both neurons and astrocytes.

In more rostral regions of the CNS, multiple discrete sources of OPCs have been identified (Spassky et al. 1998). These include the floor of the third ventricle, which generates a subset of OPCs that migrate down the optic nerve (Gao and Miller 2006) and the medial ganglion eminence (Perez Villages et al. 1999; Spassky et al. 1998; Timsit et al. 1995) that contributes oligodendrocytes to specific domains of the CNS. Not all these regions are adjacent to an obvious source of an inductive signal such as the floor plate or notochord, and at least in the optic system, the source of inductive signals appears to be retinal ganglion cell axons (Gao and Miller 2006) that express Shh (Dakubo et al. 2008) and are the ultimate target of these OPCs. Our current understanding of the origins of OPCs in the developing CNS remains incomplete, and it seems likely based on cell tracing studies that many regions of the neural tube generate OPCs (Kessaris et al. 2006) but do so in response to specific local cues in a temporal and spatially regulated manner.

1.2.3 *Environmental Factors That Dictate the Location of OPCs*

The localized appearance of OPCs at defined stages in the developing CNS provided insights into environmental cues mediating their initial appearance. The ability to induce oligodendrocytes in dorsal and lateral regions of the developing spinal cord through notochord transplantation (Orentas and Miller 1996, 1998; Pringle et al. 1996) demonstrated the importance of the cellular environment and implicated the morphogen Sonic hedgehog in their initial appearance.

Sonic hedgehog is a member of the hedgehog family that contains three different members termed Sonic, Indian, and Desert hedgehogs (Ingham and McMahon 2001). The hedgehog pathway is critical in a wide range of morphogenetic events including patterning of the digits, lungs, and neural axis. Each member of the hedgehog family is functionally similar but has a distinct distribution in the embryo (Ingham and McMahon 2001). In the nervous system, the most highly expressed member of the hedgehog family is Sonic hedgehog (Shh) that is responsible for contributing to the patterning of the neural tube in both rostral–caudal and dorsoventral axis (Ericson et al. 1997; Jessell 2000). In caudal regions of the embryo during early development, Shh is expressed in the notochord where it contributes to the induction of ventral midline derived neural cells including the floor plate and motor neurons (Jessell 2000). In many cases, there is a connection between the types of cells induced by Shh and their subsequent expression of Shh. For example, the floor plate is induced by high concentrations of Shh generated by the adjacent notochord but then contributes to the further patterning of adjacent ventral spinal cord cell populations through the release of Shh (Jessell 2000). At high concentrations, Shh released by the notochord and floor plate induces adjacent ventral midline cells of the neural tube to develop into motor neurons and subsequently OPCs (Jessell 2000; Rowitch 2004). It has been proposed that the graded signaling of Shh from the notochord and overlying floor plate establishes a series of ventral to dorsal domains in the ventricular zone in which precursor cells become specified to distinct cell fates depending on the specific repertoire of transcription factors that are induced (Jessell 2000; Rowitch 2004). The downstream signaling of the hedgehog pathway has been extensively studied (Jiang and Hui 2008). In general, in the absence of hedgehog, ligands signaling through the smoothened receptor are inhibited by a surface molecule termed patched. The binding of hedgehog to patched releases this inhibition, and smoothened signaling is initiated. Downstream of this interaction, in the absence of hedgehog, the Gli proteins are cleaved to generate a Gli repressor that blocks the transcription of hedgehog-responsive genes, while in the presence of hedgehog, the activation of smoothened leads to a switch from a repressor to an activator (Jiang and Hui 2008). The activation of the Gli signaling pathway has been described in neural stem cells and OPCs, and although the details of the hedgehog signaling pathway in the oligodendrocyte lineage have yet to be fully resolved (Hu et al. 2009), it seems likely that they are similar to those described in other mammalian cell systems.

The inductive effects of hedgehog signaling are countered by inhibitory cues that suppress the initial appearance of OPCs. Several factors have been implicated in inhibiting OPC appearance including members of the bone morphogenetic protein family (BMPs) and the Wnts. The appearance and development of OPCs in dorsal spinal cord are significantly delayed compared to ventral regions. One mechanism that inhibits OPC development in dorsal regions is the localized expression of BMPs (Gomes et al. 2003; Grinspan et al. 2000). Bone morphogenetic proteins are members of the transforming growth factor beta superfamily and signal through SMAD intracellular pathways (Massague 1998). In the spinal cord, BMPs are primarily expressed in dorsal midline regions of the spinal cord until mid-embryogenesis and in vitro studies have shown that exposure of developing neural cells to BMP4 or 7 inhibits the appearance of oligodendrocytes and enhances the appearance of astrocytes (Grinspan et al. 2000; Miller et al. 2004). Signaling through the BMP pathway may not simply inhibit the induction of oligodendrocytes. Since in vitro OPCs are bipotential cells capable of giving rise to type 2 astrocytes or oligodendrocytes, one possibility is that BMPs influence OPCs to give rise to type 2 astrocytes rather than oligodendrocytes and this has been clearly shown in a series of in vitro studies (Grinspan et al. 2000; Miller et al. 2004). Consistent with this hypothesis, perturbation of the BMP receptor 1 in OPCs in vivo results in a transient reduction in the number of immature oligodendrocytes and an enhancement in the number of astrocytes (Gomes et al. 2003). Remarkably, this cellular imbalance is rapidly corrected during subsequent development (Samanta et al. 2007). Direct evidence for an antagonism between BMP and Shh signaling comes from in vitro studies in which the induction of oligodendrocytes by Shh can be directly inhibited by BMPs (Miller et al. 2004). It seems likely that in normal development the inhibition of oligodendrocyte differentiation by BMPs effectively outcompetes the inductive effects of Shh, and in regions where Shh is operative, additional local inhibitors of BMP signaling such as noggin are also present. The induction of OPCs in dorsal regions of the spinal cord may also reflect localized blockade of BMP signaling. The generation of OPCs in dorsal spinal cord is independent of Shh but stimulated by fibroblast growth factor (FGF) (Cai et al. 2005). In vitro studies suggest that in neural precursor cells treatment with FGF blocks BMP-stimulated nuclear translocation of phosphorylated Smads and subsequent development of OPCs (Bilican et al. 2008). Not only are BMPs important during development for oligodendrogenesis, it is becoming clear they regulate neural cell responses and fate after demyelination or spinal cord injury (Cheng et al. 2007; Fuller et al. 2007).

Additional signaling pathways may regulate OPC development. One such pathway that has received considerable attention is the Wnt signaling system that may contribute to the initial patterning of the spinal cord. Multiple studies implicated Wnts in the control of dorsal cell proliferation and more recently in neural tube patterning (Ulloa and Marti 2010). Explant studies using the chick spinal cord demonstrated expression of Wnt in dorsal regions of the developing spinal cord at the time that OPCs were being generated ventrally and that inhibition of Wnt signaling in the ex vivo model resulted in a dramatic increase in the number of OPC cells (Shimizu et al. 2005) suggesting that Wnt had an inhibitory effect on OPC development. Subsequent studies suggest that Wnt may have a more profound effect on OPC maturation than induction. Cell culture studies implicated Wnt signaling through β-catenin in the regulation of OPC maturation that was independent of the induction of OPCs or the control of their proliferation or cell fate determination (Feigenson et al. 2009). These functional characteristics distinguish Wnt signaling from other inhibitors of OPC development such as BMPs. Genetic manipulations resulting in constitutive activation of Wnt/β-catenin signaling specifically in cells of the oligodendrocyte lineage resulted in similar numbers of OPC compared to wild-type littermate controls but delayed the development of mature oligodendrocytes and myelination. Somewhat unexpectedly, these effects were transient and the histology of the adult animals was relatively normal (Feigenson et al. 2009). Confirmation of a critical role for Wnt in the timing of OPC development and identification of a persistent role for Wnt/β-catenin in the adult followed the identification of the transcription factor Tcf4 as a major mediator of oligodendrocyte Wnt signaling (Fancy et al. 2009). Conditional activation of the β-catenin pathway or perturbation of APC, an inhibitor of Wnt, resulted in a significant delay in the timing of oligodendrocyte maturation during development and a delay in remyelination in the adult CNS (Fancy et al. 2009). Thus far all the data suggest Wnt signaling in oligodendrocytes or OPCs is mediated through the canonical pathway, and whether there is a role for the noncanonical pathway in mediating the effects of Wnt on OPCs and oligodendrocytes remains to be determined.

Multiple factors may initiate the induction of OPCs in the developing CNS. While Shh is clearly important in early OPC development, OPCs also develop in cultures derived from Shh null animals indicating the existence of alternative pathways for their induction. The best defined of these is through FGF signaling pathway. As discussed above, FGF signaling may induce a blockade of BMP signaling (Bilican et al. 2008) and promote OPC induction. Later in development, OPCs express multiple FGF receptors that initiate a variety of cellular responses (Fortin et al. 2005). The relative roles of FGF in OPC induction are likely to be more pronounced in rostral regions of the CNS that contain multiple anatomically distinct areas of OPC appearance (Kessaris et al. 2006; Spassky et al. 1998). Axonal signals such as the neuregulins have been implicated in OPC induction; however, their role remains somewhat controversial. While in vitro the absence of NRG-1 significantly inhibited the appearance of OPCs (Vartanian et al. 1999) and suggested involvement of distinct ErbB receptors (Park et al. 2001; Sussman et al. 2005), subsequent studies using genetic manipulations in vivo have failed to provide any evidence of a role for neuregulins or their receptors in the development of CNS myelination (Brinkmann et al. 2008). It seems likely that while not essential in vivo for OPC development and CNS myelination, distinct isoforms of NRG and their receptors contribute significantly to normal OPC development and further studies are needed to clarify their precise roles.

1.3 Control of Oligodendrocyte Precursor Differentiation in the CNS

1.3.1 Regulation of Neonatal OPC Differentiation

The number of OPCs that are generated in germinal zones of the developing animal is small compared to the number of oligodendrocytes in the adult CNS. This difference is a consequence of the extensive proliferation of OPCs during subsequent development. A large number of studies have been focused on understanding the molecular control of OPC proliferation. While the identification of mitogenic signals is critical for understanding how the appropriate cohort of oligodendrocytes is generated, equally important is an understanding of the cellular and molecular cues that lead to the cessation of proliferation and differentiation of OPCs.

A wide range of growth factors have been shown to promote the proliferation of OPCs. The best characterized of these is platelet-derived growth factor (Richardson et al. 1988). Overexpression of PDGF results in increased numbers of OPCs during development, while conditional knockout of the PDGFαR in OPCs results in a dramatic reduction in their numbers (Calver et al. 1998). Furthermore, the proliferative response of OPCs to PDGF can be significantly modulated by signaling through the CXCR2 chemokine receptor (Robinson et al. 1998). Many other growth factors including FGF and IGF-1 promote OPC proliferation in vitro, as do several neurotrophins such as NT3 and BDNF. Whether all the growth factors that have been shown to promote OPC proliferation in vitro have specific or overlapping roles in vivo is not currently clear.

A critical step in the generation of the appropriate number of oligodendrocytes during development is the conversion of proliferative OPCs to differentiated, non-migratory oligodendrocytes. How this transition is accomplished is not well understood. One attractive hypothesis is that intrinsic timers or clocks in OPCs control the timing at the level of individual cells and that this is clonally related (Temple and Raff 1986). Classical in vitro studies using clonal analysis of rat optic nerve OPCs in which sister cells were separated and grown in isolation revealed that they generated clones of similar sizes and the cells within a clone differentiated synchronously (Temple and Raff 1986). These studies raised the possibility that OPC differentiation might be regulated by cell intrinsic timers. How such timers might operate is uncertain but may be related the number of cell divisions a cell undergoes (Raff et al. 1988; Temple and Raff 1986). What sets such a clock is also unknown, and it seems likely that this intrinsic control is modulated by local environmental cues. For example, in high-density mixed cell cultures of developing rat spinal cord, clonally derived OPCs do not differentiate synchronously, but rather virtually every clone contains both OPCs and differentiated oligodendrocytes (Zhang and Miller 1995), which may reflect the immediate environment of individual OPCs. Subsequent studies in the optic nerve have, however, provided molecular substrates for inductive timers of OPC differentiation including regulators of the cell cycle, such as p21 and p57. How such an intrinsic mechanism actually controls cell differentiation is uncertain. It has been proposed that mitogenic signaling cascades may be diminished with each cycle such that, ultimately, the progenitor cell drops out of the cell cycle and differentiates constitutively. Support for this hypothesis comes from studies in which removal of growth factors drives precocious differentiation of OPCs, and promoting their continued proliferation through stimulation with both PDGF and FGF inhibits their differentiation (Bogler et al. 1990).

Recent studies provide insights into the molecular mechanism regulating neonatal OPC differentiation. Analyses of the expression of small noncoding microRNAs (miRNA) demonstrated differentiation-associated changes in expression levels for several miRNAs in the oligodendrocyte lineage. Single miRNAs can regulate the expression of multiple genes, and perturbation of their processing in OPCs has been shown to block differentiation into oligodendrocytes (Dugas et al. 2010; Zhao et al. 2010). Specifically, a distinct miRNA (miR-219) appears to be a critical regulator of oligodendrocyte differentiation and is both necessary and sufficient to promote OPC differentiation (Dugas et al. 2010; Zhao et al. 2010). One of the mechanisms by which miR-219 may operate is through the repression of genes that are required to maintain OPC proliferation such as PDGFαR. The inhibition of proliferation coupled with the upregulation of specific transcription factors essential for the onset of oligodendrocyte differentiation and myelination (Emery et al. 2009; Howng et al. 2010) is likely the central regulator of intrinsic control of OPC differentiation. The coordinated control of the expression of multiple genes can be affected by epigenetic regulation, and there is strong evidence for a role of epigenetic factors in controlling oligodendrocyte differentiation. Culture studies have suggested that HDAC1 is a critical modulator of genes that regulate oligodendrocyte differentiation. Manipulation of HDAC1 results in altered timing of oligodendrocyte appearance both in culture and in vivo (Li and Richardson 2009; Marin-Husstege et al. 2002). The precise signaling cascades that regulate this phenomenon are currently being resolved.

1.3.2 The Adult Oligodendrocyte Precursor

During development, not all OPCs differentiate, and the adult CNS contains a significant number of cells that have the morphological phenotype of OPCs and express OPC markers including PDFGαR and NG2. These adult progenitors were originally described in cell cultures where they had a longer cell cycle and reduced migratory capacity (Wolswijk and Noble 1989). The development of animals with reporters driven from promoters that are largely restricted to OPC such as NG2 or PDGFαR has provided remarkable insights into this interesting cell type (Rivers et al. 2008; Tripathi et al. 2010). For example, there are unexpectedly high numbers of adult OPCs in the CNS, and they are located in both gray and white matter. The cells have a unique morphology and appear to have a distinct physiological profile. Several studies have indicated that adult OPCs are electrically active (De Biase et al. 2010; Tripathi et al. 2011) and suggest that they may contribute to monitoring or regulating electrical activity particularly in white matter. Not only are adult OPCs numerous, but they are also capable of continued proliferation as might be anticipated for precursors cells. It was generally thought that in the absence of injury or disease there was relatively little turnover or de novo generation of oligodendrocytes in the adult CNS. The use of inducible reporters has, however, indicated that a substantial fraction of the oligodendrocytes in the adult corpus callosum are generated from precursor cells after the major period of cell division. Whether these newly formed oligodendrocytes represent replacement or additional cells is currently unclear. While a primary function of the adult OPCs may be to provide a resource to replenish cells lost to injury or that are required during the normal life span of the animal, there appears to be too many of these cells for that to be their only function. Likewise, the ability of these cells to sense electrical activity argues against a purely precursor function. Somewhat surprisingly, deletion of the majority of adult OPCs and blockade of their replacement do not appear to result in a major functional deficit, so the ultimate function of these cells remains unclear.

1.4 Control of OPC Migration in the Developing CNS

1.4.1 Signals Regulating OPC Migration

The restricted location of OPC's origins in the embryonic CNS combined with the widespread distribution of oligodendrocytes in the adult CNS reflects the extensive migratory capacity of OPCs. Unlike the exquisite guidance of distinct neuronal cells or axonal growth cones to their targets, the migration of OPCs is regulated in a more general way even though in some cases similar molecular mechanisms appear to be operative. Culture studies demonstrate that PDGF is a strong chemoattractant for OPCs and is implicated in facilitating the widespread dispersal of these cells. It seems likely, however, that PDGF is not directing OPCs to particular locations. It is relatively ubiquitously distributed throughout the CNS and consequently appears to act as a general dispersal cue or simply to promote motility (Frost et al. 2009). Several other signals have been implicated in regulating OPC migration. These encompass a wide range of different classes of molecules, and it seems likely they all converge on a common target that modulates basic cell motility mechanisms at the level of the cytoskeleton. Such signals include neurotransmitter receptors such as AMPA and kainate receptors for glutamate (Karadottir and Attwell 2007). These receptors associate with alpha (v) b-3 integrin receptors (Gudz et al. 2006) as part of a complex that also contains proteolipid protein (PLP) (Gudz et al. 2006). Functionally, PLP is important in the signaling pathway since OPCs with mutations in PLP fail to show AMPA-stimulated response. In addition, OPCs express the chemokine receptors CXCR2 and CXCR4. Stimulation of CXCR4 with the ligand CXCL12 promotes OPC migration that is blocked by CXCR4 antagonists (Dziembowska et al. 2005). A common linkage between these different stimuli is the modulation of intracellular calcium, suggesting that it is an important modulator of OPC motility. The downstream targets of Ca^{2+} activation and the mechanisms by which Ca^{2+} is activated appear to be important in determining OPC responses since activation of the chemokine receptor CXCR2, another G-protein-coupled receptor that fluxes Ca^{2+}, inhibits rather than enhances OPC migration.

1.4.2 Guidance of OPC Migration

The ability to specifically identify OPCs in culture and in vivo has allowed their dispersal and migration to be studied in culture as well as in fixed and living tissue (Jarjour and Kennedy 2004; Tsai and Miller 2002; Tsai et al. 2003, 2009). These studies indicate that while the dispersal of OPCs is not totally random, highly defined preexisting pathways of OPC migration do not appear to exist. Rather the migration of OPCs is guided by more general chemoattractants and chemorepellents (reviewed in Jarjour and Kennedy 2004) that direct OPCs to presumptive white matter regions.

In the spinal cord, OPCs initially arise in the ventral midline and subsequently migrate widely to the peripheral white matter. This dispersal of OPCs is largely the result of the localized expression of the chemorepellent netrin-1. Netrin-1 is expressed at the ventral midline at the time that OPCs initially arise and in vitro OPCs respond to a gradient of netrin-1 by orienting away from high concentrations and migrating toward lower concentrations (Jarjour et al. 2003; Tsai et al. 2003). In animals lacking netrin-1, the dispersal of spinal cord OPCs is inhibited (Jarjour et al. 2003; Tsai et al. 2003), and the cells remain concentrated at their source. The response of OPCs to netrin-1 depends on the receptors DCC and UNC5H1 (Jarjour et al. 2003; Spassky et al. 2002; Tsai et al. 2003), and blocking DCC signaling negates netrin-1 directional cues. The dispersal of OPCs from the ventral midline appears to be important to position cells to receive appropriate proliferative and differentiated cues (Tsai et al. 2009) that are localized in the developing white matter. What roles netrin-1 plays in OPC guidance in other regions of the CNS is less clear. Localization studies suggest that it acts as a chemorepellent to promote migration of OPCs that arise in the brain along the optic nerve toward the retina, and this is consistent with some in vitro studies (Sugimoto et al. 2001) but not others (Spassky et al. 2002). Whether these differences reflect the responses of OPCs at different stages of development or whether there are distinct subpopulations of OPCs who have distinct responses to dispersal cues is unclear.

Other guidance cues such as the semaphorins have been shown to influence OPC migration. The semaphorins are a large family of guidance molecules that signal through receptors composed of neuropilins and plexins (Raper 2000). Oligodendrocyte precursors express neuropilins (Spassky et al. 2002; Sugimoto et al. 2001) and are either attracted or repelled by semaphorins depending on the specific semaphoring family member. For example, Sema 3A appears to be a chemorepellent for OPCs, while Sema 3F is a chemoattractant (Spassky et al. 2002). The importance of semaphorin-guided migration during development is currently unclear although it has been implicated in myelin repair (Williams et al. 2007).

1.4.3 Targeting OPCs to Presumptive Myelinating Regions

The final localization of oligodendrocytes and the formation of myelin sheaths are tightly regulated. One of the most striking examples of localization of myelinating oligodendrocytes in the adult CNS is at the junction of the optic nerve and the retina. Retinal ganglion cell axons project from cell bodies located in the retina through the optic nerve head along the optic nerve to the visual centers of the brain. The portion of the retinal ganglion cell axons traversing the optic nerve is uniformly myelinated, while the region in the retina is unmyelinated. The abrupt transition from myelinated to unmyelinated regions of the same axons occurs at the lamina cribrosa (ffrench-Constant et al. 1988) in species such as rat, mouse, and human. How this distinct patterning is established is unclear, but it clearly reflects the defined distribution of oligodendrocytes. In other CNS regions, OPCs need to be targeted to axon-rich regions at the appropriate stage of development to ensure normal myelination (Tsai et al. 2009). Several mechanisms have been implicated in patterning the final distribution of oligodendrocytes in the adult CNS. One simple concept is that the cessation of migration is a consequence of OPC differentiation. While OPCs are highly migratory once they differentiate, they become multiprocessed and stop migrating. Thus local signals that drove OPC differentiation would, by default, inhibit migration. Currently no such localized differentiation signal has been described. Alternatively, localized cues may directly inhibit OPC migration allowing for localized proliferation and differentiation. One such signal is the chemokine CXCL1 signaling through the CXCR2 receptor (Tsai et al. 2002).

In the developing spinal cord, the final localization of OPCs to presumptive white matter reflects the arrest of OPC migration mediated by stop signals. Localized expression of the chemokine CXCL1 by a subset of astrocytes in presumptive white matter inhibits the continued dispersal of OPCs (Tsai et al. 2002). In animals lacking CXCR2, OPCs are generated normally in the germinal zones and disperse throughout the spinal cord. These cells, however, fail to stop in proximal white matter regions but rather coalesce at the periphery of the cord. While the final distribution eventually normalizes, the perturbed development results in long-term functional deficits (Padovani-Claudio et al. 2006). What regulates the local expression of the chemokine in astrocytes is currently unclear. In the setting of demyelination/remyelination, however, CXCL1/CXCR2 signaling appears to be important in mediating repair (Kerstetter et al. 2009; Liu et al. 2010). Other chemokines and their receptors such as CXCR4 and its ligand SDF1 have been suggested to be involved in regulating OPC development (Dziembowska et al. 2005; Patel et al. 2010).

Several studies implicate components of the extracellular matrix in regulating the final localization of oligodendrocytes. The most compelling data comes from studies on tenascin-C. Exposure of OPCs to a tenascin-C substrate in vitro inhibits their continued migration (Frost et al. 1996), and the expression of tenascin-C by a subset of astrocytes at the junction of the optic nerve and retina is coincident with the boundary of myelination (Bartsch et al. 1994) suggesting that this patterning of myelination is a consequence of the inhibition of OPC migration into the retina (ffrench-Constant et al. 1988). Further analysis of tenascin-C null animals revealed a similar sharp demarcation between myelinated and nonmyelinated segments of retinal ganglion cell axons (Garcion et al. 2001) implying that tenascin-C is not a sole regulator of OPC migration in this region. These animals did, however, reveal a role for tenascin-C in regulating the rate of migration of OPCs in early development in other regions of the CNS (Garcion et al. 2001).

1.5 Control of Myelination During Development of the CNS

Several factors are likely to control the differentiation of oligodendrocytes and the onset of CNS myelination (Fig. 1.2). Traditionally it was believed that the trigger to initiate myelination was when the axon reached a specific diameter. This simple

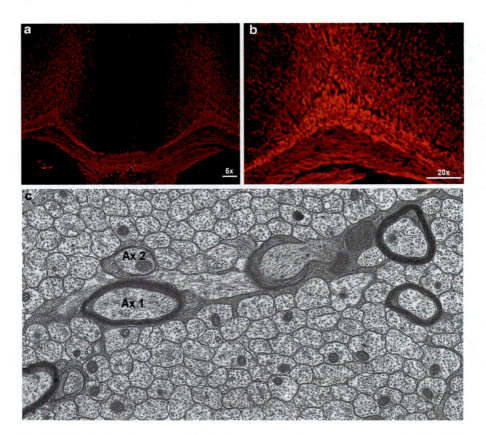

Fig. 1.2 Developing oligodendrocytes in the intact central nervous system can be identified through expression of myelin basic protein (MBP). (a) Low magnification of a coronal section through P16 mouse brain labeled with anti-MBP. The corpus callosum, an axon tract that traverses the hemispheres, is intensely labeled with antibodies indicative of its white matter characteristics while lighter but significant MBP expression is seen in the developing cortex. (b) Higher magnification of the corpus callosum and overlying cortex showing the extensive expression of MBP at P16 in the mouse brain. (c) An electron micrograph of myelination in developing mouse brain. The oligodendrocyte process has generated four myelinated axons seen in cross section and has elaborated smaller processes to multiple other axons. Each of the axons undergoing myelination from this oligodendrocyte process is at different stages of completion. The large axon Ax1 is virtually fully myelinated while myelination is just initiated in the smaller axon (Ax2) above it. $Bar = 25$ mm in **a** and **b**, **c** = ×20,000

notion is probably incorrect since in the CNS axons with a wide range of diameters are myelinated and some of these are very small diameter. Furthermore, myelination itself alters axonal diameter; thus myelinated axons are larger because they are myelinated. It seems likely that the trigger to form myelin has at least three components (1) The axon has to achieve a level of maturation at which myelination can be initiated; this may be reflected or at least correlated with an increase in axon diameter. (2) The oligodendrocyte has to be competent to engage the myelination machinery. This is likely to be linked to the precursor exiting from the cell cycle and probably a rearrangement of its cytoskeleton. (3) There is a release of inhibitory cues that block myelination.

The timing of myelination in regions of the CNS such as the spinal cord that contain multiple axonal tracks likely reflects the interplay between maturing OPCs and distinct axonal populations. For example, in rodents, axons in ventral spinal cord are myelinated around the time of birth; axons of the dorsally located cortical spinal track are not myelinated until the second or third postnatal week. Although signals such as neuregulin-1 have been identified that modulate levels of myelination, at least in the periphery (Vartanian et al. 1997), the axonal competence cues for myelination in the CNS are unknown. Likewise, instructive signals that activate myelination programs in mature oligodendrocytes remain to be defined.

The signals that regulate the timing of myelination by oligodendrocytes are beginning to be described. Clearly oligodendrocyte differentiation precedes the onset of myelination, but it is unclear whether there is a direct linkage between the onset of differentiation and the activation of the myelination program or whether these are independently controlled events. One hypothesis is that following differentiation oligodendrocytes become dependent on survival signals derived from adjacent axons (Barres and Raff 1994) and in the absence of such signals they die while in their presence OPCs survive and myelinate. Recent studies have provided clear evidence that the onset of myelination is regulated by signaling between oligodendrocytes and their adjacent axons.

Several molecules have been implicated in mediating axon–OPC interactions including the transmembrane glycoprotein LINGO-1 (Mi et al. 2004). LINGO-1 is a leucine-rich repeat (LRR) molecule whose expression is restricted to the CNS. Originally thought to be part of the complex that regulated axonal outgrowth (Mi et al. 2004), LINGO-1 was subsequently identified as a key regulator of oligodendrocyte differentiation (Mi et al. 2005, 2007) that is expressed on the surface of axons and OPCs. Early in development, LINGO-1 is expressed abundantly and its expression decreases prior to myelination (Mi et al. 2005). Functional studies both in vitro and during development suggest that a primary role for LINGO-1 is to inhibit the differentiation of oligodendrocytes and retard the onset of myelination. Blocking LINGO-1 signaling enhances the maturation of oligodendrocytes and accelerates the appearance of CNS myelin (Mi et al. 2005) in vitro. In vivo, the appearance of myelin is accelerated in LINGO-1 null animals (Mi et al. 2005) suggesting that expression of LINGO-1 is important for the modulation for the timing of myelination. Recent studies demonstrate that inhibiting LINGO expression enhances remyelination in multiple experimental models of demyelination (Mi et al. 2009). The precise signaling pathways that mediate LINGO-1 signaling are not totally resolved although the expression of LINGO-1 on axons is modulated by NGF (Lee et al. 2007) and may involve FYN kinase signaling.

Other regulators of the timing of myelination include expression of the surface glycoprotein neural cell adhesion molecule (NCAM) that has been shown to mediate interactions and adhesion between adjacent cells (Acheson et al. 1991). Prior to myelination, both oligodendrocytes and their target axons express a form of NCAM that contains a high level of charged polysialic acid residues. If those residues are removed, biologically or through the use of a specific endonuclease (endoN), myelination occurs; if by contrast high levels of PSA remain, oligodendrocyte maturation and myelination are inhibited. Whether this represents a specific regulation of oligodendrocyte maturation or is a reflection of a more general influence of PSA expression is unclear since removal of PSA also alters the migratory behavior of OPCs in vitro.

The Notch signaling pathway has also been implicated in regulation of myelination. In vitro studies identified a role for Notch signaling in regulating OPC differentiation. OPCs express Notch receptors on their surface, and activation of those receptors inhibits their differentiation (Wang et al. 1998). These studies suggested that blocking Notch signaling would promote OPC differentiation and in studies on in vitro myelination. Since Notch and its ligands are expressed at high levels in the setting of demyelination, recent studies have focused largely on defining a role for this pathway in myelin repair (Fancy et al. 2010). The current level of understanding is unclear. Notch is expressed on multiple cell types and whether the positive effects seen in models of remyelination reflect a direct effect on the OPCs remains uncertain. There are likely additional pathways that control the onset of CNS myelination. These include those mediated by myelin debris. In vitro when OPCs are grown on a substrate of CNS myelin, they fail to differentiate into oligodendrocytes (Miller 1999). While this mechanism may contribute to the genesis of a population of persistent OPCs in the adult CNS, it may be more critical in the regulation of myelin repair in the setting of demyelination. Subsequent studies showed that the removal of myelin debris is important for successful remyelination and begun to define signaling through the Fyn–Rho pathway as an integral component of this phenomenon. An additional pathway that may be important for regulation of OPC differentiation in the setting of injury is the signaling mediated by the extracellular matrix molecule hyaluronan. Hyaluronan was increased in the setting of CNS demyelination and under experimental conditions inhibits the differentiation of OPCs (Back et al. 2005). The relative importance of the different mechanisms controlling OPC differentiation has yet to be determined, but it seems likely that the contribution of each will differ depending on the location, subpopulation of OPCs involved, and the context of the individual situation.

1.6 Conclusions

Our understanding of the biology of the oligodendrocyte lineage has expanded dramatically over the past 3 decades. Even so there remain many unanswered questions as to how CNS myelin is generated. One of the striking features is the large number of molecular pathways that have been implicated in regulating the appearance, proliferation migration, and differentiation of oligodendrocytes. Much of this information has been generated through the use of in vitro models, and while the importance of some pathways has been reinforced through the use of animal models, others have not. One possible explanation is that when placed in culture, neural precursors and OPCs begin to express an array of cell surface molecules and receptors not expressed in the setting of development, and thus this data is not directly relevant to normal development. It seems more likely, however, that there is substantial duplication in the molecular pathways mediating early oligodendrocyte development and that deletion of a single pathway in vivo does not therefore result in a profound phenotype. Several reasons may account for the diversity in OPC promoting signals. One possibility is that there are multiple populations of OPC that have overlapping but distinct arrays of growth factor receptors. Such diversity may reflect the timing of OPC appearance or spatial restriction of origins with a different profile of local cues. Consistent with this hypothesis, the limited populations of OPCs that develop from dorsal spinal cord depend on signaling by FGFs, while the majority of OPC that develop from ventral regions depend upon signaling by Shh and not FGF. Likewise, the ultimate fate of OPCs derived from different regions differs. In the forebrain, the earliest OPCs that are generated from the most ventral regions are subsequently lost as a result of competition with later more dorsally derived OPCs (Kessaris et al. 2006; Richardson et al. 2006).

The most poorly understood aspect of the oligodendrocyte lineage is the generation and function of those OPCs that escape differentiation during normal development and establish "adult OPCs." Whether these cells represent a pool of precursors that can be recruited to generate oligodendrocytes in the adult CNS or are an independent functionally unique population of cells will likely be resolved with the application of cell-specific targeting approaches.

References

Abney ER, Bartlett PP, Raff MC (1981) Astrocytes, ependymal cells, and oligodendrocytes develop on schedule in dissociated cell cultures of embryonic rat brain. Dev Biol 83:301–310

Acheson A, Sunshine JL, Rutishauser U (1991) NCAM polysialic acid can regulate both cell-cell and cell-substrate interactions. J Cell Biol 114:143–153

Back SA, Tuohy TM, Chen H, Wallingford N, Craig A et al (2005) Hyaluronan accumulates in demyelinated lesions and inhibits oligodendrocyte progenitor maturation. Nat Med 11:966–972

Bansal R, Stefansson K, Pfeiffer SE (1992) Proligodendroblast antigen (POA), a developmental antigen expressed by A007/O4-positive oligodendrocyte progenitors prior to the appearance of sulfatide and galactocerebroside. J Neurochem 58:2221–2229

Baracskay KL, Kidd GJ, Miller RH, Trapp BD (2007) NG2-positive cells generate A2B5-positive oligodendrocyte precursor cells. Glia 55:1001–1010

Barres BA, Raff MC (1994) Control of oligodendrocyte number in the developing rat optic nerve. Neuron 12:935–942

Barres BA, Raff MC, Gaese F, Bartke I, Dechant G, Barde YA (1994) A crucial role for neurotrophin-3 in oligodendrocyte development. Nature 367:371–375

Bartsch U, Faissner A, Trotter J, Dorries U, Bartsch S et al (1994) Tenascin demarcates the boundary between the myelinated and non-myelinated part of retinal ganglion cell axons in the developing and adult mouse. J Neurosci 14:4756–4768

Bilican B, Fiore-Heriche C, Compston A, Allen ND, Chandran S (2008) Induction of Olig2 precursors by FGF involves BMP signalling blockade at the Smad level. PLoS One 3:e2863

Bogler O, Wren D, Barnett SC, Land H, Noble M (1990) Cooperation between two growth factors promotes extended self-renewal and inhibits differentiation of oligodendrocyte-type-2 astrocyte (O-2A) progenitor cells. Proc Natl Acad Sci USA 87:6368–6372

Brinkmann BG, Agarwal A, Sereda MW, Garratt AN, Muller T et al (2008) Neuregulin-1/ErbB signaling serves distinct functions in myelination of the peripheral and central nervous system. Neuron 59:581–595

Cai J, Qi Y, Hu X, Tan M, Liu Z et al (2005) Generation of oligodendrocyte precursor cells from mouse dorsal spinal cord independent of on Nkx6 regulation and Shh signaling. Neuron 45:41–53

Calver AR, Hall AC, Yu WP, Walsh FS, Heath JK et al (1998) Oligodendrocyte population dynamics and the role of PDGF in vivo. Neuron 20:869–882

Cheng X, Wang Y, He Q, Qiu M, Whittemore SR, Cao Q (2007) Bone morphogenetic protein signaling and olig1/2 interact to regulate the differentiation and maturation of adult oligodendrocyte precursor cells. Stem Cells 25:3204–3214

Dakubo GD, Beug ST, Mazerolle CJ, Thurig S, Wang Y, Wallace VA (2008) Control of glial precursor cell development in the mouse optic nerve by sonic hedgehog from retinal ganglion cells. Brain Res 1228:27–42

De Biase LM, Nishiyama A, Bergles DE (2010) Excitability and synaptic communication within the oligodendrocyte lineage. J Neurosci 30:3600–3611

Dugas JC, Cuellar TL, Scholze A, Ason B, Ibrahim A et al (2010) Dicer1 and miR-219 Are required for normal oligodendrocyte differentiation and myelination. Neuron 65:597–611

Dziembowska M, Tham TN, Lau P, Vitry S, Lazarini F, Dubois-Dalcq M (2005) A role for CXCR4 signaling in survival and migration of neural and oligodendrocyte precursors. Glia 50:258–269

Emery B, Agalliu D, Cahoy JD, Watkins TA, Dugas JC et al (2009) Myelin gene regulatory factor is a critical transcriptional regulator required for CNS myelination. Cell 138:172–185

Ericson J, Briscoe J, Rashbass P, Heyningen V, Jessell TM (1997) Graded sonic hedgehog signaling and the specification of cell fate in the ventral neural tube. Cold Spring Harb Symp Quant Biol 62:451–466

Fancy SP, Baranzini SE, Zhao C, Yuk DI, Irvine KA et al (2009) Dysregulation of the Wnt pathway inhibits timely myelination and remyelination in the mammalian CNS. Genes Dev 23:1571–1585

Fancy SP, Kotter MR, Harrington EP, Huang JK, Zhao C et al (2010) Overcoming remyelination failure in multiple sclerosis and other myelin disorders. Exp Neurol 225:18–23

Feigenson K, Reid M, See J, Crenshaw EB 3rd, Grinspan JB (2009) Wnt signaling is sufficient to perturb oligodendrocyte maturation. Mol Cell Neurosci 42:255–265

ffrench-Constant C, Miller RH, Burne JF, Raff MC (1988) Evidence that migratory oligodendrocyte-type-2 astrocyte (O-2A) progenitor cells are kept out of the rat retina by a barrier at the eye-end of the optic nerve. J Neurocytol 17:13–25

Fortin D, Rom E, Sun H, Yayon A, Bansal R (2005) Distinct fibroblast growth factor (FGF)/FGF receptor signaling pairs initiate diverse cellular responses in the oligodendrocyte lineage. J Neurosci 25:7470–7479

Frost E, Kiernan BW, Fassner A, ffrench-Constant C (1996) Regulation of oligodendrocyte precursor migration by extracellular matrix: evidence for substrate-specific inhibition of migration by Tenascin-C. Dev Neurosci 18:266–273

Frost EE, Zhou Z, Krasnesky K, Armstrong RC (2009) Initiation of oligodendrocyte progenitor cell migration by a PDGF-A activated extracellular regulated kinase (ERK) signaling pathway. Neurochem Res 34:169–181

Fuller ML, DeChant AK, Rothstein B, Caprariello A, Wang RZ, Hall AK, Miller RH (2007) Bone morphogenetic proteins promote gliosis in demyelinating spinal cord lesions. Annals of Neurology, 62(3):288–300. PMID: 17696121

Gao L, Miller RH (2006) Specification of optic nerve oligodendrocyte precursors by retinal ganglion cell axons. J Neurosci 29:7619–7628

Garcion E, Faissner A, ffrench-Constant C (2001) Knockout mice reveal a contribution of the extracellular matrix molecule tenascin-C to neural precursor proliferation and migration. Development 128:2485–2496

Gomes W, Mehler M, Kessler J (2003) Transgenic overexpression of BMP4 increases astroglial and decreases oligodendroglial lineage commitment. Dev Biol 255:164–177

Grinspan JB, Edell E, Carpio DF, Beesley JS, Lavy L et al (2000) Stage-specific effects of bone morphogenetic proteins on the oligodendrocyte lineage. J Neurobiol 43:1–17

Gudz TI, Komuro H, Macklin WB (2006) Glutamate stimulates oligodendrocyte progenitor migration mediated via an alphav integrin/myelin proteolipid protein complex. J Neurosci 26: 2458–2466

Howng SY, Avila RL, Emery B, Traka M, Lin W et al (2010) ZFP191 is required by oligodendrocytes for CNS myelination. Genes Dev 24:301–311

Hu BY, Du ZW, Li XJ, Ayala M, Zhang SC (2009) Human oligodendrocytes from embryonic stem cells: conserved SHH signaling networks and divergent FGF effects. Development 136:1443–1452

Ingham PW, McMahon AP (2001) Hedgehog signaling in animal development: paradigms and principles. Genes Dev 15:3059–3087

Jacobson M (1978) Developmental neurobiology. Plenum, New York

Jarjour AA, Kennedy TE (2004) Oligodendrocyte precursors on the move: mechanisms directing migration. Neuroscientist 10:99–105

Jarjour AA, Manitt C, Moore SW, Thompson KM, Yuh SJ, Kennedy TE (2003) Netrin1 is a chemorepellent for oligodendrocyte precursor cells in the embryonic spinal cord. J Neurosci 23:68–72

Jessell T (2000) Neuronal specification in the spinal cord: inductive signals and transcriptional codes. Nat Rev Genet 1:20–29

Jiang J, Hui CC (2008) Hedgehog signaling in development and cancer. Dev Cell 15:801–812

Karadottir R, Attwell D (2007) Neurotransmitter receptors in the life and death of oligodendrocytes. Neuroscience 145:1426–1438

Kerstetter AE, Padovani-Claudio DA, Bai L, Miller RH (2009) Inhibition of CXCR2 signaling promotes recovery in models of multiple sclerosis. Exp Neurol 220:44–56

Kessaris N, Fogarty M, Iannarelli P, Grist M, Wegner M, Richardson W (2006) Competing waves of oligodendrocytes in the forebrain and postnatal elimination of an embryonic lineage. Nat Neurosci 9:173–179

Kondo T, Raff M (2000) Oligodendrocyte precursor cells reprogrammed to become multipotential CNS stem cells. Science 289:1754–1757

Lee X, Yang Z, Shao Z, Rosenberg SS, Levesque M et al (2007) NGF regulates the expression of axonal LINGO-1 to inhibit oligodendrocyte differentiation. J Neurosci 27:220–225

Li H, Richardson WD (2009) Genetics meets epigenetics: HDACs and Wnt signaling in myelin development and regeneration. Nat Neurosci 12:815–817

Liu L, Belkadi A, Darnall L, Hu T, Drescher C et al (2010) CXCR2-positive neutrophils are essential for cuprizone-induced demyelination: relevance to multiple sclerosis. Nat Neurosci 13:319–326

Marin-Husstege M, Muggironi M, Liu A, Casaccia-Bonnefil P (2002) Histone deacetylase activity is necessary for oligodendrocyte lineage progression. J Neurosci 22:10333–10345

Massague J (1998) TGF-beta signal transduction. Annu Rev Biochem 67:753–791

Mi S et al (2004) LINGO-1is a component of the Nogo-66 receptor/p75 signaling complex. Nat Neurosci 7:221–228

Mi S, Miller RH, Lee X, Scott ML, Shulag-Morskaya S et al (2005) LINGO-1 negatively regulates myelination by oligodendrocytes. Nat Neurosci 8(6):745–751

Mi S, Hu B, Hahm K, Luo Y, Kam Hui ES et al (2007) LINGO-1 antagonist promotes spinal cord remyelination and axonal integrity in MOG-induced experimental autoimmune encephalomyelitis. Nat Med 10:1228–1233

Mi S, Miller RH, Tang W, Lee X, Hu B et al (2009) Promotion of central nervous system remyelination by induced differentiation of oligodendrocyte precursor cells. Ann Neurol 65:304–315

Miller RH (1999) Contact with central nervous system myelin inhibits oligodendrocyte progenitor maturation. Dev Biol 216:359–368

Miller RH (2002) Regulation of oligodendrocyte development in the vertebrate CNS. Prog Neurobiol 67:451–467

Miller RH, Dinsio KJ, Wang R-Z, Geertman R, Maier CE, Hall AK (2004) Patterning of spinal cord oligodendrocyte development by dorsally derived BMP4. J Neurosci Res 76:9–19

Nishiyama A, Lin X-H, Giese N, Heldin C-H, Stallcup WB (1996) Interaction between NG2 proteoglycan and PDGF alpha-receptor on O2A progenitor cells is required for optimal response to PDGF. J Neurosci Res 43:315–330

Ono K, Bansal R, Payne J, Rutishauser U, Miller RH (1995) Early development and dispersal of oligodendrocyte precursors in the embryonic chick spinal cord. Development 121:1743–1754

Orentas DM, Miller RH (1996) The origin of spinal cord oligodendrocytes is dependent on local influences from the notochord. Dev Biol 177:43–53

Orentas DM, Miller RH (1998) Regulation of oligodendrocyte development. Mol Neurosci 18(3):247–259

Padovani-Claudio D, Lui L, Ransohoff RM, Miller RH (2006) Alterations in the oligodendrocyte lineage, myelin and white matter in adult mice lacking the chemokine receptor CXCR2. Glia 54:471–483

Park SK, Miller RH, Krane I, Vartanian T (2001) The erbB2 gene is required for the development of terminally differentiated spinal cord oligodendrocytes. J Cell Biol 154:1245–1258

Patel JR, McCandless EE, Dorsey D, Klein RS (2010) CXCR4 promotes differentiation of oligodendrocyte progenitors and remyelination. Proc Natl Acad Sci USA 107:11062–11067

Perez Villages EM, Olivier C, Spassky N, Poncet C, Cochard P et al (1999) Early specification of oligodendrocytes in the chick embryonic brain. Dev Biol 216:98–113

Pringle NP, Richardson WD (1993) A singularity of PDGF alpha-receptor expression in the dorsoventral axis of the neural tube may define the origin of the oligodendrocyte lineage. Development 117:525–533

Pringle NP, Yu W, Guthrie S, Roelink H, Lumsden A et al (1996) Determination of neuroepithelial cell fate: induction of the oligodendrocyte lineage by ventral midline cells and sonic hedgehog. Dev Biol 177:30–42

Raff MC (1989) Glial cell diversification in the rat optic nerve. Science 243:1450–1455

Raff MC, Mirsky R, Fields KL, Lisak RP, Dorfman SH et al (1978) Galactocerebroside is a specific cell surface antigenic marker for oligodendrocytes in culture. Nature 274:813–816

Raff MC, Miller RH, Noble M (1983) A glial progenitor cell that develops in vitro into an astrocyte or an oligodendrocyte depending on culture medium. Nature 303:390–396

Raff MC, Abney ER, Miller RH (1984) Two glial cell lineages diverge prenatally in rat optic nerve. Dev Biol 106:53–60

Raff MC, Lillien LE, Richardson WD, Burne JF, Noble MD (1988) Platelet-derived growth factor from astrocytes drives the clock that times oligodendrocyte development in culture. Nature 333:562–565

Raper JA (2000) Semaphorins and their receptors in vertebrates and invertebrates. Curr Opin Neurobiol 10:88–94

Reynolds BA, Weiss S (1992) Generation of neurons and astrocytes from isolated cells of the adult mammalian central nervous system. Science 255:1707–1710

Richardson WD, Pringle N, Mosley MJ, Westermark B, Dubois-Dalcq M (1988) A role for platelet-derived growth factor in normal gliogenesis in the central nervous system. Cell 53:309–319

Richardson WD, Kessaris N, Pringle N (2006) Oligodendrocyte wars. Nat Rev Neurosci 7:11–18

Rivers LE, Young KM, Rizzi M, Jamen F, Psachoulia K et al (2008) PDGFRA/NG2 glia generate myelinating oligodendrocytes and piriform projection neurons in adult mice. Nat Neurosci 11:1392–1401

Robinson S, Tani M, Streiter RM, Ransohoff RM, Miller RH (1998) The chemokine growth related oncogene-a promotes spinal cord precursor proliferation. J Neurosci 18:10457–10463

Rowitch D (2004) Glial specification in the vertebrate neural tube. Nat Rev Neurosci 5:409–419

Samanta J, Burke GM, McGuire T, Pisarek AJ, Mukhopadhyay A et al (2007) BMPR1a signaling determines numbers of oligodendrocytes and calbindin-expressing interneurons in the cortex. J Neurosci 27:7397–7407

Shimizu T, Kagawa T, Wada T, Muroyama Y, Takada S, Ikenaka K (2005) Wnt signaling controls the timing of oligodendrocyte development in the spinal cord. Dev Biol 282:397–410

Sommer I, Schachner M (1981) Monoclonal antibodies (O1 to O4) to oligodendrocyte cell surfaces: an immunocytological study in the central nervous system. Dev Biol 83:311–327

Spassky N, Goujet-Zalc C, Parmantier E, Olivier C, Martinez S et al (1998) Multiple Restricted Origin of Oligodendrocytes. J Neurosci 18:8331–8343

Spassky N, de Castro F, Le Bras B, Heydon K, Queraud-LeSaux F et al (2002) Directional guidance of oligodendroglial migration by class 3 semaphorins and netrin-1. J Neurosci 22: 5992–6004

Sugimoto Y, Taniguchi M, Yagi T, Akagi Y, Nojyo Y, Tamamaki N (2001) Guidance of glial precursor cell migration by secreted cues in the developing optic nerve. Development 128: 3321–3330

Sussman CR, Vartanian T, Miller RH (2005) The ErbB4 neuregulin receptor mediates suppression of oligodendrocyte maturation. J Neurosci 25:5757–5762

Temple S, Raff MC (1986) Clonal analysis of oligodendrocyte development in culture: evidence for a developmental clock that counts cell divisions. Cell 44:773–779

Timsit S, Martinez S, Allinquant B, Peyron F, Puelles L, Zalc B (1995) Oligodendrocytes originate in a restricted zone of the embryonic ventral neural tube defined by DM-20 mRNA expression. J Neurosci 15:1012–1024

Tripathi RB, Rivers LE, Young KM, Jamen F, Richardson WD (2010) NG2 glia generate new oligodendrocytes but few astrocytes in a murine experimental autoimmune encephalomyelitis model of demyelinating disease. J Neurosci 30:16383–16390

Tripathi RB, Clarke LE, Burzomato V, Kessaris N, Anderson PN et al (2011) Dorsally and ventrally derived oligodendrocytes have similar electrical properties but myelinate preferred tracts. J Neurosci 31:6809–6819

Tsai H-H, Miller RH (2002) Glial cell migration directed by axon guidance cues. Trends Neurosci 25:173–175

Tsai H, Frost E, Robinson S, Ransohoff R, Sussman C et al (2002) The chemokine receptor CXCR2 controls positioning of oligodendrocyte precursors in developing spinal cord by arresting their migration. Cell 110:373–383

Tsai H-H, Tessier-Lavinge M, Miller RH (2003) Netrin 1 mediates spinal cord oligodendrocyte precursor dispersal. Development 130:2095–2105

Tsai HH, Macklin WB, Miller RH (2009) Distinct modes of migration position oligodendrocyte precursors for localized cell division in the developing spinal cord. J Neurosci Res 87(15):3320–3330

Ulloa F, Marti E (2010) Wnt won the war: antagonistic role of Wnt over Shh controls dorso-ventral patterning of the vertebrate neural tube. Dev Dyn 239:69–76

Vartanian T, Goodearl A, Viehover A, Fischbach G (1997) Axonal neuregulin signal cells of the oligodendrocyte lineage through activation of HER4 and Schwann cells through HER2 and HER3. J Cell Biol 137:211–220

Vartanian T, Fischbach G, Miller R (1999) Failure of spinal cord oligodendrocyte development in mice lacking neuregulin. Proc Natl Acad Sci USA 96:731–735

Wang S, Sdrulla AD, diSibio G, Bush G, Nofziger D et al (1998) Notch receptor activation inhibits oligodendrocyte differentiation. Neuron 21:63–75

Warf BC, Fok-Seang J, Miller RH (1991) Evidence for the ventral origin of oligodendrocyte precursors in the rat spinal cord. J Neurosci 11:2477–2488

Williams A, Piaton G, Aigrot MS, Belhadi A, Theaudin M et al (2007) Semaphorin 3A and 3F: key players in myelin repair in multiple sclerosis? Brain 130:2554–2565

Wolswijk G, Noble M (1989) Identification of an adult-specific glial progenitor cell. Development 105:387–400

Zhang H, Miller RH (1995) Asynchronous differentiation of clonally related spinal cord oligodendrocytes. Mol Cell Neurosci 6:16–31

Zhao X, He X, Han X, Yu Y, Ye F et al (2010) MicroRNA-mediated control of oligodendrocyte differentiation. Neuron 65:612–626

Zhu X, Bergles DE, Nishiyama A (2008) NG2 cells generate both oligodendrocytes and gray matter astrocytes. Development 135:145–157

Chapter 2
Demyelination and Remyelination in Multiple Sclerosis

Lars Bø, Margaret Esiri, Nikos Evangelou, and Tanja Kuhlmann

2.1 Introduction

The defining trait of multiple sclerosis (MS) histopathology is the presence of spatially separate focal areas of demyelination (called MS lesions, or plaques) of different age and inflammatory activity in the central nervous system (CNS). Although the nature and cause of the initial MS lesion change are not known, recent years have seen major advances in our understanding of MS pathology and pathogenesis. The aim of this review is to give a brief account of the histopathology of demyelination and remyelination in MS, with an emphasis on some issues of current interest in MS pathology research, including MS lesion staging, pathological heterogeneity, gray matter pathology, and axonal loss.

L. Bø
National Competence Center for Multiple Sclerosis, Department of Neurology,
Haukeland University Hospital, Bergen, Norway

Department of UK Clinical Medicine, University of Bergen, Bergen, Norway

M. Esiri
Neuropathology Department, West Wing, John Radcliffe Hospital,
Oxford Radcliffe NHS Trust, Oxford OX3 9DU, UK

Department of Clinical Neurology, Oxford University, Oxford, UK

N. Evangelou (✉)
Division of Clinical Neurology, Nottingham University Hospital, University of Nottingham,
C Floor, South Block, Queen's Medical Centre, Nottingham NG7 2UH, UK
e-mail: nikos.evangelou@nottingham.ac.uk

T. Kuhlmann
Institute of Neuropathology, University Hospital Münster,
Domagkstr. 19, 48149 Münster, Germany

2.2 Demyelination in MS

2.2.1 MS Lesion Classification

Lesion staging is important in order to distinguish the sequence of events in MS pathogenesis. Several staging systems have been proposed (van der Valk and De Groot 2000), partly reflecting an uncertainty about the nature of initial lesion changes. Despite the recent advances of imaging studies, MRI still lacks pathological specificity, and hence the evolution of MS lesions has to be deducted from pathological studies. Unfortunately, animal studies are also of limited value in lesion classification as MS is a disease specific to humans and the lesion evolution can only be indirectly extrapolated from experimental autoimmune encephalomyelitis (EAE) models. Any MS lesion staging system is thus a hypothesis that attempts to reconstruct a temporal sequence of lesion evolution from still images from different lesions. Complicating this issue is the MS pathological heterogeneity, clinical heterogeneity, and variability of MS lesion pathology dependent upon lesion location (Bo et al. 2003a; Confavreux and Vukusic 2006; Lucchinetti et al. 2000; Peterson et al. 2001; Revesz et al. 1994). The choice of MS classification system will to some extent reflect the research area of interest, for example, the role of chronic inflammation in MS pathogenesis or the mechanisms of early demyelination. MS lesions have therefore been classified according to multiple variables, including lesion location and distribution, pattern and extent of inflammation, presence of myelin/myelin degradation products in macrophages, extent of remyelination, pattern of oligodendrocyte loss, and presence of complement deposition (Bo et al. 1994, 2003b; Frohman et al. 2006; Gay et al. 1997; Lassmann et al. 1998; Lucchinetti et al. 2000; Sanders et al. 1993; Trapp et al. 1998; van der Valk and De Groot 2000).

2.2.2 MS Lesion Location

The location of MS lesions is critical in any therapy aiming to repair individual lesions, such as cell transplantation. Any treatment employed will likely to be met with partial response unless it can be used in gray and white matter, in the brain as well as the cord. Furthermore it is important to note that only a proportion of pathologically proven lesions are visualized with our current MR techniques and this is especially important for the gray matter (Seewann et al. 2011). MS lesions may occur anywhere in the CNS parenchyma, the predilection sites being the optic nerve and chiasm, periventricular white matter, subpial cerebral and cerebellar cortex, brain stem, and cervical spinal cord. Gray matter demyelination has been recognized to be widespread in MS, and especially in the spinal cord, it appears to be even more extensive than white matter demyelination (Fig. 2.1) (Bo et al. 2003b; Gilmore et al. 2009; Kutzelnigg et al. 2005; Sanders et al. 1993; Vercellino

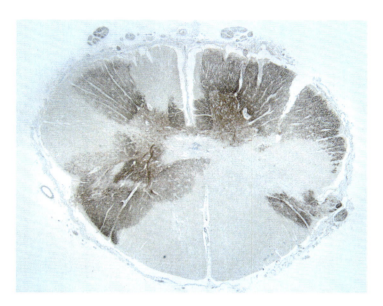

Fig. 2.1 Spinal cord demyelination. The spinal cord is a predilection site for MS lesions, and in many advanced cases, only islands of myelinated tissue remain in both the white and gray matter. Immunolabeled for MBP

et al. 2005). A proportion of chronic MS patients have a pattern of subpial cerebral demyelination in all cerebral regions; this has been termed general subpial demyelination (GSD (Bo et al. 2003b; Vercellino et al. 2005)). GSD probably represents the extreme end of a spectrum, rather than a distinct subgroup. In some GSD patients, the percentage of demyelinated area in cerebral cortex may approach 70 % (Bo et al. 2003b). The GSD patients do not have an increased percentage of demyelination in white matter, indicating that the pathogenesis in some MS cases may have a specificity for gray matter myelin (Bo et al. 2003b; Vercellino et al. 2005). Cortical GM lesions may be divided into three subtypes depending on lesion location: (1) combined WM/GM lesions, (2) lesions located entirely within GM, and (3) subpial lesions (Peterson et al. 2001).

2.2.3 Inflammation

White matter MS lesions are usually divided into active, chronic active, or chronic inactive, based upon pattern and extent of inflammation (Bo et al. 1994; Trapp et al. 1998; van der Valk and De Groot 2000). In active lesions, there is macrophage infiltration throughout the lesion; in chronic active lesions, there is macrophage infiltration at the lesion border but little infiltration at the lesion center; and in chronic inactive lesions, there is little infiltration throughout the lesion. In chronic

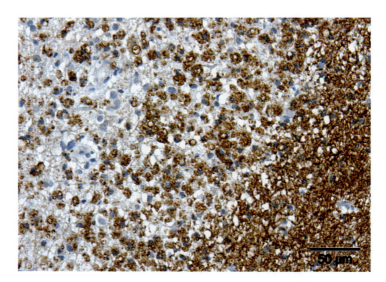

Fig. 2.2 Actively demyelinating MS lesion. Myelin basic protein (MBP) is a major constituent of the myelin sheath of oligodendrocytes in the central nervous system. Macrophages phagocytosing myelin debris can be easily detected in actively demyelinating lesions

MS, chronic active lesions are the most common lesion subtype in the brain (Bø, unpublished observation), whereas most of the lesions in the spinal cord are inactive (Tallantyre et al. 2009). The chronic persistence of inflammation in MS lesions may be clinically relevant as it may mediate lesion growth and chronic axonal loss.

2.2.4 Myelin/Myelin Degeneration Products in Macrophages

The degeneration of myelin and myelin degeneration products in macrophages after myelin phagocytosis follows a predictable time course, which may give information about the age and activity level of individual MS lesions/lesion areas. Major myelin proteins, such as PLP and MBP (Fig. 2.2), can be detected immunohistochemically in macrophages for 2–3 days, while the smaller in size myelin protein MOG, which accounts for only 0.05 % of the total myelin proteins, or CNP (Fig. 2.3a) is only detectable for approximately 1 day after phagocytosis (van der Goes et al. 2005). In one of the proposed clasification systems, immunopositivity for minor myelin proteins within macrophages is considered to signify early active lesions, minor myelin proteins late active lesions and all other lesins were considered inactive (Bruck et al. 1995 Davie et al. 1994). It is possible that the sensitivity of immunohistochemistry for minor myelin proteins in macrophages also may depend upon the rate of ongoing myelin degradation, however, so that minor myelin protein degradation in "slow burning" demyelination may go undetected. As for histochemical myelin lipid

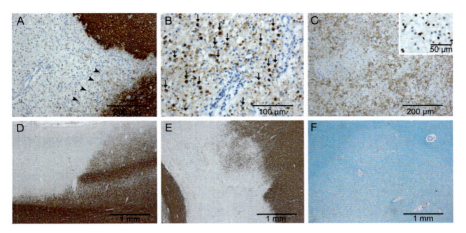

Fig. 2.3 Remyelination. In early MS lesion stages (**a–c**), remyelination is a common phenomenon, whereas remyelination is often limited or absent in chronic MS lesions (**d, e**). In early MS lesions with ongoing demyelination characterized by macrophages containing myelin degradation products (*arrow heads* in **a**), oligodendrocytes are frequently preserved even in completely demyelinated lesion areas (*arrows* in **b**). Remyelinating lesion areas are characterized by thin irregularly formed myelin sheaths (**c**); furthermore numerous olig2-expressing oligodendroglial lineage cells are present (*insert* in **c**). In chronic MS, demyelinated lesions are localized not only in the white matter but also in the cortex (**d**). Remyelination is frequently limited to the lesion border (**e**); however, approximately 20 % of the chronic MS lesions are completely remyelinated (so-called shadow plaques) (**f**). Immunohistochemistry for CNPase (**a** and **c**), immunohistochemistry for NogoA (**b**), immunohistochemistry for MBP (**d** and **e**), Luxol-Fast-Blue staining (**f**), immunohistochemistry for Olig2 (*insert* in **c**)

stains, Luxol Fast Blue is thought to persist for a few days within macrophages, while oil red O may persist a few months (Davie et al. 1994).

2.2.5 Pathological Heterogeneity

An interindividual heterogeneity in the pathology of white matter MS lesions has been described in a material of biopsies and autopsies from MS patients with acute MS or MS with a short disease duration (Lucchinetti et al. 2000). Four pathology subtypes were described (1) lesions with oligodendrocyte loss within the demyelinated area, no signs of complement activation; (2) lesions similar to (1), but with signs of complement activation; (3) lesions with a lesional and perilesional loss of oligodendrocytes by apoptosis, no sign of BBB loss, and no sign of complement activation; and (4) lesions characterized by lesional and perilesional extensive loss of oligodendrocytes, but no signs of oligodendrocyte apoptosis. Type 4 lesions are very rare. In individual patients, all active lesions had the same pattern; no intraindividual heterogeneity was observed. In a later report, early lesions with elements in common both with the type 3 and type 2 lesion pattern were described in seven MS patients with a short disease duration (Barnett and Prineas 2004). These changes were detected in areas of only slight myelin pathology, which may have represented a very early lesion stage. In the great majority of chronic MS patients, complement and immunoglobulin deposits were found similar to pattern 2 lesions (Breij et al. 2008).

The examination of autopsy specimens cannot predict the clinical classification of MS. In all types of MS, the pathological hallmarks of inflammation, demyelination, and axonal loss are present. In PPMS, fewer demyelinating gray and white matter lesions are detected, but the degree of axonal loss seems to be similar to SPMS. This raises the possibility that axonal loss is, to a significant degree, independent of demyelinating lesions or that axons in PPMS are more vulnerable to damage even with a modest degree of inflammation (Tallantyre et al. 2009).

2.3 The Cellular Components of the MS Lesion

2.3.1 *Oligodendrocytes and Myelin*

Oligodendrocyte loss has been postulated by Barnett (Barnett and Prineas 2004) to be the initial event in lesion formation. Certainly in the center of chronic inactive lesions, oligodendrocytes are in general not detected, but in other lesion types, there is a large degree of variability of oligodendrocyte numbers, as high numbers of oligodendrocytes may be present both in active lesions and at the inflammatory edge of chronic active lesions. A recent study indicates that this variability may be mainly interindividual, as in a subgroup of MS patients a high number of oligodendrocytes were retained in lesion areas (Patrikios et al. 2006). The failure to detect oligodendrocytes in chronic lesions may be due to technical factors, as PLP-immunopositive oligodendrocytes were detected in chronic lesions using highly sensitive immunohistochemical methods (Chang et al. 2002). In the chronic MS lesions, oligodendrocyte processes were extended to axons but did not seem to be able to initiate the formation of myelin. This suggests that there are molecules expressed by demyelinating axons in MS lesions that inhibit the initiation of myelination. PSA-NCAM, an adhesion molecule expressed by axons in lesion areas, may have such a function. The extent of oligodendrocyte loss in gray matter lesions is similar to that in white matter lesions. The mechanisms of oligodendrocyte death in MS are not well characterized. In active lesions, cells with dense pyknotic nuclei are frequently observed, indicating an apoptotic process. Oligodendrocyte nuclear fragmentation is rare, however, and cells with dense "apoptotic" nuclei are not immunostained with anti-caspase 3 antibodies, suggesting an apoptosis through non-caspase-dependent mechanisms. In some MS patients, a pattern of "dying back oligodendrogliopathy" has been described, where the initial change observed is in the most distal part of the oligodendrocyte, the periaxonal process, suggesting a critical role of disturbed oligodendrocyte function in the initiation of demyelination

(Rodriguez and Scheithauer 1994). Oligodendrocytes in MS lesion areas express MHC class I which may make them susceptible to damage by CD8 positive T cells (Hoftberger et al. 2004) with vesicular dissolution of myelin, prior to phagocytosis by macrophages.

2.3.2 Astrocytes

In the majority of chronic inactive white matter MS lesions, there is prominent astrogliosis, with a dense meshwork of glial fibrillary acidic protein (GFAP)-positive processes. In active lesions, large hypertrophic astrocytes may be present, which may appear to be in close proximity or even contain the cell bodies of one or more oligodendrocytes. This process seems to be different from phagocytosis. The reactive astrocytes of MS lesions frequently contain myelin or myelin degradation products. Astrogliosis is also detected in diffusely abnormal white matter areas with concomitant microglial activation and BBB abnormality. The cause of the general astrocyte activation outside of lesions is unclear, possibly related to inflammatory mediators, diffusing from lesions, stimulating this proliferation. In contrast, in purely cortical lesions, astroglial changes are small or absent. Astroglial scar has been thought to impair the recruitment of OPC in MS lesions, although direct evidence is lacking.

2.3.3 Microglia/Macrophages

Demyelination in MS generally is thought to be mediated by resident microglial cells, or infiltrating monocytes/macrophages. The relative contribution of these two cell types is not known. In confocal microscopy studies, it has been demonstrated that macrophages that are close to myelin sheaths extend processes to the myelin sheaths. By electron microscopy, it has been observed that outer lamellae are separated from the myelin sheaths and attached to coated pits on the surface of macrophages. Myelin was observed to enter the cytoplasm of the macrophages through elongated pinocytic vesicles (Prineas and Connell 1978). Ultimately the myelin sheath may become loosened from the axon and macrophage processes may be interspersed between the axon and the myelin sheath. Myelin fragments have been observed in coated pits in macrophages, and macrophages display a "capping" of Fc receptors, indicating a receptor-mediated endocytic process. Complement activation products and IgG are polarized to the macrophage/myelin interface at areas of demyelination, suggesting an opsonization of myelin by complement and IgG (Breij et al. 2008; Prineas and Graham 1981). Other investigators have described a vesicular dissolution of myelin sheaths prior to the phagocytosis by macrophages (Guo and Gao 1983; Lassmann 1983). Lipid-laden macrophages accumulate in perivascular spaces in the lesions and also in the normal-appearing white matter (NAWM). The macrophages may be transported to regional lymph nodes in the neck, as lipid-laden macrophages have been observed in cervical lymph nodes.

The morphology of microglia changes toward the edge of active and chronic active lesions, from cells with small elongated cell bodies, rod-shaped nuclei, and thin, highly branched processes to larger rounded cells with thick and less branched processes, to that of large lipid-laden macrophages within the lesions (Revesz et al. 1994). Lipid macrophages in MS lesions may be recruited from microglia or from blood-borne monocytes; the relative contribution of these cell populations is not known. Although MHC class II expression has been detected on astrocytes, the majority of studies indicate that macrophages/monocytes, microglia, and perivascular macrophages are the main cell types expressing MHC class II in MS lesions and thus being able to present antigen to CD4 T cells. Activated microglia have been shown to express a variety of proinflammatory cytokines, noticeably IL-1, TNF-α, IL-6, IL-12, and IL-23. The cytokine profile of MS lesion macrophages indicates that they constitute a mixture between a proinflammatory and anti-inflammatory phenotype, with anti-inflammatory cells being the majority.

2.3.4 Blood Vessels

The BBB is damaged early in the pathogenesis of MS lesions, and there is leakage of plasma proteins throughout all MS lesion stages (Barnes et al. 1991; Kwon and Prineas 1994). Vascular inflammation does not seem to require myelin as periphlebitis has been elegantly demonstrated in the retina that lacks myelin (Green et al. 2010). The leakage of plasma proteins may be due to a combination of increased pinocytic transport and to the disruption of intercellular tight junctions (Brown 1978; Plumb et al. 2002). Similarly, there is increased transport of cells through the vessel wall, by migration of leukocytes through the endothelium and by migration between endothelial cells. The blood vessel wall is thickened in old MS lesions by the deposition of collagen, and the perivascular spaces (Virchow-Robin spaces) are widened. The expression of adhesion molecules on parenchymal blood vessels is increased, including the expression of ICAM-1. In the majority of WM MS lesions, there is a central small blood vessel, suggesting that the vascular compartment determines the direction of lesion spread (Dawson's finger). As antigen presentation by perivascular macrophages seems sufficient to initiate brain parenchymal inflammation, the mechanisms of lesion spread could be through perivascular traffic of perivascular macrophages that present myelin autoantigens.

2.3.5 Lymphocytes

In white matter MS lesions, the majority of infiltrating T cells are lymphocytes, distributed in perivascular infiltrates, but also scattered through the parenchyma. CD8-positive T cells are more prevalent at the lesion border than at the lesion center, while the opposite was true for CD4-positive cells; in all locations, CD8 T cells

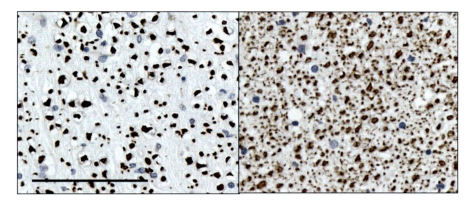

Fig. 2.4 Axonal loss. Neurofilament stain of the lateral corticospinal tract of a patient with MS (*left*) demonstrates extensive axonal loss compared to control tissue (*right*). *Scale bar* 100 μm

were predominant (Bo et al. 2003a). There is also a small minority of gamma delta T cells present in lesions; this may be pathogenetically important, as these may lyse cells expressing heat shock proteins, such as lesion oligodendrocytes. Distinct T cell clones, some of which were CD8 positive, were detected in lesion and NAWM areas in several anatomical regions in individual MS patients, indicating that T cells in different lesions in these patients responded to common antigens. T cells may be cleared from the lesions by transport along perivascular spaces or through apoptosis (Ozawa et al. 1994). B cells are present in lower numbers in the brain parenchyma, as are plasma cells. B cells are seen more commonly in chronic lesions particularly when active (Esiri 1977). Granulocytes are rare, except in very acute cases with destructive lesions (Marburg 1906).

2.3.6 Axons

Axonal loss occurs in all demyelinated MS lesions and is considered the main cause of irreversible chronic disease progression in MS (Fig. 2.4). The extent of ongoing axonal loss correlates with extent of macrophage infiltration. The rate of axonal loss is thus highest in early lesion stages, where there is intense inflammation, and is also high at the inflammatory edge of chronic active lesions (Bitsch et al. 2000; Ferguson et al. 1997; Trapp et al. 1998). In areas with few or no perivascular infiltrates or macrophages, in chronic/inactive MS lesions, there is a low-grade axonal loss, suggesting that persistent demyelination mediates axonal pathology and loss independent of inflammation (Bitsch et al. 2000; Ferguson et al. 1997; Trapp et al. 1998). Significant axonal loss has also been detected in NAWM (Evangelou et al. 2000a), and a recent study supports the assumption that significant degree of axonal loss is due to Wallerian degeneration even in early MS cases (Dziedzic et al. 2010).

A strong correlation of regional lesion load with axon numbers in the corresponding projection areas in the corpus callosum indicates a substantial role for lesional axonal transection to diffuse axonal loss in the NAWM (Evangelou et al. 2000b).

Axon pathology is observed in lesions both as axonal caliber changes and as transected axons form end-bulbs or ovoids (Trapp et al. 1998). In transected axons, there is an accumulation of amyloid precursor protein (APP) and non-phosphorylated neurofilament (Ferguson et al. 1997; Trapp et al. 1998). In acute MS, in response to demyelination, sodium channels are distributed along demyelinated axons, thereby providing an anatomical basis for continuous nerve impulse conduction in demyelinated areas. In chronic MS lesions, continuous sodium channel expression in axons is lost. The sodium channel Nav1.6, seen in healthy white matter at the nodes of Ranvier, was present in chronic lesions at approximately one-third of the axons, and then in a patchy manner (Black et al. 2007; Craner et al. 2003). The Na+/K+ ATPase is necessary for the ion balance of functional axons. It is retained in acutely demyelinated axons but frequently absent in chronic MS lesions, indicating that the majority of axons in chronic demyelinated lesions are not functionally active (Young et al. 2008). The epidemiological evidence for gender differences in disability accumulation is conflicting; hence a recent comparison of axonal loss in acute lesions that failed to show any gender differences in the APP-positive spheroids in either acute of chronic lesions is of interest.

2.4 Gray Matter Pathology

In chronic MS patients, extensive cortical subpial MS lesions are frequent, and the percentage demyelinated area is similar in gray matter and white matter. GM demyelination is extensive also in cerebellar cortex; this may be a cause of cerebellar dysfunction in MS (Kutzelnigg et al. 2007). Extent of gray matter and white matter demyelination is weakly correlated, if at all (Bo et al. 2003b, 2007; Kutzelnigg et al. 2007). Episodic and anterograde memory is frequently affected in MS. This could be due to hippocampal MS pathology (Figs. 2.3 and 2.5). Extensive demyelination has been detected in the hippocampus in MS patients with substantial alterations in the cholinergic neurotransmitter system in the MS hippocampus, which were different from those in AD hippocampus (Geurts et al. 2007; Papadopoulos et al. 2009). There is increased neuronal apoptosis in cortical MS lesions, and significant neuronal death has also been demonstrated in the thalamus in MS. Specific subpopulations of neurons are vulnerable in MS. In primary motor cortex, parvalbumin interneurons within layer 2 were significantly reduced, with no concurrent change in the number of calretinin-positive neurons. In the lateral geniculate nucleus, there was disproportionate pathology of small neurons, which correlated with axonal loss in the optic nerve in MS patients (Evangelou et al. 2001). Cortical demyelinated lesions are not specific for MS; intracortical

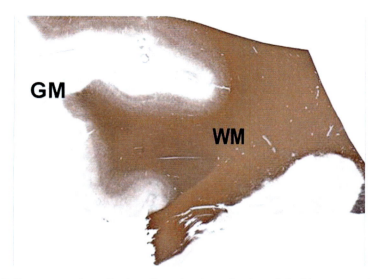

Fig. 2.5 Gray matter demyelination. Extensive demyelination of the hippocampus has been observed in a number of studies and has been implicated as a significant cause of the memory symptoms MS patients report frequently. In this tissue block, subpial demyelination is seen throughout the cortical ribbon

and leukocortical lesions were also detected in progressive multifocal leukoencephalopathy (PML). Subpial lesions were not observed in PML, however (Moll et al. 2008). Cortical lesions using conventional MR scanners are largely undetectable in vivo because of the low sensitivity of MRI (5 %) for purely gray matter plaques (Geurts et al. 2005). Extensive cortical pathology is not associated with increased focal or diffuse WM pathology, indicating that the extent or distribution of WM abnormalities cannot be used to identify patients with extensive GM demyelination (Bo et al. 2007). The presence and extent of remyelination in MS cortex are not easy to characterize. In a direct comparison of remyelination of WM and GM MS lesions of the same patients, GM remyelination was consistently more extensive (Albert et al. 2007). Until recently it was accepted that the pathology of gray matter demyelination differs from white matter lesions in that there is less inflammation in the GM. In fact, in purely cortical lesions, no significant increase in T lymphocyte infiltration, no BBB damage, and no complement activation were detected (Bo et al. 2003a; Brink et al. 2005; van Horssen et al. 2007). The GM part of combined lesions (type 1) has levels of inflammation intermediate between that of white matter and purely gray matter lesions. Recent evidence suggests that in some early MS patients, cortical demyelination was frequently inflammatory (Lucchinetti et al. 2011).

During inflammation, white matter neurons are destroyed, but there is evidence of neurogenesis in a subgroup of chronic subcortical white matter lesions (Chang et al. 2008). The subventricular zone is a possible source of those neurons as an increase of progenitor neurons was found in the SVZ when bordering demyelinating lesions.

2.4.1 Meningeal Inflammation

A low-grade meningeal inflammation is frequently detected in chronic MS (Black et al. 2007; Dawson 1916; Gay et al. 1997). Recently structures similar to B cell follicles containing germinal centers were detected in meningeal inflammatory infiltrates (Serafini et al. 2004). Meningeal B cell follicles were present in approximately 40 % of secondary progressive MS (SPMS) patients, but not in primary progressive MS (PPMS) cases. Follicle-positive SPMS cases have been reported to show a more rapid disease development, with a younger age at disease onset, irreversible disability and death, and more extensive cortical pathology, indicating a detrimental effect of meningeal B cell follicles in MS pathogenesis (Magliozzi et al. 2007; Serafini et al. 2004). On the contrary, the presence of meningeal inflammation (predominantly T cells) does not seem to be correlated with the occurrence of cortical lesions (Kooi et al. 2009). Recent publications offer conflicting evidence regarding the presence of Epstein–Barr virus in MS brains and specifically in meningeal B cells. Although clearly there is epidemiological evidence in support of EBV playing a pathogenetic role, the pathological evidence is still weak (Lassmann et al. 2011).

2.5 Normal-Appearing White Matter

The concept of NAWM was proposed in the years when conventional MR techniques failed to show abnormalities outside WM plaques. Now it accepted that at least in chronic MS there is probably a low-grade inflammation in MS brain white matter, outside of established lesions (Kutzelnigg et al. 2005). This inflammation is variable among individuals and consists of nodules of activated microglia and perivascular infiltrates of lymphocytes/monocytes. In LFB-stained sections, large WM areas of diffusely lighter staining are present outside of lesions, suggestive of edema, remyelination, gliosis, axonal degeneration, or combinations of these. Some authors refer to those abnormalities that are increasingly easier to detect with highfield MRI scanners and different sequences as diffusely abnormal white matter. In MS NAWM and normal-appearing gray matter (NAGM), there is a slow ongoing loss of axons, as evidenced with APP or nonphosphorylated neurofilament-positive axonal end-bulbs. White matter interneurons are lost in MS, but in a subpopulation of patients, there is an increase in white matter neuron number at the lesion edge. Vascular fibrinogen leakage in the NAWM has been demonstrated by immunohistochemistry, indicating that the BBB is damaged also outside of demyelinated lesions.

2.6 Remyelination in MS

The thickness of the myelin sheaths depends on the axon diameter and is tightly regulated under physiological conditions. Remyelinated axons are characterized by uniformly thin and shortened internodes. The gold standard to detect remyelination is electron microscopy, and in MS, remyelinated axons were first described in 1965 (Perier and Gregoire 1965; Prineas and Connell 1979). By light microscopy, remyelination can be detected using conventional lipophilic stains such as Luxol Fast Blue or by immunohistochemistry using antibodies directed against myelin proteins such as myelin basic protein (MBP), proteolipid protein (PLP), or CNPase (Fig. 2.3). Lesion areas with advanced remyelination display a paler staining intensity due to thinner myelin sheaths and reduced numbers of axons (Fig. 2.3d).

Remyelination is frequently found at the lesion border, starts early during the formation of lesions, and is also present in lesions with active demyelination (Fig. 2.3) (Goldschmidt et al. 2009; Lucchinetti et al. 1999; Prineas et al. 1993a, b; Raine and Wu 1993). The comparison of lesions derived from patients with either a short or a long disease duration suggests that remyelination is more frequent and more extensively observed in early MS lesions stages (Goldschmidt et al. 2009). In average, about 10–20 % of chronic lesions are completely remyelinated (so-called shadow plaques) (Barkhof et al. 2003; Patani et al. 2007). However, remyelinated lesion areas may be more vulnerable to repeated demyelinating activity compared to NAWM (Bramow et al. 2010; Prineas et al. 1993b). Histological analyses do not allow longitudinal studies, and the lack of imaging techniques specifically measuring remyelination makes it difficult to determine exactly when remyelination occurs in individual lesions and how long it continues. However, the identification of differentiating oligodendroglial lineage cells in early disease stages and the lack of myelinating oligodendrocytes in chronic MS support the hypothesis that remyelination mostly occurs relatively early during lesion formation (Chang et al. 2002; Kuhlmann et al. 2008; Wolswijk 1998a, b, 2000). The extent of remyelination varies between individual lesions of the same patient and might be influenced by lesion size and lesion location (Goldschmidt et al. 2009; Patani et al. 2007). Periventricular and cerebellar lesions, for example, display a lower extent of remyelination than subcortical lesions (Goldschmidt et al. 2009). Similarly, cortical lesions show more extensive remyelination in the majority of patients compared to white matter lesions from the same patient (Albert et al. 2007). How the anatomical localization may influence remyelination is unknown; potential explanation includes differences in the microenvironment or location-dependent differences in the oligodendroglial progenitor populations and their remyelination capabilities. Additional factors might influence remyelination, such as patient-dependent factors. Patrikios et al. (2006), for example, could show that a subset of patients was more prone to remyelination than others. In rodent animal models, reduced remyelination capacity is associated with increasing age and male sex (Gilson and Blakemore 1993; Li et al. 2006; Shields et al. 1999; Sim et al. 2002); however, such a correlation has not been detected yet in MS which might be explained by the heterogeneity of the analyzed tissue samples. Since a higher proportion of remyelinated shadow plaques and increased remyelination capacity was observed in primary versus secondary progressive MS, the disease course might be an additional factor contributing to remyelination outcome (Bramow et al. 2010).

Oligodendroglial precursor cells (OPCs) are believed to be the cells responsible for remyelination. In contrast to OPCs, transplanted mature oligodendrocytes are not able to remyelinate a demyelinated lesion (Groves et al. 1993; Keirstead and Blakemore 1997; Targett et al. 1996; Zhang et al. 1999). In demyelinated lesions depleted of oligodendroglial lineage cells, OPCs occur with onset of remyelination suggesting that OPC and not mature oligodendrocytes are responsible for remyelination (Fancy et al. 2004; Levine and Reynolds 1999; Watanabe et al. 2002).

2.6.1 Proliferation, Migration, and Differentiation of OPCs in MS Lesions

Prerequisite for successful remyelination is the proliferation, migration, and differentiation of OPCs. Every of these steps might be disturbed in MS lesions. Whereas proliferation of oligodendroglial precursor cells is a frequent phenomenon in de- and remyelinating animal models, proliferating OPCs are a rare event in human MS studies (Kuhlmann et al. 2008; Schonrock et al. 1998). However, this does not exclude that earlier lineage cells, for example, subventricular neural stem cells may proliferate, migrate, differentiate, and promote remyelination (Nait-Oumesmar et al. 2007). At the lesion border, higher numbers of oligodendroglial progenitors and mature oligodendrocytes are observed (Kuhlmann et al. 2008; Prineas et al. 1993b), suggesting that migration of OPCs into the lesions might be impaired. Migration of oligodendroglial lineage cells is regulated by a complex network of short- and long-range migration cues which are either secreted such as growth factors (e.g., FGF, PDGF), guidance molecules (netrins, certain semaphorins), and chemokines (e.g., CXCL1) or contact-mediated, for example, extracellular matrix molecules (for review see Jarjour and Kennedy 2004). Semaphorins 3A and 3F, for example, are upregulated in close proximity to active but not inactive MS plaques (Williams et al. 2007). Furthermore, animal studies demonstrate that overexpression of Sema 3A impairs the migration of OPCs, whereas overexpression of Sema 3F accelerates OPC's recruitment and remyelination (Piaton et al. 2011; Syed et al. 2011). Not only is the expression of long-range guidance molecules changed in MS lesions, but also changes of the ECM have been reported; tenascin-C expression, for example, that impairs the migration of OPCs in vitro is reduced in acute lesions, whereas chronic MS lesions display tenascin-C levels comparable to the NAWM (Gutowski et al. 1999). Furthermore, in active demyelinating lesions, upregulation of the migration promoting factors fibronectin and vitronectin has been reported (Gutowski et al. 1999; Sobel et al. 1995). In summary, these data indicate that guidance molecules are dynamically regulated in MS lesions and shift from a migration promoting to less favorable environment with lesion chronicity. However, since

OPCs are present even in chronic MS lesions, impairment of migration might not be the major cause for remyelination failure in chronic MS lesions.

Several publications indicate that the differentiation of oligodendrocytes is disturbed in chronic MS lesions (Chang et al. 2002; Wolswijk 1998a; 2002). In the majority of early MS lesions, differentiating progenitor cells are found (Kuhlmann et al. 2008), whereas in chronic MS lesions, OPCs are present but do not remyelinate despite close contact to axons (Chang et al. 2002; Wolswijk 1998a). This suggests that inhibitory signals, lack of remyelination promoting factors, or a combination of both mechanisms prevent successful remyelination. A complex and timely interaction of extracellular signals and intracellular transcription factors is required to initiate and perform remyelination successfully. This process can be disturbed by activation of inhibitory signaling pathways on many different levels. In experimental animal studies, a number of inhibitory pathways have been identified in recent years; however, whether these signaling cascades and factors also contribute to remyelination failure in MS is mostly unknown. Here, we describe a selected number of factors, such as the Notch/Jagged, Wnt, hyaluronan, and PSA-NCAM, which have been shown to might be relevant for MS.

2.6.2 Inhibitory Pathways

PSA-NCAM is an adhesion molecule that belongs to the immunoglobulin superfamily and plays a role in a number of developmental processes such as axonal pathfinding, nerve branching, cell migration as well as synaptic plasticity. During development, downregulation of axonal PSA-NCAM precedes myelination, whereas in demyelinated but not remyelinated CNS lesion areas, PSA-NCAM is reexpressed, suggesting that loss of PSA-NCAM may be prerequisite for (re-)myelination (Charles et al. 2000, 2002).

In and around MS lesions, Jagged1 is upregulated on astrocytes whereas oligodendroglial lineage cells express its receptors Notch1 and the downstream transcription factor HES5 that inhibits the differentiation of oligodendrocytes (John et al. 2002). In contrast, no Jagged1 was detected in remyelinating lesions indicating that the activation of the canonical Notch–Jagged pathway may contribute to remyelination failure in chronic MS lesions. This hypothesis was questioned by the finding that ablation of Notch1 in PLP-expressing oligodendrocytes had no effect on remyelination in demyelinating animal models (Zhang et al. 2009). However, inactivation of Notch1 in olig1-expressing oligodendroglial lineage cells resulted in accelerated remyelination indicating that Notch1 is critical early during the differentiation process (Zhang et al. 2009). Interestingly, Notch1 has not only inhibitory effects on oligodendrocytes. Binding of axonal contactin to Notch1 leads to the activation of a noncanonical pathway resulting into differentiation of oligodendrocytes and myelination via the nuclear translocation of the NICD/Deltex complex (Hu et al. 2003). In chronic MS lesions, TIP30 has been detected, a known inhibitor of the translocation of NICD to the nucleus, suggesting that not only activation of the canonical Notch signaling pathway but also inhibition of the noncanonical pathway might contribute to remyelination failure in chronic MS lesions (Nakahara et al. 2009).

The glycosaminoglycan hyaluronan is another factor secreted by astrocytes that may contribute to remyelination failure in MS (Marret et al. 1994). Hyaluronan accumulates in chronic demyelinated and remyelinated MS lesions, but expression is most intense in demyelinated lesion areas. Furthermore, degraded hyaluronan inhibits oligodendroglial differentiation and remyelination in vitro and in vivo (Back et al. 2005; Sloane et al. 2010), and this effect may be mediated via Toll-like receptor 2, a known binding partner of hyaluronan.

The Wnt pathway is another signaling cascade contributing to oligodendroglial differentiation (Shimizu et al. 2005). TCF7L2, a transcription factor activated by the Wnt/β-catenin pathway, as well as other members of the Wnt/β-catenin signaling cascade expressed in active MS lesions suggest that a dysregulation of this pathway may contribute to remyelination failure (Fancy et al. 2009). This was supported by experimental animal studies demonstrating that this pathway is activated during myelination and remyelination. Mice with only one copy of the Wnt pathway inhibitor APCs suffer from delayed remyelination (Pohl et al. 2011). Interestingly, mice lacking TCF7L2 show as well-impaired remyelination, similar to mice lacking histone deacetylases 1 and 2 (Ye et al. 2009). Lack of HDAC 1 and 2, enzymes that regulate histone acetylation and transcription by DNA packaging, is among others associated with stabilization and nuclear translocation of β-catenin. Therefore the hypothesis evolved that TCF7L2, depending on its binding partners, either promotes or inhibits oligodendroglial differentiation and remyelination (Aarli et al. 1975). Furthermore, in animal studies, histone acetylation increases with age and correlates with impaired remyelination; exposure of young animals to inhibitors of histone deacetylases is associated with delayed remyelination suggesting that changes in histone acetylation modulate remyelination capacities of OPCs (Marin-Husstege et al. 2002; Shen et al. 2005, 2008). In the aged human CNS as well as in the NAWM of patients with chronic MS, an increased histone acetylation has been observed, whereas in early MS lesions, a significantly reduced number of oligodendrocytes with acetylated histone 3 was found, indicating that in MS histone acetylation may as well influences remyelination capabilities (Pedre et al. 2011).

2.6.3 Remyelination and Inflammation

As described above, remyelination is a frequent phenomenon in early but not in chronic MS lesions. This change in remyelination capacity is associated with changes in the extent and composition of inflammation suggesting that certain factors present in the inflammatory "milieu" early during lesion formation may support remyelination. Inflammatory cells in MS patients, for example, express not only cytokines and chemokines but also neurotrophic factors such as BDNF and CNTF

(Kerschensteiner et al. 1999; Stadelmann et al. 2002). Increased density of macrophages and microglia at the lesion border correlated significantly with more extensive remyelination in tissue samples from patients with a long disease duration (Patani et al. 2007). However, this can indicate that either macrophages/microglia are prerequisite for remyelination or inflammatory cells prevent further formation of new myelin sheaths. In animal studies, dampening of the inflammatory response or lack of certain cytokines leads to impaired remyelination (Arnett et al. 2001; Chari et al. 2006; Kotter et al. 2005). In contrast, mice deficient for CXCR2, a chemokine receptor, display accelerated remyelination, and certain inflammatory factors are cytotoxic for oligodendroglial lineage cells at least in vitro (Liu et al. 2010). These combined data demonstrate that inflammation and remyelination are characterized by complex interactions and that most likely not a single but the spatiotemporal interplay of many factors determines the outcome.

2.6.4 Remyelination Promoting Therapies in MS

So far, no directly neuroprotective treatments exist. However, few compounds are under development aiming at promotion of remyelination. One such compound is an antibody directed against the leucine-rich repeat and Ig domain containing NOGO receptor interacting protein 1 (LINGO1). Lingo1 is a transmembrane receptor that is expressed on neurons and oligodendrocytes. Inhibition of Lingo1 promotes oligodendroglial differentiation and myelination in vitro and in vivo as well as remyelination in different animal models. In MS lesions, Lingo1 could be detected on neurons, astrocytes, and macrophages/microglial cells but not on oligodendrocytes, and levels of Lingo1 are decreased in brain samples from MS patients compared to controls. However, a phase 1 clinical trial evaluating the safety of anti-Lingo1 is currently under way.

2.7 Conclusion

We have advanced a lot in our knowledge of the pathology of MS, especially in relation to MS plaques. We are still far from understanding the mechanisms of ongoing deterioration observed in progressive MS. Not surprisingly, we lack medicines that can arrest the progressive phase of the disease, the cause of disability to the majority of MS patients. While our efforts continue to explore the initial events, and possibly the trigger for the formation of MS plaques, increasingly mechanisms of neurodegeneration and remyelination (failure) are becoming the focus of research in MS.

Acknowledgements Prof. Esiri has been supported by the Oxford Biomedical Research Centre and the National Institute of Health Research.

References

Aarli JA, Aparicio SR, Lumsden CE, Tonder O (1975) Binding of normal human IgG to myelin sheaths, glia and neurons. Immunology 28:171–185

Albert M, Antel J, Bruck W, Stadelmann C (2007) Extensive cortical remyelination in patients with chronic multiple sclerosis. Brain Pathol 17:129–138

Arnett HA, Mason J, Marino M, Suzuki K, Matsushima GK, Ting JP (2001) TNF alpha promotes proliferation of oligodendrocyte progenitors and remyelination. Nat Neurosci 4:1116–1122

Back SA, Tuohy TM, Chen H, Wallingford N, Craig A, Struve J, Luo NL, Banine F, Liu Y, Chang A, Trapp BD, Bebo BF Jr, Rao MS, Sherman LS (2005) Hyaluronan accumulates in demyelinated lesions and inhibits oligodendrocyte progenitor maturation. Nat Med 11:966–972

Barkhof F, Bruck W, De Groot CJ, Bergers E, Hulshof S, Geurts J, Polman CH, van der Valk P (2003) Remyelinated lesions in multiple sclerosis: magnetic resonance image appearance. Arch Neurol 60:1073–1081

Barnes D, Munro PM, Youl BD, Prineas JW, McDonald WI (1991) The longstanding MS lesion. A quantitative MRI and electron microscopic study. Brain 114(Pt 3):1271–1280

Barnett MH, Prineas JW (2004) Relapsing and remitting multiple sclerosis: pathology of the newly forming lesion. Ann Neurol 55:458–468

Bitsch A, Schuchardt J, Bunkowski S, Kuhlmann T, Bruck W (2000) Acute axonal injury in multiple sclerosis. Correlation with demyelination and inflammation. Brain 123(Pt 6):1174–1183

Black JA, Newcombe J, Trapp BD, Waxman SG (2007) Sodium channel expression within chronic multiple sclerosis plaques. J Neuropathol Exp Neurol 66:828–837

Bo L, Mork S, Kong PA, Nyland H, Pardo CA, Trapp BD (1994) Detection of MHC class II-antigens on macrophages and microglia, but not on astrocytes and endothelia in active multiple sclerosis lesions. J Neuroimmunol 51:135–146

Bo L, Vedeler CA, Nyland H, Trapp BD, Mork SJ (2003a) Intracortical multiple sclerosis lesions are not associated with increased lymphocyte infiltration. Mult Scler 9:323–331

Bo L, Vedeler CA, Nyland HI, Trapp BD, Mork SJ (2003b) Subpial demyelination in the cerebral cortex of multiple sclerosis patients. J Neuropathol Exp Neurol 62:723–732

Bo L, Geurts JJ, van der Valk P, Polman C, Barkhof F (2007) Lack of correlation between cortical demyelination and white matter pathologic changes in multiple sclerosis. Arch Neurol 64: 76–80

Bramow S, Frischer JM, Lassmann H, Koch-Henriksen N, Lucchinetti CF, Sorensen PS, Laursen H (2010) Demyelination versus remyelination in progressive multiple sclerosis. Brain 133:2983–2998

Breij EC, Brink BP, Veerhuis R, van den Berg C, Vloet R, Yan R, Dijkstra CD, van der Valk P, Bo L (2008) Homogeneity of active demyelinating lesions in established multiple sclerosis. Ann Neurol 63:16–25

Brink BP, Veerhuis R, Breij EC, van der Valk P, Dijkstra CD, Bo L (2005) The pathology of multiple sclerosis is location-dependent: no significant complement activation is detected in purely cortical lesions. J Neuropathol Exp Neurol 64:147–155

Brown WJ (1978) The capillaries in acute and subacute multiple sclerosis plaques: a morphometric analysis. Neurology 28:84–92

Bruck W, Porada P, Poser S et al. (1995) Monocyte/macrophage differentiation in early multiple sclerosis lesions. Ann Neurol 38:788–796

Chang A, Tourtellotte WW, Rudick R, Trapp BD (2002) Premyelinating oligodendrocytes in chronic lesions of multiple sclerosis. N Engl J Med 346:165–173

Chang A, Smith MC, Yin X, Fox RJ, Staugaitis SM, Trapp BD (2008) Neurogenesis in the chronic lesions of multiple sclerosis. Brain 131:2366–2375

Chari DM, Zhao C, Kotter MR, Blakemore WF, Franklin RJ (2006) Corticosteroids delay remyelination of experimental demyelination in the rodent central nervous system. J Neurosci Res 83:594–605

Charles P, Hernandez MP, Stankoff B, Aigrot MS, Colin C, Rougon G, Zalc B, Lubetzki C (2000) Negative regulation of central nervous system myelination by polysialylated-neural cell adhesion molecule. Proc Natl Acad Sci USA 97:7585–7590

Charles P, Reynolds R, Seilhean D, Rougon G, Aigrot MS, Niezgoda A, Zalc B, Lubetzki C (2002) Re-expression of PSA-NCAM by demyelinated axons: an inhibitor of remyelination in multiple sclerosis? Brain 125:1972–1979

Confavreux C, Vukusic S (2006) Natural history of multiple sclerosis: a unifying concept. Brain 129:606–616

Craner MJ, Lo AC, Black JA, Waxman SG (2003) Abnormal sodium channel distribution in optic nerve axons in a model of inflammatory demyelination. Brain 126:1552–1561

Davie CA, Hawkins CP, Barker GJ, Brennan A, Tofts PS, Miller DH, McDonald WI (1994) Serial proton magnetic resonance spectroscopy in acute multiple sclerosis lesions. Brain 117(Pt 1): 49–58

Dawson JW (1916) The histology of multiple sclerosis. Trans R Soc Edinb 50:517–740

Dziedzic T, Metz I, Dallenga T, Konig FB, Muller S, Stadelmann C, Bruck W (2010) Wallerian degeneration: a major component of early axonal pathology in multiple sclerosis. Brain Pathol 20:976–985

Esiri MM (1977) Immunoglobulin-containing cells in multiple-sclerosis plaques. Lancet 2:478

Evangelou N, Esiri MM, Smith S, Palace J, Matthews PM (2000a) Quantitative pathological evidence for axonal loss in normal appearing white matter in multiple sclerosis. Ann Neurol 47:391–395

Evangelou N, Konz D, Esiri MM, Smith S, Palace J, Matthews PM (2000b) Regional axonal loss in the corpus callosum correlates with cerebral white matter lesion volume and distribution in multiple sclerosis. Brain 123(Pt 9):1845–1849

Evangelou N, Konz D, Esiri MM, Smith S, Palace J, Matthews PM (2001) Size-selective neuronal changes in the anterior optic pathways suggest a differential susceptibility to injury in multiple sclerosis. Brain 124:1813–1820

Fancy SP, Zhao C, Franklin RJ (2004) Increased expression of Nkx2.2 and Olig2 identifies reactive oligodendrocyte progenitor cells responding to demyelination in the adult CNS. Mol Cell Neurosci 27:247–254

Fancy SP, Baranzini SE, Zhao C, Yuk DI, Irvine KA, Kaing S, Sanai N, Franklin RJ, Rowitch DH (2009) Dysregulation of the Wnt pathway inhibits timely myelination and remyelination in the mammalian CNS. Genes Dev 23:1571–1585

Ferguson B, Matyszak MK, Esiri MM, Perry VH (1997) Axonal damage in acute multiple sclerosis lesions. Brain 120(Pt 3):393–399

Frohman EM, Racke MK, Raine CS (2006) Multiple sclerosis – the plaque and its pathogenesis. N Engl J Med 354:942–955

Gay FW, Drye TJ, Dick GW, Esiri MM (1997) The application of multifactorial cluster analysis in the staging of plaques in early multiple sclerosis. Identification and characterization of the primary demyelinating lesion. Brain 120(Pt 8):1461–1483

Geurts JJ, Bo L, Pouwels PJ, Castelijns JA, Polman CH, Barkhof F (2005) Cortical lesions in multiple sclerosis: combined postmortem MR imaging and histopathology. AJNR Am J Neuroradiol 26:572–577

Geurts JJ, Bo L, Roosendaal SD, Hazes T, Daniels R, Barkhof F, Witter MP, Huitinga I, van der Valk P (2007) Extensive hippocampal demyelination in multiple sclerosis. J Neuropathol Exp Neurol 66:819–827

Gilmore CP, Donaldson I, Bo L, Owens T, Lowe J, Evangelou N (2009) Regional variations in the extent and pattern of grey matter demyelination in multiple sclerosis: a comparison between the cerebral cortex, cerebellar cortex, deep grey matter nuclei and the spinal cord. J Neurol Neurosurg Psychiatry 80:182–187

Gilson J, Blakemore WF (1993) Failure of remyelination in areas of demyelination produced in the spinal cord of old rats. Neuropathol Appl Neurobiol 19:173–181

Goldschmidt T, Antel J, Konig FB, Bruck W, Kuhlmann T (2009) Remyelination capacity of the MS brain decreases with disease chronicity. Neurology 72:1914–1921

Green AJ, McQuaid S, Hauser SL, Allen IV, Lyness R (2010) Ocular pathology in multiple sclerosis: retinal atrophy and inflammation irrespective of disease duration. Brain 133: 1591–1601

Groves AK, Barnett SC, Franklin RJ, Crang AJ, Mayer M, Blakemore WF, Noble M (1993) Repair of demyelinated lesions by transplantation of purified O-2A progenitor cells. Nature 362:453–455

Guo YP, Gao SF (1983) Concentric sclerosis. Clin Exp Neurol 19:67–76

Gutowski NJ, Newcombe J, Cuzner ML (1999) Tenascin-R and C in multiple sclerosis lesions: relevance to extracellular matrix remodelling. Neuropathol Appl Neurobiol 25:207–214

Hoftberger R, Aboul-Enein F, Brueck W, Lucchinetti C, Rodriguez M, Schmidbauer M, Jellinger K, Lassmann H (2004) Expression of major histocompatibility complex class I molecules on the different cell types in multiple sclerosis lesions. Brain Pathol 14:43–50

Hu QD, Ang BT, Karsak M, Hu WP, Cui XY, Duka T, Takeda Y, Chia W, Sankar N, Ng YK, Ling EA, Maciag T, Small D, Trifonova R, Kopan R, Okano H, Nakafuku M, Chiba S, Hirai H, Aster JC, Schachner M, Pallen CJ, Watanabe K, Xiao ZC (2003) F3/contactin acts as a functional ligand for Notch during oligodendrocyte maturation. Cell 115:163–175

Jarjour AA, Kennedy TE (2004) Oligodendrocyte precursors on the move: mechanisms directing migration. Neuroscientist 10:99–105

John GR, Shankar SL, Shafit-Zagardo B, Massimi A, Lee SC, Raine CS, Brosnan CF (2002) Multiple sclerosis: re-expression of a developmental pathway that restricts oligodendrocyte maturation. Nat Med 8:1115–1121

Keirstead HS, Blakemore WF (1997) Identification of post-mitotic oligodendrocytes incapable of remyelination within the demyelinated adult spinal cord. J Neuropathol Exp Neurol 56:1191–1201

Kerschensteiner M, Gallmeier E, Behrens L, Leal VV, Misgeld T, Klinkert WE, Kolbeck R, Hoppe E, Oropeza-Wekerle RL, Bartke I, Stadelmann C, Lassmann H, Wekerle H, Hohlfeld R (1999) Activated human T cells, B cells, and monocytes produce brain-derived neurotrophic factor in vitro and in inflammatory brain lesions: a neuroprotective role of inflammation? J Exp Med 189:865–870

Kooi EJ, Geurts JJ, van Horssen J, Bo L, van der Valk P (2009) Meningeal inflammation is not associated with cortical demyelination in chronic multiple sclerosis. J Neuropathol Exp Neurol 68:1021–1028

Kotter MR, Zhao C, van Rooijen N, Franklin RJ (2005) Macrophage-depletion induced impairment of experimental CNS remyelination is associated with a reduced oligodendrocyte progenitor cell response and altered growth factor expression. Neurobiol Dis 18:166–175

Kuhlmann T, Miron V, Cui Q, Wegner C, Antel J, Bruck W (2008) Differentiation block of oligodendroglial progenitor cells as a cause for remyelination failure in chronic multiple sclerosis. Brain 131:1749–1758

Kutzelnigg A, Lucchinetti CF, Stadelmann C, Bruck W, Rauschka H, Bergmann M, Schmidbauer M, Parisi JE, Lassmann H (2005) Cortical demyelination and diffuse white matter injury in multiple sclerosis. Brain 128:2705–2712

Kutzelnigg A, Faber-Rod JC, Bauer J, Lucchinetti CF, Sorensen PS, Laursen H, Stadelmann C, Bruck W, Rauschka H, Schmidbauer M, Lassmann H (2007) Widespread demyelination in the cerebellar cortex in multiple sclerosis. Brain Pathol 17:38–44

Kwon EE, Prineas JW (1994) Blood-brain barrier abnormalities in longstanding multiple sclerosis lesions. An immunohistochemical study. J Neuropathol Exp Neurol 53:625–636

Lassmann H (1983) Comparative neuropathology of chronic experimental allergic encephalomyelitis and multiple sclerosis. Schriftenr Neurol 25:1–135

Lassmann H, Raine CS, Antel J, Prineas JW (1998) Immunopathology of multiple sclerosis: report on an international meeting held at the Institute of Neurology of the University of Vienna. J Neuroimmunol 86:213–217

Lassmann H, Niedobitek G, Aloisi F, Middeldorp JM (2011) Epstein-Barr virus in the multiple sclerosis brain: a controversial issue – report on a focused workshop held in the Centre for Brain Research of the Medical University of Vienna, Austria. Brain 134:2772–2786

Levine JM, Reynolds R (1999) Activation and proliferation of endogenous oligodendrocyte precursor cells during ethidium bromide-induced demyelination. Exp Neurol 160:333–347

Li WW, Penderis J, Zhao C, Schumacher M, Franklin RJ (2006) Females remyelinate more efficiently than males following demyelination in the aged but not young adult CNS. Exp Neurol 202:250–254

Liu L, Darnall L, Hu T, Choi K, Lane TE, Ransohoff RM (2010) Myelin repair is accelerated by inactivating CXCR2 on nonhematopoietic cells. J Neurosci 30:9074–9083

Lucchinetti C, Bruck W, Parisi J, Scheithauer B, Rodriguez M, Lassmann H (1999) A quantitative analysis of oligodendrocytes in multiple sclerosis lesions. A study of 113 cases. Brain 122 (Pt 12):2279–2295

Lucchinetti C, Bruck W, Parisi J, Scheithauer B, Rodriguez M, Lassmann H (2000) Heterogeneity of multiple sclerosis lesions: implications for the pathogenesis of demyelination. Ann Neurol 47:707–717

Lucchinetti CF, Popescu BF, Bunyan RF, Moll NM, Roemer SF, Lassmann H, Bruck W, Parisi JE, Scheithauer BW, Giannini C, Weigand SD, Mandrekar J, Ransohoff RM (2011) Inflammatory cortical demyelination in early multiple sclerosis. N Engl J Med 365:2188–2197

Magliozzi R, Howell O, Vora A, Serafini B, Nicholas R, Puopolo M, Reynolds R, Aloisi F (2007) Meningeal B-cell follicles in secondary progressive multiple sclerosis associate with early onset of disease and severe cortical pathology. Brain 130:1089–1104

Marburg O (1906) Die sogenannte "akute multiple sklerose" (encephalomyelitis periaxialis scleroticans). J Psychiatr Neurol 27: 217–312

Marin-Husstege M, Muggironi M, Liu A, Casaccia-Bonnefil P (2002) Histone deacetylase activity is necessary for oligodendrocyte lineage progression. J Neurosci 22:10333–10345

Marret S, Delpech B, Delpech A, Asou H, Girard N, Courel MN, Chauzy C, Maingonnat C, Fessard C (1994) Expression and effects of hyaluronan and of the hyaluronan-binding protein hyaluronectin in newborn rat brain glial cell cultures. J Neurochem 62:1285–1295

Moll NM, Rietsch AM, Ransohoff AJ, Cossoy MB, Huang D, Eichler FS, Trapp BD, Ransohoff RM (2008) Cortical demyelination in PML and MS: similarities and differences. Neurology 70:336–343

Nait-Oumesmar B, Picard-Riera N, Kerninon C, Decker L, Seilhean D, Hoglinger GU, Hirsch EC, Reynolds R, Baron-Van Evercooren A (2007) Activation of the subventricular zone in multiple sclerosis: evidence for early glial progenitors. Proc Natl Acad Sci USA 104:4694–4699

Nakahara J, Kanekura K, Nawa M, Aiso S, Suzuki N (2009) Abnormal expression of TIP30 and arrested nucleocytoplasmic transport within oligodendrocyte precursor cells in multiple sclerosis. J Clin Invest 119:169–181

Ozawa K, Suchanek G, Breitschopf H, Bruck W, Budka H, Jellinger K, Lassmann H (1994) Patterns of oligodendroglia pathology in multiple sclerosis. Brain 117(Pt 6):1311–1322

Papadopoulos D, Dukes S, Patel R, Nicholas R, Vora A, Reynolds R (2009) Substantial archaeocortical atrophy and neuronal loss in multiple sclerosis. Brain Pathol 19:238–253

Patani R, Balaratnam M, Vora A, Reynolds R (2007) Remyelination can be extensive in multiple sclerosis despite a long disease course. Neuropathol Appl Neurobiol 33:277–287

Patrikios P, Stadelmann C, Kutzelnigg A, Rauschka H, Schmidbauer M, Laursen H, Sorensen PS, Bruck W, Lucchinetti C, Lassmann H (2006) Remyelination is extensive in a subset of multiple sclerosis patients. Brain 129:3165–3172

Pedre X, Mastronardi F, Bruck W, Lopez-Rodas G, Kuhlmann T, Casaccia P (2011) Changed histone acetylation patterns in normal-appearing white matter and early multiple sclerosis lesions. J Neurosci 31:3435–3445

Perier O, Gregoire A (1965) Electron microscopic features of multiple sclerosis lesions. Brain 88:937–952

Peterson JW, Bo L, Mork S, Chang A, Trapp BD (2001) Transected neurites, apoptotic neurons, and reduced inflammation in cortical multiple sclerosis lesions. Ann Neurol 50:389–400

Piaton G, Aigrot MS, Williams A, Moyon S, Tepavcevic V, Moutkine I, Gras J, Matho KS, Schmitt A, Soellner H, Huber AB, Ravassard P, Lubetzki C (2011) Class 3 semaphorins influence oligodendrocyte precursor recruitment and remyelination in adult central nervous system. Brain 134:1156–1167

Plumb J, McQuaid S, Mirakhur M, Kirk J (2002) Abnormal endothelial tight junctions in active lesions and normal-appearing white matter in multiple sclerosis. Brain Pathol 12:154–169

Pohl HB, Porcheri C, Mueggler T, Bachmann LC, Martino G, Riethmacher D, Franklin RJ, Rudin M, Suter U (2011) Genetically induced adult oligodendrocyte cell death is associated with poor myelin clearance, reduced remyelination, and axonal damage. J Neurosci 31:1069–1080

Prineas JW, Connell F (1978) The fine structure of chronically active multiple sclerosis plaques. Neurology 28:68–75

Prineas JW, Connell F (1979) Remyelination in multiple sclerosis. Ann Neurol 5:22–31

Prineas JW, Graham JS (1981) Multiple sclerosis: capping of surface immunoglobulin G on macrophages engaged in myelin breakdown. Ann Neurol 10:149–158

Prineas JW, Barnard RO, Kwon EE, Sharer LR, Cho ES (1993a) Multiple sclerosis: remyelination of nascent lesions. Ann Neurol 33:137–151

Prineas JW, Barnard RO, Revesz T, Kwon EE, Sharer L, Cho ES (1993b) Multiple sclerosis. Pathology of recurrent lesions. Brain 116(Pt 3):681–693

Raine CS, Wu E (1993) Multiple sclerosis: remyelination in acute lesions. J Neuropathol Exp Neurol 52:199–204

Revesz T, Kidd D, Thompson AJ, Barnard RO, McDonald WI (1994) A comparison of the pathology of primary and secondary progressive multiple sclerosis. Brain 117(Pt 4):759–765

Rodriguez M, Scheithauer B (1994) Ultrastructure of multiple sclerosis. Ultrastruct Pathol 18:3–13

Sanders V, Conrad AJ, Tourtellotte WW (1993) On classification of post-mortem multiple sclerosis plaques for neuroscientists. J Neuroimmunol 46:207–216

Schonrock LM, Kuhlmann T, Adler S, Bitsch A, Bruck W (1998) Identification of glial cell proliferation in early multiple sclerosis lesions. Neuropathol Appl Neurobiol 24:320–330

Seewann A, Vrenken H, Kooi EJ, van der Valk P, Knol DL, Polman CH, Pouwels PJ, Barkhof F, Geurts JJ (2011) Imaging the tip of the iceberg: visualization of cortical lesions in multiple sclerosis. Mult Scler 17:1202–1210

Serafini B, Rosicarelli B, Magliozzi R, Stigliano E, Aloisi F (2004) Detection of ectopic B-cell follicles with germinal centers in the meninges of patients with secondary progressive multiple sclerosis. Brain Pathol 14:164–174

Shen S, Li J, Casaccia-Bonnefil P (2005) Histone modifications affect timing of oligodendrocyte progenitor differentiation in the developing rat brain. J Cell Biol 169:577–589

Shen S, Sandoval J, Swiss VA, Li J, Dupree J, Franklin RJ, Casaccia-Bonnefil P (2008) Age-dependent epigenetic control of differentiation inhibitors is critical for remyelination efficiency. Nat Neurosci 11:1024–1034

Shields SA, Gilson JM, Blakemore WF, Franklin RJ (1999) Remyelination occurs as extensively but more slowly in old rats compared to young rats following gliotoxin-induced CNS demyelination. Glia 28:77–83

Shimizu T, Kagawa T, Wada T, Muroyama Y, Takada S, Ikenaka K (2005) Wnt signaling controls the timing of oligodendrocyte development in the spinal cord. Dev Biol 282:397–410

Sim FJ, Zhao C, Penderis J, Franklin RJ (2002) The age-related decrease in CNS remyelination efficiency is attributable to an impairment of both oligodendrocyte progenitor recruitment and differentiation. J Neurosci 22:2451–2459

Sloane JA, Batt C, Ma Y, Harris ZM, Trapp B, Vartanian T (2010) Hyaluronan blocks oligodendrocyte progenitor maturation and remyelination through TLR2. Proc Natl Acad Sci USA 107:11555–11560

Sobel RA, Chen M, Maeda A, Hinojoza JR (1995) Vitronectin and integrin vitronectin receptor localization in multiple sclerosis lesions. J Neuropathol Exp Neurol 54:202–213

Stadelmann C, Kerschensteiner M, Misgeld T, Bruck W, Hohlfeld R, Lassmann H (2002) BDNF and gp145trkB in multiple sclerosis brain lesions: neuroprotective interactions between immune and neuronal cells? Brain 125:75–85

Syed YA, Hand E, Mobius W, Zhao C, Hofer M, Nave KA, Kotter MR (2011) Inhibition of CNS remyelination by the presence of semaphorin 3A. J Neurosci 31:3719–3728

Tallantyre EC, Bo L, Al-Rawashdeh O, Owens T, Polman CH, Lowe J, Evangelou N (2009) Greater loss of axons in primary progressive multiple sclerosis plaques compared to secondary progressive disease. Brain 132:1190–1199

Targett MP, Sussman J, Scolding N, O'Leary MT, Compston DA, Blakemore WF (1996) Failure to achieve remyelination of demyelinated rat axons following transplantation of glial cells obtained from the adult human brain. Neuropathol Appl Neurobiol 22:199–206

Trapp BD, Peterson J, Ransohoff RM, Rudick R, Mork S, Bo L (1998) Axonal transection in the lesions of multiple sclerosis. N Engl J Med 338:278–285

van der Goes A, Boorsma W, Hoekstra K, Montagne L, de Groot CJ, Dijkstra CD (2005) Determination of the sequential degradation of myelin proteins by macrophages. J Neuroimmunol 161:12–20

van der Valk P, De Groot CJ (2000) Staging of multiple sclerosis (MS) lesions: pathology of the time frame of MS. Neuropathol Appl Neurobiol 26:2–10

van Horssen J, Brink BP, de Vries HE, van der Valk P, Bo L (2007) The blood-brain barrier in cortical multiple sclerosis lesions. J Neuropathol Exp Neurol 66:321–328

Vercellino M, Plano F, Votta B, Mutani R, Giordana MT, Cavalla P (2005) Grey matter pathology in multiple sclerosis. J Neuropathol Exp Neurol 64:1101–1107

Watanabe M, Toyama Y, Nishiyama A (2002) Differentiation of proliferated NG2-positive glial progenitor cells in a remyelinating lesion. J Neurosci Res 69:826–836

Williams A, Piaton G, Aigrot MS, Belhadi A, Theaudin M, Petermann F, Thomas JL, Zalc B, Lubetzki C (2007) Semaphorin 3A and 3F: key players in myelin repair in multiple sclerosis? Brain 130:2554–2565

Wolswijk G (1998a) Chronic stage multiple sclerosis lesions contain a relatively quiescent population of oligodendrocyte precursor cells. J Neurosci 18:601–609

Wolswijk G (1998b) Oligodendrocyte regeneration in the adult rodent CNS and the failure of this process in multiple sclerosis. Prog Brain Res 117:233–247

Wolswijk G (2000) Oligodendrocyte survival, loss and birth in lesions of chronic-stage multiple sclerosis. Brain 123(Pt 1):105–115

Wolswijk G (2002) Oligodendrocyte precursor cells in the demyelinated multiple sclerosis spinal cord. Brain 125:338–349

Ye F, Chen Y, Hoang T, Montgomery RL, Zhao XH, Bu H, Hu T, Taketo MM, van Es JH, Clevers H, Hsieh J, Bassel-Duby R, Olson EN, Lu QR (2009) HDAC1 and HDAC2 regulate oligodendrocyte differentiation by disrupting the beta-catenin-TCF interaction. Nat Neurosci 12:829–838

Young EA, Fowler CD, Kidd GJ, Chang A, Rudick R, Fisher E, Trapp BD (2008) Imaging correlates of decreased axonal Na+/K+ ATPase in chronic multiple sclerosis lesions. Ann Neurol 63:428–435

Zhang SC, Ge B, Duncan ID (1999) Adult brain retains the potential to generate oligodendroglial progenitors with extensive myelination capacity. Proc Natl Acad Sci USA 96:4089–4094

Zhang Y, et al. (2009) Notch1 signaling plays a role in regulating precursor differentiation during CNS remyelination. Proc Natl Acad Sci USA 106:19162–19167

Chapter 3
Microglial Function in MS Pathology

Trevor J. Kilpatrick and Vilija G. Jokubaitis

3.1 Repair in MS

Multiple sclerosis is traditionally defined as an inflammatory demyelinative disease of the central nervous system (CNS), although numerous studies have now demonstrated that myelin loss is associated with axonal damage (Bjartmar et al. 2000; Ferguson et al. 1997; Irvine and Blakemore 2006, 2008; Trapp et al. 1998). It is now also well established that remyelination, the process of reinvestment of myelin sheaths around denuded axons, is often extensive in both acute and chronic active MS plaques (Lassmann et al. 1997; Patani et al. 2007; Prineas et al. 1989, 1993), although it is usually incomplete.

From first principles, it is apparent that in order to optimise repair, it is necessary not only to enhance remyelination but also to inhibit ongoing pathogenesis, as well as to limit axonal degeneration, an injury that is probably, at least in part, contributed to by a deficiency of myelinogenic trophic support. Although it has been convenient to conceptually separate pathogenesis and repair, the two processes are inextricably linked. Indeed, destruction and repair often occur contemporaneously. Discrete zones of demyelination and remyelination are sometimes detected in anatomically discrete but approximate regions, for example, in Balo's concentric rings. The microenvironment of these adjacent regions obviously differs very significantly, raising the

---

T.J. Kilpatrick (✉)
Melbourne Neuroscience Institute, The University of Melbourne, Parkville, VIC, Australia

Multiple Sclerosis Division, Florey Neuroscience Institutes, The University of Melbourne, Parkville, VIC, Australia
e-mail: tkilpat@unimelf.edu.au

V.G. Jokubaitis
Centre for Neuroscience, The University of Melbourne, Parkville, VIC, Australia

Melbourne Brain Centre, Florey Neuroscience Institutes, The University of Melbourne, Parkville, VIC 3010, Australia

question as to how inflammation influences the capacity for repair but also how it can be modified to enhance that capacity.

It is well established that remyelination is the province of cycling oligodendrocyte progenitor cells (OPCs) (Blakemore and Patterson 1978; Irvine and Blakemore 2008) that are present within demyelinated lesions (Scolding et al. 1998; Wolswijk 2002). These cells need to be stimulated to migrate, divide and to differentiate in order to assume their reparative role. In the context of a demyelinative injury, OPCs must therefore be subject to external influences, some of which will be provided by the inflammatory milieu and some by neural cells, whether as a sequel to degeneration or as part of a reactive response. Recent evidence suggests that the nature of this neural cell response, in particular that provided by microglia, is a critical determinant of the extent to which repair occurs. It is, however, also recognised that microglia can potentially subserve a Janus activity in that they have been posited to contribute firstly to pathogenesis, previously thought to be the exclusive province of adaptive immunity, and secondly, to the inhibition of repair. These latter two phenomena are obviously potentially linked, although the former is directed to the mature, myelinating oligodendrocyte as well as to the axons that they ensheath, whereas the latter targets progenitor cells and possibly contributes to a hostile environment for axonogenesis.

3.2 Patterns of Microglial Activation in the Context of Disease Heterogeneity

3.2.1 Pathology

Conflicting views concerning the pathogenesis of MS have emerged and embedded in this debate are disparate perspectives on when and how microglia might exert each of their putative functions to influence disease course. Lassmann et al. identified four distinct patterns of demyelination, where only a single pattern was ever identified in any given individual (Lucchinetti et al. 1996, 2000). The four patterns of demyelination were initially reported to share the common immunopathological features of microglial activation and T-lymphocyte infiltration; however, distinguishing features were reported to include macrophage-mediated myelin destruction in pattern I; activated complement and IgG deposition in pattern II; oligodendroglial apoptosis in pattern III; and extensive loss of oligodendrocytes with lack of remyelination in pattern IV. This classification implies that MS comprises a heterogeneous group of pathologies with a common phenotypic outcome in which microglia are one of the effector cells responsible for disease pathogenesis (Fig. 3.1).

An alternative view has emerged from the examination of very early MS lesions, in some cases where death occurred within 24 h of symptom onset. Barnett and Prineas (2004) described that these very early lesions exhibited extensive oligodendroglial apoptosis together with microglial activation in the absence of lymphocytic

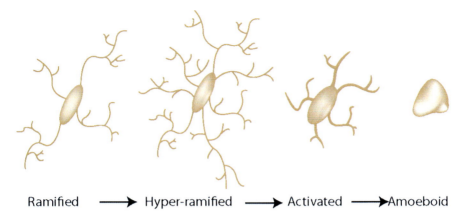

Fig. 3.1 At rest, microglia display a ramified morphology with a thin rod-like cell body. Upon activation, microglia transition to an intermediate hyper-ramified morphology, followed by an activated morphology with a hypertrophied cell body and shortened/thickened cell process. Microglia can also assume an amoeboid (macrophage-like) morphology when engaged in phagocytosis

infiltration (Barnett and Prineas 2004). These observations challenged the findings of Lassmann et al., arguing that the pathological variability described in these prior studies reflected different stages of lesion development rather than distinct pathologies, a controversy that remains essentially unresolved. These two studies also proposed differing roles for microglia in MS, with Lassmann et al. emphasising the commonality/central role of the adaptive immune response, whereas Barnett and Prineas emphasised that microglia might play a fundamental role at the earliest stage of disease activity, prior to widespread T cell infiltration. Another important feature that Barnett and Prineas identified was of oligodendroglial apoptosis, suggesting that activation of the innate immune system could be reactive, although it is unclear as to what role such activated cells might have in subsequent disease pathogenesis and, in particular, the activation of a subsequent adaptive immune response.

Further analysis from the Lassmann group has provided a clarification of their classification, which serves to accommodate some of the findings of Prineas and Barnett in relation to microglial activity, at least for so-called pattern III. The Lassmann group has chosen to interpret that early microglial activation in this pattern reflects a critical role in lesion pathogenesis, which could serve as the impetus for subsequent activation of the adaptive immune response and frank demyelination. In contrast, it has been posited by others that, in some patients, microglial activation occurs relatively late in the disease course (Weiner 2008). This activation was hypothesised to correlate with conversion from relapsing–remitting disease, driven by a peripherally invoked adaptive immune response, and which is responsive to immunomodulation, to a phase of progressive disability driven by innate immunity behind an intact blood–brain barrier, which is refractory to systemic immunotherapy.

3.2.2 Distribution of Microglia within MS Lesions and Putative Functional Roles

Detailed analysis of the distribution of microglia within extant lesions has been undertaken and provides some basis for speculation concerning their potential roles in disease pathogenesis and in the remyelinative process. Macrophages/microglia are found in large numbers in acute and chronic active lesions where, on average, they outnumber lymphocytes by 10–20 times. Microglia are also found in lower numbers in inactive, demyelinated and remyelinated lesions (Barnett et al. 2006; Brück et al. 1995).

Analysis of human tissue can only provide a snapshot of histopathology and therefore provides no direct insights into pathogenesis. Nevertheless, some intimate anatomical associations between microglia and other neural cells have been identified, suggesting that microglia might be involved in both the disruption of axons and of myelin lamellae. Such macrophages have been shown to exhibit IgG capping at the site of attachment of myelin (Prineas and Graham 1981), suggesting that immune-ligand-mediated phagocytosis could be occurring. In addition, microglial processes positive for MHC class II within active MS lesions have been shown to interact with myelin internodes, suggesting that microglia are actively involved in demyelination (Bö et al. 1994; Trapp et al. 1999). Furthermore, microglia have also been shown to interact with oligodendrocyte cell bodies, where, in one particular study, 17 % of all microglia within actively demyelinating lesion boundaries were apposed to or surrounded by oligodendroglial somata (Peterson et al. 2002). Human MS lesions and experimentally generated demyelinated plaques also contain phagocytically active microglia/macrophages that can be identified by staining for myelin degradation products or by ultrastructural examination of cell contents (Barnett and Prineas 2004; Bö et al. 1994; Brück et al. 1995; Prineas and Graham 1981).

It has been harder to obtain compelling evidence from the analysis of human tissue that microglia are actively involved in the myelin repair process. Nevertheless, histological analysis has demonstrated that myelin regeneration can occur within 2 days of the onset of MS symptoms and this correlates with the presence of oligodendrocyte progenitor cells within lesions that also contain phagocytically active microglia (Barnett and Prineas 2004). A close spatial relationship between OPCs and microglia has also been demonstrated in MS lesions (Nishiyama et al. 1997), and a recent post-mortem study of tissue from a patient with long disease course showed a high density of HLA-DR+ macrophages/microglia was positively correlated with remyelination (Patani et al. 2007). Collectively, these studies provide good circumstantial evidence that microglia have a role to play not only in pathogenesis but also in repair.

However, it remains uncertain as to whether individual microglial cells can be induced to transform from a pathogenic to a reparative phenotype or whether such transformation depends on the stimulation/activation of a separate cohort of cells.

3.3 Microglial Heterogeneity: Disparate Functions and Targets?

The question of pathological heterogeneity is pertinent not only to whether there are variable patterns between individuals but also as to whether microglial phenotype varies according to region, disease duration and/or age.

3.3.1 Developmental Origins

It is generally believed that CNS microglia are monocytic cells of mesodermal origin deriving from the yolk sac (Alliot et al. 1999; Hirasawa et al. 2005). An elegant study by Hirasawa et al. employed transgenic mice expressing a fluorescent marker enhanced green fluorescent protein (eGFP) under the control of the macrophage/microglial promoter for ionised calcium-binding adaptor molecule 1 (Iba1) to track microglial origin and development. Invasion of cells of monocytic origin into the CNS was first observed at embryonic day 10.5 (E10.5) in Iba1-eGFP mice. Microglia were shown to enter the brain parenchyma via the meninges, having been carried through the circulation from the yolk sac. A strong eGFP signal in the forebrain, spinal cord and eye was present at E11.5 and persisted into adulthood. A second, smaller microglial invasion occurred at postnatal day 6 (P6) in the supraventricular corpus callosum and cingulum. The signal from these cells diminished over the course of the second postnatal week, presumably due to migration of these cells to other brain regions. Amoeboid microglia migrate throughout the parenchyma and differentiate into a resting, ramified morphology once they have reached their final position (Cuadros and Navascués 1998; Hirasawa et al. 2005; Marín-Teva et al. 1998).

3.3.2 Regional Variation

It is also important to note that in addition to parenchymal microglia, there is a population of cells that is also resident adjacent to the CNS vasculature, known as perivascular macrophages. These cells express many of the same cell surface antigens as parenchymal microglia and the two cell types are therefore difficult to distinguish in pathological conditions. Like microglia, perivascular macrophages are bone marrow-derived and provide an interface between the relatively immune-privileged CNS environment and the circulation. These cells can interact with CNS-derived antigens and present them to circulating T-lymphocytes (Streit et al. 1999). It is interesting to note that it has been suggested that microglia and macrophages differ in their intrinsic capacity to phagocytose myelin and in the extent to which this is influenced by opsonisation (Mildner et al. 2007; Mosley and Cuzner 1996). The situation is further complicated in circumstances where monocytes invade the parenchyma. These observations raise the question as to whether these various cells have intrinsic differences in their capacity to either assist or impede the remyelinative process.

3.3.3 Sources in Disease

Initially, it had been suggested that Mac-1+/CD45hi peripheral macrophages might contribute significantly to the repertoire of inflammatory cells identified within zones of demyelination. This had been assessed by McMahon et al. by lethally irradiating mice, followed by rescue with transplanted bone marrow from GFP+ transgenic mice and then exposing the animals to cuprizone. In these animals, GFP+ cells were identified within the CNS after 2 weeks of cuprizone challenge, although they were significantly outnumbered by parenchymal microglia, such that at 3 weeks they accounted for only some 8 % of the inflammatory cells within the corpus callosum. It was reported, on the basis of HRP staining, that the blood–brain barrier was maintained in these irradiated animals: subtle deficits may not, however, have been identified using this technique. In fact, recent work using chimeric mice failed to identify any evidence of microglial progenitor recruitment from the circulation in CNS neurodegenerative disease, at least when the blood–brain barrier is intact, suggesting that microgliosis is induced by local expansion in these models (Ajami et al. 2007; Mildner et al. 2007).

3.4 Interrogating Functional Roles for Microglia in Health and Disease

3.4.1 Surveillance

A transgenic mouse strain expressing eGFP behind the promoter region for the microglia/macrophage-specific chemokine receptor CX3CR1 (fractalkine receptor) has further enriched our understanding of microglial phenotype in both a normal physiological context and in response to acute injury (Davalos et al. 2005; Nimmerjahn et al. 2005). In particular, transcranial two-photon microscopy has allowed the visualisation of superficial microglia within the first 200 μm of the cortical surface. It was revealed that microglial cell bodies are evenly distributed throughout the cortex, with cell somata lying 50–60 μm apart. Microglial cell bodies were small and generally rod shaped with numerous branched processes extending from the soma (Nimmerjahn et al. 2005). Imaging for up to 10 h revealed that there was very little somatic translocation in the quiescent state. In contrast, microglial processes were shown to be highly motile with extensive process extensions and retractions cycling in an apparently random manner with speeds of up to 4.1 μm/min. Interestingly, when neighbouring microglial cell processes came into contact, they would repel each other, thereby indicating that each cell has a defined region of surveillance. The average protrusion lifespan was about 4 min. Bulbous microglial processes were observed actively engulfing tissue detritus during routine surveillance. These observations lead the investigators to conclude that, at rest, the extracellular microenvironment is completely surveyed once every few hours (Nimmerjahn et al. 2005).

Work by Nimmerjahn et al. has identified that microglia responded immediately to a laser-induced disruption of the blood–brain barrier (BBB). Cells within 90 μm of the injury site were shown to extend processes towards the site of damage and to retract processes oriented in the opposite direction. Whilst process morphology did not change, the cell soma was shown to hypertrophy. Interestingly, the average speed of process extension was the same in the response to injury as it was in the quiescent state at rest (Nimmerjahn et al. 2005).

Using these same transgenic mice, Davalos et al. (2005) independently confirmed the findings of the Nimmerjahn study. In addition, it was found that a single microglial cell could respond to two acute injuries within very short distances of each other by extending hypertrophied branches to both sites without somatic translocation. In the Davalos study, it was shown that processes which extended to an injury site could fuse together to encapsulate the damaged tissue. It was further shown that injury-mediated branch motility was regulated by extracellular adenosine triphosphate (ATP) released by astrocytes in damaged tissue, acting via microglial purinergic P2 receptor.

The above two studies have revolutionised our understanding of microglial function in both normal physiological and acute pathological states, revealing that these cells are very quick to respond to injury, and indicating that, even at rest, microglia are highly active cells responsible for maintenance of parenchymal homeostasis.

3.4.2 Experimental Models of Demyelination

Independent of the interpretations concerning microglial function that can be gleaned from cross-sectional studies of human pathological samples, it is generally agreed that direct analysis of their function in the context of demyelination requires the use of either animal models or detailed in vitro study.

There are several experimental tools that are used in a laboratory context to study demyelination and remyelination. These can be broadly divided into two groups, immune-mediated models of demyelination and toxin-mediated models. The most common immune-mediated model of demyelination, rodent experimental autoimmune encephalomyelitis (EAE), is induced by the active sensitisation of CNS tissue to myelin antigens or by the passive transfer of auto-reactive T-lymphocytes. This leads to a stereotyped, ascending paralysis correlated with microglial activation, mononuclear cell infiltration into the CNS parenchyma and the destruction of myelin, oligodendrocytes and axons. This model can be used to study factors that influence disease pathogenesis, but not disease aetiology. As in MS, the animal models of inflammatory demyelination are complex, and demyelination and remyelination are often contemporaneous, making it difficult, in isolation, to formally assess a role for microglia in the repair process.

In addition to EAE, three toxin-mediated models are routinely used to study demyelination and remyelination. On the one hand, the copper chelator cuprizone (bis(cyclohexanone)oxaldihydrazone), when delivered to mice in their feed, induces demyelination of the midline corpus callosum and, to a lesser extent, the superior cerebellar peduncles (Blakemore 1973). On the other hand, the toxins lysolecithin and ethidium bromide can be injected focally into the white matter of the central nervous system (CNS) of experimental animals, producing demyelination, oligodendrocyte death and a variable astrocyte response, but with preservation of axons. Demyelination in these models is followed by stereotyped and extensive remyelination of over 90 % of affected axons in the lesions. Toxin-induced models are particularly useful models for investigation of the role of microglia/macrophages in repair since their presence in the lesion is in response to injury without them being involved in the primary injury, although they could still be involved in secondary pathogenic events.

3.5 Microglial Activities

The spectrum of microglial activity in response to injury has been previously posited to comprise two broad components: inflammation and immunomodulation. The inflammatory response is engaged to destroy damaged, infected or invading cells, which may incur bystander damage. Immunomodulation, on the other hand, promotes cell survival and tissue repair. In general, an inflammatory response resolves once tissue is repaired (Muzio et al. 2007). It is also important to note that other important functions for microglia have also been identified, in particular phagocytic activity. Inhibition of microglial activity can, therefore, be either beneficial or deleterious, depending on the circumstance.

A variety of molecular determinants of microglial activation have been identified. A key issue will be to identify how the relevant signalling pathways act, either in isolation or in a coordinated way, to effect the upregulation of a set of cell surface antigens leading to a specific phenotypic response, whether it be inflammatory, immunomodulatory or phagocytic or a combination thereof.

A variety of receptors expressed by microglia have been identified that, when activated, can induce the migration of these cells in response to ligands within the extracellular milieu. These include the chemokine receptors CCR5 (Carbonell et al. 2005) and CXCR3 (Rappert et al. 2004), the purinergic receptors P2Y12/13 which act in a beta-1 integrin-dependent manner (Nasu-Tada et al. 2005), the B-1 bradykinin receptor (Huisman et al. 2008), the neurotensin receptor-3 (Martin et al. 2003), the vascular endothelial growth factor receptor (Forstreuter et al. 2002), the pattern recognition receptor CD36 (Stuart et al. 2007), the neurotrophin receptor TrkA (De Simone et al. 2007) and the IL-6 superfamily receptor complexes (Sugiura et al. 2000).

3.5.1 Phagocytic Activity

Microglia have long been recognised to be phagocytic: teleologically, this activity has evolved in order to remove pathogens as part of a pro-inflammatory response. This is known to be mediated via activation of the toll-like receptors (TLRs). However, as indicated previously, in the context of autoimmune disease, microglial activity can be directed to the phagocytosis of a variety of endogenous structures including myelin lamellae, axons, apoptotic cells and myelin debris, resulting in a range of influences upon either disease pathogenesis or repair.

Several studies have used toxin-induced models of demyelination in the rat to study the role of microglial phagocytosis. Phagocytically active macrophages can be recognised in tissue sections by the presence of myelin degradation products and lysosomal lipids within their cytoplasm. It is important that such studies dissect out the influence that phagocytosis is having upon disease severity in that particular circumstance and to identify, if possible, the physical structures that are being targeted. As a corollary, it is necessary to understand the molecular mechanisms by which phagocytosis of the various targets is expedited if focused strategies to modify phagocytosis in a targeted way are to be developed for the treatment of human disease.

3.5.1.1 Targets of Phagocytic Activity

Myelin

In experimental models of MS, myelin lamellae have been observed to be located in coated pits on the macrophage surface. Coated pits that express the protein clathrin are known to be the hallmark of receptor-mediated phagocytosis. Myelin opsonised with anti-myelin antibodies is also more readily phagocytosed (Trotter et al. 1986).

Axons

Both experimental models of MS and in vitro data suggest that microglia can be directly responsible for axonal transection (Fordyce et al. 2005). Additionally, microglia in vitro have been shown to lethally injure neurons in culture via a process that requires cell proximity and which is potentially mediated via nitric oxide (Gibbons and Dragunow 2006) and also which could be potentiated by stimulating Kv1.3 channel activity (Fordyce et al. 2005).

Apoptotic Cells

Until recently, only phosphatidylserine (PS) had been identified as a recognition ligand on apoptotic cells. The PS molecule is known to be relocated from the internal to the external surface of the plasma membrane of apoptotic cells and has now been identified to be recognised by either Tim4 (Miyanishi et al. 2007) or by the secreted protein, growth arrest-specific gene 6 (Gas6). In turn, Gas6 has the capacity to bind to the TAM (Tyro 3, Axl and Mer) family of receptor tyrosine kinase receptors on microglia and through a modified type II phagocytic response, acting via Rac and Vav in a microtubule-dependent way, can induce phagocytosis (Grommes et al. 2008). It is also important to note that Gas6 not only induces phagocytosis but can also induce an anti-inflammatory response by microglia (Binder et al. 2008; Grommes et al. 2008) (see below).

Recently, other important molecular mediators of apoptotic phagocytosis and of tissue homeostasis have been identified. First, it has been demonstrated that calreticulin (an obligate endoplasmic reticulum protein that has been shown to be expressed on the cell surface during apoptosis) serves as a second general recognition ligand by binding and acting upon the LDL-receptor-related protein on the engulfing cell. Second, the triggering receptor expressed on microglial cells-2 (TREM2) has been identified as an innate immunoreceptor which acts to induce phagocytosis via phosphorylation of DNAX-activating protein of 12bDa (DAP12) and by inducing cytoskeletal reorganisation. Interestingly, knockdown of TREM2 in microglia has been shown not only to inhibit the phagocytosis of neural cells but also to potentiate a pro-inflammatory response, with upregulated expression of the genes encoding for tumour necrosis factor-alpha (TNFα) and nitric oxide (NO) synthase. A beneficial effect of TREM2 has recently been identified in EAE, given that mice given myeloid cells which overexpressed TREM2 exhibited less severe clinical disease as well as reduced axonal damage and demyelination but with increased lysosomal and phagocytic activity (Takahashi et al. 2007). On the other hand, antibody-mediated blockade of TREM2 during the effector phase of the disease resulted in exacerbated disease (Piccio et al. 2007). The TREM2/DAP12 complex has also been shown to be important in man such that either TREM2 or DAP12 deficiency results in an inflammatory neurodegenerative disease in the 4th/5th decade of life (Bianchin et al. 2006). The ligand responsible for activating this system is yet to be identified (Takahashi et al. 2005).

Myelin Debris

Macrophage depletion in young rats has been shown to result in impairment of remyelination, in part due to delayed phagocytosis of myelin debris (Kotter et al. 2005). To examine this issue further, either myelin debris, liver cell membranes or PBS were added to ethidium bromide-induced demyelinating lesions in the rat, and the capacity to remyelinate was assessed. The authors found that whilst non-CNS debris inhibited efficient remyelination relative to PBS-treated controls, myelin debris had an even greater inhibitory effect on myelin regeneration, and this effect was found to be independent of OPC and macrophage recruitment. The authors demonstrated that the remyelination block was at the level of oligodendrocyte maturation: oligodendrocytes were able to make contact with axons and to ensheath them but were unable to compact the myelin (Kotter et al. 2006).

In summary, microglia appear to play a critical phagocytic role that leads to the promotion of oligodendrocyte maturation and remyelination, possibly by removing a negative regulator. A candidate negative regulator would be the leucine-rich repeat and Ig domain-containing Nogo receptor interacting protein (LINGO), which has been shown to be expressed by oligodendrocytes (Mi et al. 2005). It is also possible, although untested, that impaired axonal regeneration could be contributed to by uncleared myelin debris (Vargas and Barres 2007).

The molecular mechanisms that regulate the capacity of microglia to remove myelin debris are beginning to be determined. Firstly, leukaemia inhibitory factor (LIF), a neuropoietic cytokine already known to be induced in animal models of central demyelination, has been shown to stimulate myelin uptake by macrophages to enhance myelin clearance. This effect is accompanied by activation of the JAK/STAT signalling pathway (Hendriks et al. 2008). Both the complement component receptor CR3 (Mac1) (Rzepecka et al. 2009) and the scavenger receptor—AI/II interact with calreticulin and have been implicated in mediating the phagocytosis of myelin by microglia and macrophages (Reichert and Rotshenker 2003), a process that appears dependent upon co-expression of MAC-2 (Rotshenker et al. 2008). Recently, it has also been identified that AI/II signals via the TAM receptor family member, Mer, during apoptotic cell uptake by murine macrophages (Todt et al. 2008). In addition, Axl, another TAM family member, appears to play a role in myelin debris clearance. In Axl null mice subjected to cuprizone-induced demyelination, there is not only a reduction in numbers of microglia but also a reduction in their phagocytic activity. This is associated with both a subsequent delay in the development of mature oligodendrocytes and increased axonal damage in the remyelinative phase (Hoehn et al. 2008).

3.5.1.2 Remodelling

An additional activity ascribed to microglia is that of synaptic stripping. Trapp et al. have demonstrated that microglia activated in response to an inflammatory stimulus can remove up to 45 % of axosomatic connections within the cortex, a process that occurs independent of phagocytosis and which has been deemed to be neuroprotective. Many of the cortical neurons that were stripped of their synapses were of GABAergic phenotype. It has been proposed that loss of inhibitory GABAergic inputs leads to an increased capacity for NMDA receptor activation which stimulates anti-apoptotic neuroprotective signalling pathways regulated by activation of cAMP-response-element-binding (CREB) protein (Trapp et al. 2007).

It has been identified that the C3R on microglia could be an important component of this response, given that both the complement components C1q and C3 are expressed by neural cells, including neuronal synapses and given that mice lacking C1q and C3 are deficient in synaptic pruning (Stevens et al. 2007). Interestingly, TNFα has been reported to inhibit phagocytosis mediated by C3 (CR3) by downregulating CR3 levels (Bruck et al. 1992). The major histocompatibility antigen class I antigen could also have a role, given that it has been shown to be implicated in mediating synapse removal after motor neuron injury (Cullheim and Thams 2007).

3.5.2 Inflammatory Activity

3.5.2.1 Molecular Determinants of Microglial Activation

Factors such as interferon-gamma (IFNγ) produced by key participants in the adaptive immune response (including NK cells, CD4+Th1 cells and CD8+T cells) appear to be key determinants of microglial activation. In fact, it has been hypothesised that a positive feedback loop might be operative, leading to enhanced activation of both microglia and dendritic cells with time, potentially explaining why systemically applied immunomodulatory therapy that works predominantly in the periphery eventually becomes ineffective. Microglial cells have been shown to exhibit increased secretion of proteolytic enzymes, potentiated oxidative stress and enhanced phagocytic activity in response to IFNγ. Microarray analyses have also revealed increases in expression of genes encoding chemokines and MHC proteins by activated microglia (Dheen et al. 2007).

A variety of other environmental factors have been implicated in potentiating a pathogenic microglial response. These include members of the TLR family which are activated via exposure to damage-associated molecular patterns (DAMPs), although they can also have roles in inducing tolerance (Fischer and Ehlers 2008). Macrophage colony stimulating factor also appears to be a significant positive regulator of microglial activity, given that its overexpression induces microglial proliferation and pro-inflammatory cytokine expression (Mitrasinovic et al. 2001).

3.5.2.2 Consequences of Pro-inflammatory Activity in Central Demyelination

A seminal study in clarifying the role of microglia in demyelination made use of a transgenic mouse strain where the thymidine kinase of herpes simplex virus (HSVTK) was inserted under the regulatory control of the alpha chain of CR3, namely, CD11b (Heppner et al. 2005). In this context, inducible suppression of microglia by activation of HSVTK ameliorated EAE disease severity. Furthermore, microglial paralysis in ex vivo slice cultures led to the suppression of reactive oxygen species, pro-inflammatory cytokine and chemokine release, implicating these cells in disease pathogenesis and again advocating a pathogenic role for microglia in demyelination, at least in the effector phase of the disease (Heppner et al. 2005).

3.5.2.3 Molecular Basis for the Pathogenic Effect

Microglia can potentiate inflammation and contribute to the pathogenic process in a variety of ways. First, they serve as antigen-presenting cells and do so via MHC class II molecules which present processed antigens to CD4-positive T-lymphocytes. The MHC class II molecule is upregulated by activated microglia in pathogenic states (Kim and de Vellis 2005), and this serves to amplify the immune response via the process of epitope spreading.

Microglia also have the capacity to secrete a wide variety of signalling molecules including inflammatory cytokines such as TNFα, interleukin-1 (IL-1) as well as leukotrienes, complement components, reactive oxygen intermediates and proteolytic and lipolytic enzymes. These molecules can exert harmful effects by acting either directly or indirectly to induce neural cell damage (Kim and de Vellis 2005; Neumann et al. 2009), although some can also have beneficial effects, depending upon the context in which they are expressed (see below).

The free radical nitric oxide (NO) has been directly implicated in inducing oligodendrocyte death (Gibbons and Dragunow 2006; Merrill et al. 1993) and in promoting MS pathogenesis (Parkinson et al. 1997). Microglia that produce nitric oxide and inducible nitric oxide synthase (iNOS, the enzyme responsible for NO synthesis) have been found in active lesions but not in either chronic MS lesions or normal tissue, supporting the notion that microglia-derived NO is, in part, responsible for disease pathogenesis (Bagasra et al. 1995; De Groot et al. 1997; Liu et al. 2001; Oleszak et al. 1998).

3.5.3 Immunomodulation

Microglia can secrete anti-inflammatory cytokines that could either support damaged cells during a demyelinative insult or serve to enhance repair. Recent work has validated that microglia have an essential role to play in the immunomodulatory effects exerted by β-interferon, a cytokine used for therapeutic benefit in MS. This was assessed by studying the influence of either generalised or cell lineage-specific deletion of the interferon type 1 receptor (IFNAR), which is responsible for mediating β-interferon-induced signalling, in mice subjected to EAE (Prinz et al. 2008). Mice with a generalised deletion of IFNAR performed significantly worse than their wild-type counterparts, with increased myelin destruction, corresponding to increased macrophage invasion of the CNS parenchyma. Interestingly, conditional deletion of IFNAR from myeloid cells (in particular from CD11b-positive microglia, monocytes and granulocytes, but not from either T- or B-lymphocytes) aggravated disease severity. These experiments thus established that CD11b-positve macrophages are the target of endogenously produced interferon: by extrapolation, it is likely that β-interferon as a therapeutic agent exerts a significant component of its effects via these cells (Prinz et al. 2008).

Recent data also indicate that the conversion of macrophages to an immunomodulatory phenotype involves not only IFNAR in isolation but also other receptors, in particular the TAM subfamily of receptor tyrosine kinases. This is potentially of direct relevance to the pathogenesis of MS, as the TAMs and their cognate ligand Gas6 have recently been identified as having a protective effect during the demyelinative stage of cuprizone-induced demyelination (Binder et al. 2008). Over the course of a 6-week insult, expression of Gas6, Axl and Mer mirrored microglial activation and infiltration within the corpus callosum. Gas6 knockout mice displayed a significantly greater degree of demyelination, which was not only commensurate with loss of mature oligodendroglia but also an increase of microglia/macrophage numbers (Binder et al. 2008). Complementary in vitro studies established that exogenous Gas6 both potentiated oligodendrocyte survival and modulated the expression of pro-inflammatory cytokines by lipopolysaccharide (LPS)-stimulated microglia.

The IL-6 family of cytokines can also regulate microglial activity. All members of this family, including LIF, CNTF and IL-6, signal through a common gp130 receptor subunit, with specificity of ligand binding driven by an array of alternative receptor subunits which associate with gp130 to form heterologous receptor complexes. Hendriks et al. (2008) have reported that LIF can inhibit the production of oxygen radicals and TNFα by macrophages (Hendriks et al. 2008). Kerr and Patterson have also reported that LIF can induce microglia to produce IGF-1, another trophic factor that is capable of inducing oligodendrocyte survival (Gveric et al. 1999; Hinks and Franklin 1999; Kerr and Patterson 2005) (see Sect. 3.5.4). Interestingly, it has been shown that partial phosphorylation of the transcription factor, STAT-3, by CNTF in microglia results in enhanced survival of neurons in mixed cultures, whereas application of IL-6, which leads to more robust phosphorylation of STAT-3, results in potentiated neuronal loss. This suggests that gp130-mediated signalling can contribute to either pathogenesis or repair, depending on circumstance, predicated on the availability of particular ligands within the family and upon the expression of key co-receptors such as the non-signal transducing GPI-linked CNTF receptor-alpha, which is required for CNTF binding.

Other cytokines could also exert anti-inflammatory roles: for example, IL-10, which can also activate STAT-3, inhibits the production of TNFα and nitric oxide by macrophages.

It has also been reported that retinoic acid (RA), a vitamin A metabolite, suppresses the activation of microglial cells and inhibits the expression of TNFα and NO (Diab et al. 2004; Xu and Drew 2006). These effects are mediated through two subfamilies of nuclear receptors, the retinoic acid receptors (RARs) and the retinoid X receptors (RXRs). It has also been shown that RA enhances the expression of the gene encoding transforming growth factor-β1 (TGFβ$_1$), a factor which, by increasing IκB expression, can inhibit nuclear translocation of NFκB, a key transcription factor implicated in the pathogenic effect of microglia (Dheen et al. 2005).

3.5.4 Repair

It is not only possible to modify the inflammatory response for beneficial effect, but the activity of the so-called pro-inflammatory cytokines can be redirected.

These redirected activities include the killing of pathogenic cells, enhancement of the recruitment and maturation of OPCs, and the provision of neuroprotection in order to enhance repair.

Pathogenic events in both MS and in the animal model EAE are self-limiting, and microglia could exert an important role in this regulatory activity. During an EAE-induced demyelinating insult, NO production by microglia/macrophages has been shown to limit the expansion of auto-reactive T-lymphocytes and to aid in EAE resolution (Juedes and Ruddle 2001). Additionally, auto-reactive T cells undergo apoptosis via an activation-induced mechanism (Pender 1999; Pender et al. 1991). These apoptotic cells are then subject to microglial/macrophage phagocytosis (Chan et al. 2003) that may be IFNγ dependent (Chan et al. 2001).

There is also a cornucopia of evidence from both human pathological studies and experimental models that remyelination often occurs in the presence of inflammation (Morell et al. 1998; Raine and Wu 1993; Totoiu et al. 2004; Wolswijk 2002). A more limited body of work suggests that the two processes are actually linked (Foote and Blakemore 2005) such that some aspects of the inflammatory response appear to be required in order to enhance remyelination. Interesting data have emerged from the study of the *taiep* rat, a myelin mutant that over the course of 1 year progressively loses myelin, eventually displaying features of chronic demyelination in the presence of astrogliosis, but in the absence of acute inflammation. This model therefore enables dedicated analysis of the factors that influence remyelination. This was achieved by inducing the death of endogenous OPCs by exposure to X-irradiation, after which the rats were subjected to transplantation of myelinogenic OPCs. It was shown that axons were still receptive to remyelination, with extensive remyelination occurring at the site of transplantation, but with limited repair distal to the transplantation site. Furthermore, remyelination was highly correlated with acute inflammation and microglial activation at the injection site. Once again, these observations speak to the importance of identifying the molecular players responsible for this selective response.

Kotter et al. attempted to interrogate the role that the innate immune system might play in expediting remyelination in the lysolecithin model by delivering liposomes containing clodronate (dichloromethylene biphosphonate), which has been reported to deplete macrophage subpopulations. The intervention resulted in a significant reduction in oligodendrocyte-induced myelination, probably reflecting inhibited OPC recruitment (Kotter et al. 2001). A reduction in the gene encoding the macrophage scavenger receptor-B was identified, but it was concluded that the effect was independent of the degree of extracellular myelin debris clearance/phagocytosis (Kotter et al. 2005). Interestingly, macrophage depletion induced changes in the mRNA expression of IGF-1 and TGFα but not platelet-derived growth factor-A (PDGF-A) or fibroblast growth factor-2 (FGF-2) which, together with EGF and NT-3, are factors known to influence oligodendrocyte development and which have previously been identified to be secreted by macrophages in vitro.

A role for TLR-mediated inflammation in the induction of remyelination induced by either transplanted or endogenous progenitor cells has been posited. This has been studied by administering the toll-like receptor-2 (TLR-2) ligand zymosan to retinae that have been seeded with transplanted OPCs, allowing for interrogation of the influence of OPC-induced myelin without the confounder of myelin debris. In the absence of zymosan, the investigators found that transplanted OPCs could myelinate unmyelinated axons within the nerve fibre layer. However, the addition of zymosan resulted in significantly increased myelination relative to controls, and this was highly correlated both with reactive gliosis and with the recruitment of microglia/macrophages into the retina. The authors further found that increased myelination in zymosan-treated retinae was associated with a 44-fold increase in $TGF\beta_1$ expression and a 2.4-fold increase in IL-1β expression (Setzu et al. 2006). Similarly, the addition of LPS into the corpus callosum, which signals through TLR-4, was shown to enhance the recruitment of endogenous OPCs into sites of demyelination and to enhance expression of the myelin protein, proteolipid protein in ethidium bromide-induced demyelinated lesions (Glezer et al. 2006). It is important to note, however, that this effect is contextual, given that LPS injected into immature rodents can induce oligodendrocyte loss and hypomyelination (Lehnardt et al. 2002) and that LPS injected into the spinal cord can cause demyelination (Felts et al. 2005).

Recent experimental studies, predominantly in cuprizone-induced demyelination, have also unveiled a reparative role for a number of molecules including TNFα, iNOS, MHCII and osteopontin, which had previously been best known for their roles in disease pathogenesis (Arnett et al. 2001, 2003; Linares et al. 2006; Selvaraju et al. 2004). These factors have now been shown to influence repair by promoting the recruitment and/or maturation of oligodendroglia. In some instances, they could also provide a neuroprotective effect to expedite a viable axonal substrate upon which remyelination can proceed.

Even IL-12, a pro-inflammatory cytokine, has been reported to exert beneficial effects after spinal cord injury (Yaguchi et al. 2008), including enhanced proliferation of OPC around the injured site, as well as enhanced remyelination. In the context of spinal cord injury, exogenous IL-12 administered on the day of trauma was reported to increase the numbers of microglia and dendritic cells, as well as levels of the neurotrophin, brain-derived neurotrophic factor (BDNF). An endogenous role for IL-12 of this nature remains unexplored, although microglia are known to produce both the IL-12 cytokine and its receptor, suggesting the possibility of an autocrine loop.

The TNFα has been shown to promote the proliferation of OPCs and remyelination in cuprizone-mediated demyelination (Arnett et al. 2001). Mice in which TNFα was deleted showed impaired remyelination and a reduction of cycling OPCs that are normally responsible for this repair process. The TNFα is known to signal via two receptors, the TNFα receptor 1 (TNFR1) and the TNFα receptor 2 (TNFR2/p75), and so the authors sought to identify which receptor was responsible for these effects. Conditional deletion of TNFR1 did not have an effect on either demyelination or remyelination relative to controls; however, TNFR2 deletion showed the same inhibition of OPC proliferation and remyelination as the TNFα knockout animals, illustrating that proliferative and protective effects of TNFα signal through the TNFR2, and that deleterious effects may be mediated via TNFR1 (Arnett et al. 2001). Similarly, mice carrying a deletion for either MHCII or iNOS have been shown to suffer from impaired remyelination and more severe demyelination, respectively, in cuprizone-mediated demyelination relative to wild-type mice. These observations argue that like TNFα, MHCII and NO potentially have dual roles in demyelinating disorders whereby they are not only implicated in pathogenesis but could also potentiate remyelination and/or oligodendroglial survival in the context of repair (Arnett et al. 2002, 2003). The potential role of MHCII is of particular interest as the remyelination phase in cuprizone-induced demyelination has been shown to be independent of T cells, as shown by the use of RAG−/− mice (Arnett et al. 2001). It has also been hypothesised that MHCII might have an independent signalling role in microglia (Arnett et al. 2003) but the exact nature of this and how it relates to the modulation of oligodendroglial function remains uncertain.

Interleukin-1β (IL-1β) is a pro-inflammatory cytokine that is produced mainly by microglia and macrophages which can promote astrocyte proliferation (Giulian and Lachman 1985) and in turn can regulate the expression of TNFα, thus mediating deleterious signalling pathways in degenerative contexts (Lee et al. 1995). However, using the cuprizone model of demyelination, it has been demonstrated that IL-1β knockout mice demyelinate in a similar fashion to wild types, even though IL-1β had previously been reported to induce the production of IL-6, TNFα and NO. However, IL-1β knockout mice displayed impaired remyelination, associated with the inhibition of oligodendroglial differentiation (Mason et al. 2001). Interestingly, it has been shown that in culture, IL-1β can inhibit proliferation of late oligodendroglial progenitors and promote their differentiation (Vela et al. 2002), although others had previously reported that it could be cytotoxic to oligodendrocytes (Merrill 1991).

Osteopontin is a pro-inflammatory cytokine that is highly expressed in MS lesions where it is secreted by activated microglia/macrophages and astrocytes. It contributes to tissue injury through the inhibition of T cell apoptosis and the promotion of the secretion other pro-inflammatory cytokines (Chabas et al. 2001; Hur et al. 2007). On the other hand, osteopontin is upregulated during remyelination following toxin-mediated demyelination (Selvaraju et al. 2004; Zhao et al. 2008). In vitro cultures of oligodendrocytes prepared from osteopontin knockout mice have been shown to express less MBP than wild-type oligodendrocytes (Selvaraju et al. 2004). However, in vivo experiments demonstrate that osteopontin is functionally redundant in the remyelinative process as osteopontin-null mice are capable of successful remyelination (Zhao et al. 2008).

The above results complement a number of other studies that have now demonstrated a key role for macrophage-derived IGF-1, during remyelination (Diemel et al. 2004; Hinks and Franklin 1999; Kiefer et al. 1998; Kotter et al. 2005). It has been observed that IGF-1 can promote oligodendrocyte differentiation and myelin sheath formation (Hinks and Franklin 2000). In lysolecithin-induced demyelination, macrophage depletion resulted in remyelination impairment due to a transient delay in OPC recruitment as well as complex changes in the expression profile of IGF-1, but most importantly a reduction in IGF-1 levels during the early stages of remyelination.

Hinks and Franklin studied remyelination in young (<3 months old) and aged (>9 months old) rats exposed to lysolecithin. They found that the process was delayed in the older animals and that this correlated with impaired activation of both macrophages and mRNA encoding PDGF-A, IGF-1 and TGFβ1 (Hinks and Franklin 2000). These findings raised the possibility that macrophage activation could be responsible for the production of PDGF-A to induce the proliferation of oligodendrocyte progenitors and for the production of IGF-1 and TGFβ1 which could potentiate the maturation of these cells.

The capacity of microglia to remodel the extracellular matrix via secretion of proteinases such as matrix metalloproteinase-9 (MMP-9) could also be an important determinant of the extent of repair. Interestingly, mice lacking MMP-9 have been shown to exhibit reduced remyelination after lysolecithin-induced demyelination, due to inhibited clearance of NG2, a proteoglycan that retards the differentiation of oligodendrocytes (Larsen et al. 2003).

It remains to be determined the extent to which microglia influence OPC migration. The chemokine receptor CXCR2 and its ligand CXCL1 have been shown to be critical in controlling the position of OPCs in the developing spinal cord (Tsai et al. 2002). Filipovic et al. (2003) found the CXCL1 to be expressed by activated microglia, including at the border of MS lesions, although neither CXCL1 nor CXCR2 was found to be expressed by OPCs in lesional tissues.

Taken together, the above data highlight a role for microglia in creating a myelogenic signalling environment not only through the clearance of myelin debris but also via the secretion of soluble trophic mediators, proteases and chemokines. This speaks both to the complexity of microglial activity and to the oversimplification of formally separating microglial function into "detrimental" inflammatory and "protective" immunomodulatory roles.

3.6 Conclusion

It is now well established that microglia can exert a number of distinct activities that serve to modulate either the pathogenesis of central demyelination or its repair. As indicated in this review, these activities include pro-inflammatory effects, the induction of phagocytosis of myelin lamellae, myelin debris or apoptotic cells and the generation of factors that either directly or indirectly potentiate repair. Although at face value these represent distinct processes, evidence is emerging to suggest that these disparate activities are, in some instances, linked. It has, for example, been identified that microglia stimulated to become actively phagocytic are also induced to secrete a modified cytokine profile that promotes neural cell repair. The molecular events that regulate these distinct processes are rapidly being identified, and recent work is suggesting that there could be key molecular associations between important signalling pathways (e.g. those initiated by the TAM receptors and the interferon receptor-alpha) that are responsible for integrating these various microglial functions in a coordinated way. It is also important to note that the influence that microglia exert may be contextual. This is well illustrated by the fact that many molecules produced by microglia have been shown to be pathogenic in the active demyelinative phase but can also, in other contexts, promote repair. The switch in these influences is almost certainly dependent upon a number of factors, but a key determinant is likely to be variance in the receptor profile of the target cell population. This complexity provides particular challenges in the development of therapeutic strategies that are designed to modulate microglial activity. It is likely that these approaches will be most relevant and useful if the modulation is pathway specific rather than broad. As a corollary, successful application is likely to be predicated on an understanding of the receptor expression profile of the target tissue, which will ultimately depend upon the dedicated application of key technologies such as positron emission tomography, necessitating an integrated approach to therapeutic development.

References

Ajami B, Bennett JL, Krieger C, Tetzlaff W, Rossi FM (2007) Local self-renewal can sustain CNS microglia maintenance and function throughout adult life. Nat Neurosci 10(12):1538–1543

Alliot F, Godin I, Pessac B (1999) Microglia derive from progenitors, originating from the yolk sac, and which proliferate in the brain. Brain Res Dev Brain Res 117(2):145–152

Arnett HA, Mason J, Marino M, Suzuki K, Matsushima GK, Ting JP (2001) TNF alpha promotes proliferation of oligodendrocyte progenitors and remyelination. Nat Neurosci 4(11):1116–1122

Arnett HA, Hellendall RP, Matsushima GK, Suzuki K, Laubach VE, Sherman P, Ting JP (2002) The protective role of nitric oxide in a neurotoxicant-induced demyelinating model. J Immunol 168(1):427–433

Arnett HA, Wang Y, Matsushima GK, Suzuki K, Ting JP (2003) Functional genomic analysis of remyelination reveals importance of inflammation in oligodendrocyte regeneration. J Neurosci 23(30):9824–9832

Bagasra O, Michaels FH, Zheng YM, Bobroski LE, Spitsin SV, Fu ZF, Tawadros R, Koprowski H (1995) Activation of the inducible form of nitric oxide synthase in the brains of patients with multiple sclerosis. Proc Natl Acad Sci USA 92(26):12041–12045

Barnett MH, Prineas JW (2004) Relapsing and remitting multiple sclerosis: pathology of the newly forming lesion. Ann Neurol 55(4):458–468

Barnett MH, Henderson AP, Prineas JW (2006) The macrophage in MS: just a scavenger after all? Pathology and pathogenesis of the acute MS lesion. Mult Scler 12(2):121–132

Bianchin MM, Lima JE, Natel J, Sakamoto AC (2006) The genetic causes of basal ganglia calcification, dementia, and bone cysts: DAP12 and TREM2. Neurology 66(4):615–616, author reply 615–616

Binder MD, Cate HS, Prieto AL, Kemper D, Butzkueven H, Gresle MM, Cipriani T, Jokubaitis VG, Carmeliet P, Kilpatrick TJ (2008) Gas6 deficiency increases oligodendrocyte loss and microglial activation in response to cuprizone-induced demyelination. J Neurosci 28(20):5195–5206

Bjartmar C, Kidd GJ, Mörk S, Rudick R, Trapp BD (2000) Neurological disability correlates with spinal cord axonal loss and reduced N-acetyl aspartate in chronic multiple sclerosis patients. Ann Neurol 48(6):893–901

Blakemore WF (1973) Demyelination of the superior cerebellar peduncle in the mouse induced by cuprizone. J Neurol Sci 20(1):63–72

Blakemore WF, Patterson RC (1978) Suppression of remyelination in the CNS by X-irradiation. Acta Neuropathol 42(2):105–113

Bö L, Mörk S, Kong PA, Nyland H, Pardo CA, Trapp BD (1994) Detection of MHC class II-antigens on macrophages and microglia, but not on astrocytes and endothelia in active multiple sclerosis lesions. J Neuroimmunol 51(2):135–146

Bruck W, Bruck Y, Friede RL (1992) TNF-alpha suppresses CR3-mediated myelin removal by macrophages. J Neuroimmunol 38(1–2):9–17

Brück W, Porada P, Poser S, Rieckmann P, Hanefeld F, Kretzschmar HA, Lassmann H (1995) Monocyte/macrophage differentiation in early multiple sclerosis lesions. Ann Neurol 38(5): 788–796

Carbonell WS, Murase S, Horwitz AF, Mandell JW (2005) Migration of perilesional microglia after focal brain injury and modulation by CC chemokine receptor 5: an in situ time-lapse confocal imaging study. J Neurosci 25(30):7040–7047

Chabas D, Baranzini SE, Mitchell D, Bernard CC, Rittling SR, Denhardt DT, Sobel RA, Lock C, Karpuj M, Pedotti R, Heller R, Oksenberg JR, Steinman L (2001) The influence of the proinflammatory cytokine, osteopontin, on autoimmune demyelinating disease. Science 294(5547):1731–1735

Chan A, Magnus T, Gold R (2001) Phagocytosis of apoptotic inflammatory cells by microglia and modulation by different cytokines: mechanism for removal of apoptotic cells in the inflamed nervous system. Glia 33(1):87–95

Chan A, Seguin R, Magnus T, Papadimitriou C, Toyka KV, Antel JP, Gold R (2003) Phagocytosis of apoptotic inflammatory cells by microglia and its therapeutic implications: termination of CNS autoimmune inflammation and modulation by interferon-beta. Glia 43(3):231–242

Cuadros MA, Navascués J (1998) The origin and differentiation of microglial cells during development. Prog Neurobiol 56(2):173–189

Cullheim S, Thams S (2007) The microglial networks of the brain and their role in neuronal network plasticity after lesion. Brain Res Rev 55(1):89–96

Davalos D, Grutzendler J, Yang G, Kim JV, Zuo Y, Jung S, Littman DR, Dustin ML, Gan WB (2005) ATP mediates rapid microglial response to local brain injury in vivo. Nat Neurosci 8(6):752–758

De Groot CJ, Ruuls SR, Theeuwes JW, Dijkstra CD, Van Der Valk P (1997) Immunocytochemical characterization of the expression of inducible and constitutive isoforms of nitric oxide synthase in demyelinating multiple sclerosis lesions. J Neuropathol Exp Neurol 56(1):10–20

De Simone R, Ambrosini E, Carnevale D, Ajmone-Cat MA, Minghetti L (2007) NGF promotes microglial migration through the activation of its high affinity receptor: modulation by TGF-beta. J Neuroimmunol 190(1–2):53–60

Dheen ST, Jun Y, Yan Z, Tay SS, Ling EA (2005) Retinoic acid inhibits expression of TNF-alpha and iNOS in activated rat microglia. Glia 50(1):21–31

Dheen ST, Kaur C, Ling EA (2007) Microglial activation and its implications in the brain diseases. Curr Med Chem 14(11):1189–1197

Diab A, Hussain RZ, Lovett-Racke AE, Chavis JA, Drew PD, Racke MK (2004) Ligands for the peroxisome proliferator-activated receptor-gamma and the retinoid X receptor exert additive anti-inflammatory effects on experimental autoimmune encephalomyelitis. J Neuroimmunol 148(1–2):116–126

Diemel LT, Wolswijk G, Jackson SJ, Cuzner ML (2004) Remyelination of cytokine- or antibody-demyelinated CNS aggregate cultures is inhibited by macrophage supplementation. Glia 45(3):278–286

Felts PA, Woolston AM, Fernando HB, Asquith S, Gregson NA, Mizzi OJ, Smith KJ (2005) Inflammation and primary demyelination induced by the intraspinal injection of lipopolysaccharide. Brain 128(Pt 7):1649–1666

Ferguson B, Matyszak MK, Esiri MM, Perry VH (1997) Axonal damage in acute multiple sclerosis lesions. Brain 120(Pt 3):393–399

Filipovic R, Jakovcevski I, Zecevic N (2003) GRO-alpha and CXCR2 in the human fetal brain and multiple sclerosis lesions. Dev Neurosci 25(2–4):279–290

Fischer M, Ehlers M (2008) Toll-like receptors in autoimmunity. Ann N Y Acad Sci 1143:21–34

Foote AK, Blakemore WF (2005) Inflammation stimulates remyelination in areas of chronic demyelination. Brain 128(Pt 3):528–539

Fordyce CB, Jagasia R, Zhu X, Schlichter LC (2005) Microglia Kv1.3 channels contribute to their ability to kill neurons. J Neurosci 25(31):7139–7149

Forstreuter F, Lucius R, Mentlein R (2002) Vascular endothelial growth factor induces chemotaxis and proliferation of microglial cells. J Neuroimmunol 132(1–2):93–98

Gibbons HM, Dragunow M (2006) Microglia induce neural cell death via a proximity-dependent mechanism involving nitric oxide. Brain Res 1084(1):1–15

Giulian D, Lachman LB (1985) Interleukin-1 stimulation of astroglial proliferation after brain injury. Science 228(4698):497–499

Glezer I, Lapointe A, Rivest S (2006) Innate immunity triggers oligodendrocyte progenitor reactivity and confines damages to brain injuries. FASEB J 20(6):750–752

Grommes C, Lee CY, Wilkinson BL, Jiang Q, Koenigsknecht-Talboo JL, Varnum B, Landreth GE (2008) Regulation of microglial phagocytosis and inflammatory gene expression by Gas6 acting on the Axl/Mer family of tyrosine kinases. J Neuroimmune Pharmacol 3(2):130–140

Gveric D, Cuzner ML, Newcombe J (1999) Insulin-like growth factors and binding proteins in multiple sclerosis plaques. Neuropathol Appl Neurobiol 25(3):215–225

Hendriks JJ, Slaets H, Carmans S, de Vries HE, Dijkstra CD, Stinissen P, Hellings N (2008) Leukemia inhibitory factor modulates production of inflammatory mediators and myelin phagocytosis by macrophages. J Neuroimmunol 204(1–2):52–57

Heppner FL, Greter M, Marino D, Falsig J, Raivich G, Hövelmeyer N, Waisman A, Rülicke T, Prinz M, Priller J, Becher B, Aguzzi A (2005) Experimental autoimmune encephalomyelitis repressed by microglial paralysis. Nat Med 11(2):146–152

Hinks GL, Franklin RJ (1999) Distinctive patterns of PDGF-A, FGF-2, IGF-I, and TGF-beta1 gene expression during remyelination of experimentally-induced spinal cord demyelination. Mol Cell Neurosci 14(2):153–168

Hinks GL, Franklin RJ (2000) Delayed changes in growth factor gene expression during slow remyelination in the CNS of aged rats. Mol Cell Neurosci 16(5):542–556

Hirasawa T, Ohsawa K, Imai Y, Ondo Y, Akazawa C, Uchino S, Kohsaka S (2005) Visualization of microglia in living tissues using Iba1-EGFP transgenic mice. J Neurosci Res 81(3):357–362

Hoehn HJ, Kress Y, Sohn A, Brosnan CF, Bourdon S, Shafit-Zagardo B (2008) Axl-/- mice have delayed recovery and prolonged axonal damage following cuprizone toxicity. Brain Res 1240:1–11

Huisman C, Kok P, Schmaal L, Verhoog P (2008) Bradykinin: a microglia attractant in vivo? J Neurosci 28(14):3531–3532

Hur EM, Youssef S, Haws ME, Zhang SY, Sobel RA, Steinman L (2007) Osteopontin-induced relapse and progression of autoimmune brain disease through enhanced survival of activated T cells. Nat Immunol 8(1):74–83

Irvine KA, Blakemore WF (2006) Age increases axon loss associated with primary demyelination in cuprizone-induced demyelination in C57BL/6 mice. J Neuroimmunol 175(1–2):69–76

Irvine KA, Blakemore WF (2008) Remyelination protects axons from demyelination-associated axon degeneration. Brain 131(Pt 6):1464–1477

Juedes AE, Ruddle NH (2001) Resident and infiltrating central nervous system APCs regulate the emergence and resolution of experimental autoimmune encephalomyelitis. J Immunol 166(8): 5168–5175

Kerr BJ, Patterson PH (2005) Leukemia inhibitory factor promotes oligodendrocyte survival after spinal cord injury. Glia 51(1):73–79

Kiefer R, Schweitzer T, Jung S, Toyka KV, Hartung HP (1998) Sequential expression of transforming growth factor-beta1 by T-cells, macrophages, and microglia in rat spinal cord during autoimmune inflammation. J Neuropathol Exp Neurol 57(5):385–395

Kim SU, de Vellis J (2005) Microglia in health and disease. J Neurosci Res 81(3):302–313

Kotter MR, Setzu A, Sim FJ, van Rooijen N, Franklin RJ (2001) Macrophage depletion impairs oligodendrocyte remyelination following lysolecithin-induced demyelination. Glia 35(3):204–212

Kotter MR, Zhao C, van Rooijen N, Franklin RJ (2005) Macrophage-depletion induced impairment of experimental CNS remyelination is associated with a reduced oligodendrocyte progenitor cell response and altered growth factor expression. Neurobiol Dis 18(1):166–175

Kotter MR, Li WW, Zhao C, Franklin RJ (2006) Myelin impairs CNS remyelination by inhibiting oligodendrocyte precursor cell differentiation. J Neurosci 26(1):328–332

Larsen PH, Wells JE, Stallcup WB, Opdenakker G, Yong VW (2003) Matrix metalloproteinase-9 facilitates remyelination in part by processing the inhibitory NG2 proteoglycan. J Neurosci 23(35):11127–11135

Lassmann H, Brück W, Lucchinetti C, Rodriguez M (1997) Remyelination in multiple sclerosis. Mult Scler 3(2):133–136

Lee SC, Dickson DW, Brosnan CF (1995) Interleukin-1, nitric oxide and reactive astrocytes. Brain Behav Immun 9(4):345–354

Lehnardt S, Lachance C, Patrizi S, Lefebvre S, Follett PL, Jensen FE, Rosenberg PA, Volpe JJ, Vartanian T (2002) The toll-like receptor TLR4 is necessary for lipopolysaccharide-induced oligodendrocyte injury in the CNS. J Neurosci 22(7):2478–2486

Linares D, Taconis M, Mana P, Correcha M, Fordham S, Staykova M, Willenborg DO (2006) Neuronal nitric oxide synthase plays a key role in CNS demyelination. J Neurosci 26(49):12672–12681

Liu JS, Zhao ML, Brosnan CF, Lee SC (2001) Expression of inducible nitric oxide synthase and nitrotyrosine in multiple sclerosis lesions. Am J Pathol 158(6):2057–2066

Lucchinetti CF, Brück W, Rodriguez M, Lassmann H (1996) Distinct patterns of multiple sclerosis pathology indicates heterogeneity on pathogenesis. Brain Pathol 6(3):259–274

Lucchinetti C, Brück W, Parisi J, Scheithauer B, Rodriguez M, Lassmann H (2000) Heterogeneity of multiple sclerosis lesions: implications for the pathogenesis of demyelination. Ann Neurol 47(6):707–717

Marín-Teva JL, Almendros A, Calvente R, Cuadros MA, Navascués J (1998) Tangential migration of ameboid microglia in the developing quail retina: mechanism of migration and migratory behavior. Glia 22(1):31–52

Martin S, Vincent JP, Mazella J (2003) Involvement of the neurotensin receptor-3 in the neurotensin-induced migration of human microglia. J Neurosci 23(4):1198–1205

Mason JL, Suzuki K, Chaplin DD, Matsushima GK (2001) Interleukin-1beta promotes repair of the CNS. J Neurosci 21(18):7046–7052

Merrill JE (1991) Effects of interleukin-1 and tumor necrosis factor-alpha on astrocytes, microglia, oligodendrocytes, and glial precursors in vitro. Dev Neurosci 13(3):130–137

Merrill JE, Ignarro LJ, Sherman MP, Melinek J, Lane TE (1993) Microglial cell cytotoxicity of oligodendrocytes is mediated through nitric oxide. J Immunol 151(4):2132–2141

Mi S, Miller RH, Lee X, Scott ML, Shulag-Morskaya S, Shao Z, Chang J, Thill G, Levesque M, Zhang M, Hession C, Sah D, Trapp B, He Z, Jung V, McCoy JM, Pepinsky RB (2005) LINGO-1 negatively regulates myelination by oligodendrocytes. Nat Neurosci 8(6):745–751

Mildner A, Schmidt H, Nitsche M, Merkler D, Hanisch UK, Mack M, Heikenwalder M, Bruck W, Priller J, Prinz M (2007) Microglia in the adult brain arise from Ly-6ChiCCR2+ monocytes only under defined host conditions. Nat Neurosci 10(12):1544–1553

Mitrasinovic OM, Perez GV, Zhao F, Lee YL, Poon C, Murphy GM Jr (2001) Overexpression of macrophage colony-stimulating factor receptor on microglial cells induces an inflammatory response. J Biol Chem 276(32):30142–30149

Miyanishi M, Tada K, Koike M, Uchiyama Y, Kitamura T, Nagata S (2007) Identification of Tim4 as a phosphatidylserine receptor. Nature 450(7168):435–439

Morell P, Barrett CV, Mason JL, Toews AD, Hostettler JD, Knapp GW, Matsushima GK (1998) Gene expression in brain during cuprizone-induced demyelination and remyelination. Mol Cell Neurosci 12(4–5):220–227

Mosley K, Cuzner ML (1996) Receptor-mediated phagocytosis of myelin by macrophages and microglia: effect of opsonization and receptor blocking agents. Neurochem Res 21(4):481–487

Muzio L, Martino G, Furlan R (2007) Multifaceted aspects of inflammation in multiple sclerosis: the role of microglia. J Neuroimmunol 191(1–2):39–44

Nasu-Tada K, Koizumi S, Inoue K (2005) Involvement of beta1 integrin in microglial chemotaxis and proliferation on fibronectin: different regulations by ADP through PKA. Glia 52(2):98–107

Neumann H, Kotter MR, Franklin RJ (2009) Debris clearance by microglia: an essential link between degeneration and regeneration. Brain 132(Pt 2):288–295

Nimmerjahn A, Kirchhoff F, Helmchen F (2005) Resting microglial cells are highly dynamic surveillants of brain parenchyma in vivo. Science 308(5726):1314–1318

Nishiyama A, Yu M, Drazba JA, Tuohy VK (1997) Normal and reactive NG2+ glial cells are distinct from resting and activated microglia. J Neurosci Res 48(4):299–312

Oleszak EL, Zaczynska E, Bhattacharjee M, Butunoi C, Legido A, Katsetos CD (1998) Inducible nitric oxide synthase and nitrotyrosine are found in monocytes/macrophages and/or astrocytes in acute, but not in chronic, multiple sclerosis. Clin Diagn Lab Immunol 5(4):438–445

Parkinson JF, Mitrovic B, Merrill JE (1997) The role of nitric oxide in multiple sclerosis. J Mol Med 75(3):174–186

Patani R, Balaratnam M, Vora A, Reynolds R (2007) Remyelination can be extensive in multiple sclerosis despite a long disease course. Neuropathol Appl Neurobiol 33(3):277–287

Pender MP (1999) Activation-induced apoptosis of autoreactive and alloreactive T lymphocytes in the target organ as a major mechanism of tolerance. Immunol Cell Biol 77(3):216–223

Pender MP, Nguyen KB, McCombe PA, Kerr JF (1991) Apoptosis in the nervous system in experimental allergic encephalomyelitis. J Neurol Sci 104(1):81–87

Peterson JW, Bö L, Mörk S, Chang A, Ransohoff RM, Trapp BD (2002) VCAM-1-positive microglia target oligodendrocytes at the border of multiple sclerosis lesions. J Neuropathol Exp Neurol 61(6):539–546

Piccio L, Buonsanti C, Mariani M, Cella M, Gilfillan S, Cross AH, Colonna M, Panina-Bordignon P (2007) Blockade of TREM-2 exacerbates experimental autoimmune encephalomyelitis. Eur J Immunol 37(5):1290–1301

Prineas JW, Graham JS (1981) Multiple sclerosis: capping of surface immunoglobulin G on macrophages engaged in myelin breakdown. Ann Neurol 10(2):149–158

Prineas JW, Kwon EE, Goldenberg PZ, Ilyas AA, Quarles RH, Benjamins JA, Sprinkle TJ (1989) Multiple sclerosis. Oligodendrocyte proliferation and differentiation in fresh lesions. Lab Invest 61(5):489–503

Prineas JW, Barnard RO, Kwon EE, Sharer LR, Cho ES (1993) Multiple sclerosis: remyelination of nascent lesions. Ann Neurol 33(2):137–151

Prinz M, Schmidt H, Mildner A, Knobeloch KP, Hanisch UK, Raasch J, Merkler D, Detje C, Gutcher I, Mages J, Lang R, Martin R, Gold R, Becher B, Brück W, Kalinke U (2008) Distinct and nonredundant in vivo functions of IFNAR on myeloid cells limit autoimmunity in the central nervous system. Immunity 28(5):675–686

Raine CS, Wu E (1993) Multiple sclerosis: remyelination in acute lesions. J Neuropathol Exp Neurol 52(3):199–204

Rappert A, Bechmann I, Pivneva T, Mahlo J, Biber K, Nolte C, Kovac AD, Gerard C, Boddeke HW, Nitsch R, Kettenmann H (2004) CXCR3-dependent microglial recruitment is essential for dendrite loss after brain lesion. J Neurosci 24(39):8500–8509

Reichert F, Rotshenker S (2003) Complement-receptor-3 and scavenger-receptor-AI/II mediated myelin phagocytosis in microglia and macrophages. Neurobiol Dis 12(1):65–72

Rotshenker S, Reichert F, Gitik M, Haklai R, Elad-Sfadia G, Kloog Y (2008) Galectin-3/MAC-2, Ras and PI3K activate complement receptor-3 and scavenger receptor-AI/II mediated myelin phagocytosis in microglia. Glia 56(15):1607–1613

Rzepecka J, Rausch S, Klotz C, Schnoller C, Kornprobst T, Hagen J, Ignatius R, Lucius R, Hartmann S (2009) Calreticulin from the intestinal nematode Heligmosomoides polygyrus is a Th2-skewing protein and interacts with murine scavenger receptor-A. Mol Immunol 46(6):1109–1119

Scolding N, Franklin R, Stevens S, Heldin CH, Compston A, Newcombe J (1998) Oligodendrocyte progenitors are present in the normal adult human CNS and in the lesions of multiple sclerosis. Brain 121(Pt 12):2221–2228

Selvaraju R, Bernasconi L, Losberger C, Graber P, Kadi L, Avellana-Adalid V, Picard-Riera N, Van Evercooren AB, Cirillo R, Kosco-Vilbois M, Feger G, Papoian R, Boschert U (2004) Osteopontin is upregulated during in vivo demyelination and remyelination and enhances myelin formation in vitro. Mol Cell Neurosci 25(4):707–721

Setzu A, Lathia JD, Zhao C, Wells K, Rao MS, Ffrench-Constant C, Franklin RJ (2006) Inflammation stimulates myelination by transplanted oligodendrocyte precursor cells. Glia 54(4):297–303

Stevens B, Allen NJ, Vazouez LE, Howell GR, Christopherson KS, Nouri N, Micheva KD, Mehalow AK, Huberman AD, Stafford B, Sher A, Litke AM, Lambris JD, Smith SJ, John SW, Barres BA (2007) The classical complement cascade mediates CNS synapse elimination. Cell 14:131(6):1034–1036

Streit WJ, Walter SA, Pennell NA (1999) Reactive microgliosis. Prog Neurobiol 57(6):563–581

Stuart LM, Bell SA, Stewart CR, Silver JM, Richard J, Goss JL, Tseng AA, Zhang A, El Khoury JB, Moore KJ (2007) CD36 signals to the actin cytoskeleton and regulates microglial migration via a p130Cas complex. J Biol Chem 282(37):27392–27401

Sugiura S, Lahav R, Han J, Kou SY, Banner LR, de Pablo F, Patterson PH (2000) Leukaemia inhibitory factor is required for normal inflammatory responses to injury in the peripheral and central nervous systems in vivo and is chemotactic for macrophages in vitro. Eur J Neurosci 12(2):457–466

Takahashi K, Rochford CD, Neumann H (2005) Clearance of apoptotic neurons without inflammation by microglial triggering receptor expressed on myeloid cells-2. J Exp Med 201(4):647–657

Takahashi K, Prinz M, Stagi M, Chechneva O, Neumann H (2007) TREM2-transduced myeloid precursors mediate nervous tissue debris clearance and facilitate recovery in an animal model of multiple sclerosis. PLoS Med 4(4):e124

Todt JC, Hu B, Curtis JL (2008) The scavenger receptor SR-A I/II (CD204) signals via the receptor tyrosine kinase Mertk during apoptotic cell uptake by murine macrophages. J Leukoc Biol 84(2):510–518

Totoiu MO, Nistor GI, Lane TE, Keirstead HS (2004) Remyelination, axonal sparing, and locomotor recovery following transplantation of glial-committed progenitor cells into the MHV model of multiple sclerosis. Exp Neurol 187(2):254–265

Trapp BD, Peterson J, Ransohoff RM, Rudick R, Mörk S, Bö L (1998) Axonal transection in the lesions of multiple sclerosis. N Engl J Med 338(5):278–285

Trapp BD, Bo L, Mork S, Chang A (1999) Pathogenesis of tissue injury in MS lesions. J Neuroimmunol 98(1):49–56

Trapp BD, Wujek JR, Criste GA, Jalabi W, Yin X, Kidd GJ, Stohlman S, Ransohoff R (2007) Evidence for synaptic stripping by cortical microglia. Glia 55(4):360–368

Trotter J, DeJong LJ, Smith ME (1986) Opsonization with antimyelin antibody increases the uptake and intracellular metabolism of myelin in inflammatory macrophages. J Neurochem 47(3):779–789

Tsai HH, Frost E, To V, Robinson S, Ffrench-Constant C, Geertman R, Ransohoff RM, Miller RH (2002) The chemokine receptor CXCR2 controls positioning of oligodendrocyte precursors in developing spinal cord by arresting their migration. Cell 110(3):373–383

Vargas ME, Barres BA (2007) Why is Wallerian degeneration in the CNS so slow? Annu Rev Neurosci 30:153–179

Vela JM, Molina-Holgado E, Arévalo-Martín A, Almazán G, Guaza C (2002) Interleukin-1 regulates proliferation and differentiation of oligodendrocyte progenitor cells. Mol Cell Neurosci 20(3):489–502

Weiner HL (2008) A shift from adaptive to innate immunity: a potential mechanism of disease progression in multiple sclerosis. J Neurol 255(Suppl 1):3–11

Wolswijk G (2002) Oligodendrocyte precursor cells in the demyelinated multiple sclerosis spinal cord. Brain 125(Pt 2):338–349

Xu J, Drew PD (2006) 9-Cis-retinoic acid suppresses inflammatory responses of microglia and astrocytes. J Neuroimmunol 171(1–2):135–144

Yaguchi M, Ohta S, Toyama Y, Kawakami Y, Toda M (2008) Functional recovery after spinal cord injury in mice through activation of microglia and dendritic cells after IL-12 administration. J Neurosci Res 86(9):1972–1980

Zhao C, Fancy SP, Ffrench-Constant C, Franklin RJ (2008) Osteopontin is extensively expressed by macrophages following CNS demyelination but has a redundant role in remyelination. Neurobiol Dis 31(2):209–217

Chapter 4
Endogenous Remyelination in the CNS

Robin J.M. Franklin, Chao Zhao, Catherine Lubetzki, and Charles ffrench-Constant

4.1 Introduction

In striking contrast to the generally inadequate attempts at regeneration that follows damage to neuronal elements, the sequela to CNS demyelination is often a robust *regenerative* process called remyelination. In this chapter, we (1) review current knowledge on the biology of remyelination, including the cells and molecular signals involved; (2) describe when remyelination occurs and when and why it fails, including the consequences of its failure; and (3) discuss approaches for enhancing endogenous remyelination therapeutically in demyelinating diseases.

R.J.M. Franklin (✉) • C. Zhao
Wellcome Trust and MRC, Cambridge Stem Cell Institute, and Department of Veterinary Medicine, University of Cambridge, Madingley Road, Cambridge CB3 0ES, UK
e-mail: rjf1000@cam.ac.uk

C. Lubetzki
Départment de Neurologie et CR-ICM, AP-HP, Inserm 975, UPMC, Hôpital de la Salpêtrière, 75013 Paris, France

C. ffrench-Constant
MRC Centre for Regenerative Medicine/ Scottish Centre for Regenerative Medicine, The University of Edinburgh, Edinburgh BioQuarter, 5 Little France Drive, Edinburgh EH16 4UU, UK

University of Edinburgh Centre for Translational Research, Centre for Inflammation Research, Queen's Medical Research Institute, The University of Edinburgh, 47 Little France Crescent, Edinburgh EH16 4TJ, UK

I.D. Duncan and R.J.M. Franklin (eds.), *Myelin Repair and Neuroprotection in Multiple Sclerosis*, DOI 10.1007/978-1-4614-2218-1_4,
© Springer Science+Business Media New York 2013

4.2 Identifying Remyelination

Remyelination is the process in which entirely new myelin sheaths are restored to demyelinated axons. The term myelin repair is also sometimes used; however, it has the connotation of a damaged but otherwise intact myelin internode being "patched up", a process for which there is no evidence and which does not emphasise the truly regenerative nature of remyelination in which there is restoration of the pre-lesion cytoarchitecture. The tissue reconstruction in remyelination is complete expect for one caveat: the striking correlation between axon diameter and myelin sheath thickness and length established during myelination is less apparent in remyelination. Instead, myelin sheath thickness and length shows little increase with increasing axonal diameter with the result that the myelin is generally thinner and shorter than would be expected for a given diameter of axon (Fig. 4.1). Although some remodelling

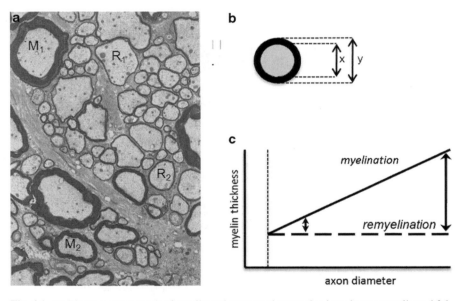

Fig. 4.1 (a) Electron micrograph of myelinated axons and axons that have been remyelinated following ethidium bromide injection into the adult rat caudal cerebellar peduncle. The myelinated axons have clearly discernible myelin sheaths whose thickness is proportional to the axon diameters (compare M_1 with M_2). The remyelinated axons are also clearly discernible on account of the relatively thin myelin sheaths, whose thinness is consistent and independent of the axon diameter (compare R_1 with R_2). Thus, remyelination is easy to identify for larger diameter axons; however, for small diameter axons, the distinction between myelinated and remyelinated becomes difficult (note the small diameter axons in this image—are they myelinated or remyelinated?) (b) The relationship between the axon diameter (x) and the myelinated axon (y) is expressed as the g ratio: the thinner the myelin sheath, the higher the g ratio, and hence remyelinated axons, unless very small diameter, have g ratios that are higher than myelinated axons. (c) In developmental myelination, as the axon diameter increases, the myelin sheath thickness increases (see M_1 with M_2 in **a**), whereas in remyelination, the myelin sheath thickness remains the same regardless of the diameter (see R_1 with R_2 in **a**). Thus, remyelination in large diameter axons is easy to distinguish from myelination—but as the axon diameter decreases, this distinction becomes more difficult such that for the smallest diameter axons it can be all but impossible (see Stidworthy et al. 2003)

of the new myelin internode occurs, the original dimensions are never attained. The relationship between axon diameter and myelin sheath is expressed as a fraction of the axonal circumference to the axon plus myelin sheath circumference, called the g ratio. The identification of abnormally thin myelin sheaths (> normal g ratio) remains the most reliable means of unequivocally identifying remyelination. This effect is obvious when large diameter axons are remyelinated but is less clear with smaller diameter axons such as those within the corpus callosum, where g ratios of remyelinated axons are indistinguishable from those of normally myelinated axons (Stidworthy et al. 2003). How is the relationship between myelin parameters and axon size established in myelination and why does it disengage in remyelination? In the PNS, axonally expressed neuregulin (NRG) 1 type III plays a pivotal role: reduced expression leads to a thinner myelin sheath (increased g ratio), while overexpression leads to a thicker than expected myelin sheath (decreased g ratio) (Michailov et al. 2004). In the CNS, however, the role of neuregulins is less clear: myelination in mice where the Nrg1 gene has been excised or double mutants for Erb3/4 in the CNS argue that Nrg1 is not necessary for CNS myelination and that other signals must also contribute to the precise relationship between axon and oligodendrocyte (Brinkmann et al. 2008). Likewise, the mechanistic basis of the increased g ratio in remyelination is not known and appears different from developmental myelination since overexpression of Nrg leads to hypermyelination in development but not during remyelination (Brinkmann et al. 2008). Similarly, activation of the Akt pathway, which results in thicker than expected myelin sheaths in development (Flores et al. 2008), does not result in thicker remyelinated sheaths following demyelination in the adult (Harrington et al. 2010) (Fig. 4.2). One hypothesis is that whereas the myelinating oligodendrocyte associates with a dynamically changing axon yet to achieve its full length and diameter, the remyelinating oligodendrocyte engages an axon that is comparatively static having already reached it mature size (Franklin and Hinks 1999). As a result, the remyelinating oligodendrocyte is not subjected to the same dynamic stresses encountered by the myelinating oligodendrocyte.

4.3 Remyelination Is the Normal Response to Demyelination

Remyelination can be viewed as a regenerative process sharing many common features with regenerative processes occurring in other tissues of the body and as being the default response to demyelination. This viewpoint is based on evidence from both experimentally-induced and clinical demyelination. When demyelination is induced by toxins injurious to oligodendrocytes and myelin (e.g. dietary cuprizone or direct delivery of lysolecithin or ethidium bromide), then remyelination usually proceeds to completion, albeit in an age-dependent manner (Blakemore and Franklin 2008). Similarly, there is evidence that axons undergoing primary demyelination in experimental or clinical traumatic injury undergo complete remyelination and that the persistence of chronically demyelinated axons is unusual (Lasiene et al. 2008). An exception is when demyelination is induced by or associated with the adaptive immune response, such occurs in the autoimmune-mediated condition

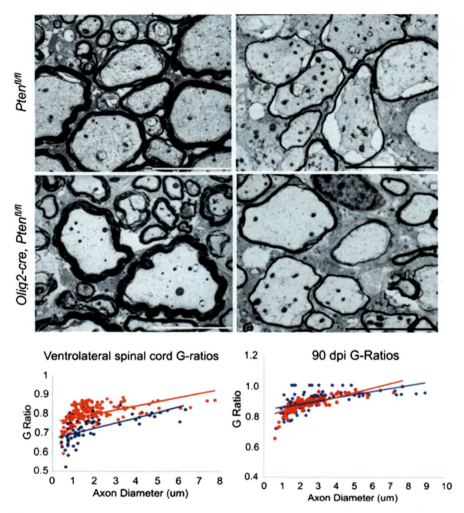

Fig. 4.2 When the Akt pathway is over-activated in mice where Pten has been conditionally deleted from oligodendrocyte lineage cells, the myelin sheath thickness is increased in relationship to axon diameter (i.e. the g ratio decreases). This is shown in the two electron micrographs on the *left* taken from the ventrolateral spinal cord white matter of adult control mice (Pten$^{fl/fl}$) and mice in which Pten is deleted from the oligodendrocyte lineage (Olig2-Cre, Pten$^{fl/fl}$). These images are confirmed below in the g-ratio plots (*blue dots*=Pten$^{fl/fl}$, *red dots*=Olig2-Cre, Pten$^{fl/fl}$). Following lysolecithin-induced demyelination on the ventrolateral white matter, the remyelinated axons in both experimental groups are the same (electron micrographs and *g*-ratio plots on *right*) (from Harrington et al. 2010)

multiple sclerosis (MS) and in its laboratory animal model experimental autoimmune encephalomyelitis (EAE). In this circumstance, remyelination is required to take place in an environment intrinsically hostile to the oligodendrocyte lineage. Thus, remyelination failure associated with MS (and EAE) can be seen not as a generic feature of remyelination biology but rather as a feature of specific disease states. However, even in MS, a disease normally associated with failed or inadequate remyelination, there is evidence that in some patients, complete remyelination occurs in a significant proportion of lesions (Patrikios et al. 2006; Patani et al. 2007; Goldschmidt et al. 2009; Piaton et al. 2009). Similarly, remyelination can be extensive in EAE, and models with significant persistent demyelination are unusual (Hampton et al. 2008; Linington et al. 1992). Remyelination appears to be especially efficient following demyelination of cerebral cortical grey matter in both experimental models (Merkler et al. 2006) and clinical disease (Albert et al. 2007), although the reason for this is unclear.

4.4 Remyelination Restores Function and Protects Axons

Remyelination restores saltatory conduction and reverses functional deficits (Liebetanz and Merkler 2006; Jeffery et al. 1999; Smith et al. 1979). Compelling evidence in support of functional restoration by remyelination has recently been provided by an unusual demyelinating condition in cats where the reversal of clinical signs is associated with spontaneous remyelination (Duncan et al. 2009).

A further and key function of remyelination is axon survival (Irvine and Blakemore 2008). Axonal and neuronal loss is now recognised as the major cause of chronic progressive disease (Trapp and Nave 2008), and it occurs as a secondary consequence of demyelination in addition to any primary effect of inflammation. Such a hypothesis explains why patients taking immunosuppressive therapies or with apparently quiescent disease still show increasing disability and progression, as these patients will have persistent demyelination as a result of failure of remyelination even in the absence of active disease. It should be noted that remyelination is not required for resolution of symptoms in an acute relapse: this likely results from resolution of inflammation and adaptive responses of the axon to restore conduction.

Evidence that myelin is required for axon survival is based on observations of genetic mouse models and studies of human pathology (Nave and Trapp 2008). Transgenic mice lacking CNP or PLP show long-term axonal degeneration, even in the presence of myelin sheaths that are either ultrastructurally normal or show only minor abnormalities (Griffiths et al. 1998; Lappe-Siefke et al. 2003). Further analysis of the PLP mutant mice has revealed a disturbance in axoplasmic transport in the absence of PLP (Edgar et al. 2004) and has led to the identification of myelin-associated sirtuin 2 as a potential mediator of long-term axonal stability (Werner et al. 2007). Myelin is also important for axon survival in humans, as patients with Pelizaeus–Merzbacher disease (PMD) caused by mutations in PLP show axon loss (Garbern et al. 2002), and studies of MS autopsy tissue show that axon preservation is seen in those areas where remyelination has occurred (Kornek et al. 2000). Axon degeneration has recently been observed as a consequence of genetically induced oligodendrocyte-specific ablation, even in Rag 1-deficient mice that have no functional lymphocytes (Pohl et al. 2011), thus providing compelling evidence that axon survival is dependent on intact oligodendrocytes and that axon degeneration in chronically demyelinated lesions can occur independently of inflammation. These observations imply that remyelination therapies will promote axon sparing in MS by the production of an oligodendrocyte-derived trophic factor signal to the axon.

4.5 The Mechanisms of Remyelination

4.5.1 Oligodendrocyte Precursor Cells Are the Main Source of New Myelin-Forming Oligodendrocytes

Remyelination involves the generation of new mature oligodendrocytes since (1) there is a greater number of oligodendrocytes within an area of remyelination compared to the equivalent area before myelination (Prayoonwiwat and Rodriguez 1993) and (2) remyelination occurs within areas depleted of oligodendrocytes (Sim et al. 2002b). In the vast majority of cases, the new oligodendrocytes that mediate remyelination are derived from a population of adult CNS stem/precursor cells, most often referred to as adult oligodendrocyte precursor (or progenitor) cells and sometimes called NG2 cells (in this chapter, OPC will generally refer to these cells). These multiprocessed proliferating cells are widespread throughout the CNS, occurring in both white matter and grey matter at a density similar to that of microglial cells (5–8 % of the cell population) (Horner et al. 2000; Richardson et al. 2011; Dawson et al. 2003). Adult OPCs are derived from their developmental forebears, and the two cells share many similarities, although the adult cell has a longer basal cell cycle time and slower rate of migration (Wolswijk and Noble 1989). Relevant to remyelination, the adult OPC can be induced to proliferate and migrate like perinatal cells in vitro by the growth factors PDGF and FGF (Wolswijk and Noble 1992), both of which are significantly upregulated during remyelination (Hinks and Franklin 1999).

Evidence obtained using Cre-lox fate mapping in transgenic mice following experimental demyelination has shown that OPCs produce the vast majority of remyelinating oligodendrocytes (Zawadzka et al. 2010; Tripathi et al. 2010) (Fig. 4.3). Remyelinating oligodendrocytes can also come from the stem and precursor cells of the adult subventricular zone (SVZ), either from the precursor cells contributing to the rostral migratory stream (RMS) (Nait-Oumesmar et al. 1999) or from the type B, GFAP-expressing stem cells of the SVZ per se (Menn et al. 2006). However, the contribution that SVZ-derived cells make relative to that of local OPCs may be relatively small, and their contribution to repair away from white matter tracts that are not close to the SVZ is likely to be negligible.

4.5.2 Remyelination Requires the Activation, Recruitment and Differentiation of Adult Oligodendrocyte Precursor Cells

In response to injury, OPCs in the vicinity undergo a switch from an essentially quiescent state to a regenerative phenotype. This activation is the first step in the

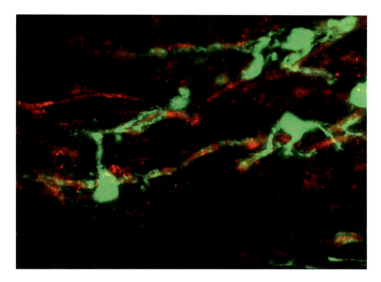

Fig. 4.3 OPCs can be labelled by a Cre-lox labelling strategy enabling the fate of these cells following lysolecithin-induced demyelination to be traced. In this confocal image, a YFP (*green*)-labelled cell is associated with a PLP+ (*red*) myelin sheath providing evidence that the labelled OPC has differentiated into a myelin sheath forming oligodendrocyte responsible for remyelination (see Zawadzka et al. 2010)

remyelination process and involves not only changes in morphology but also upregulation of several genes, many associated with the generation of oligodendrocytes during development such as the transcription factors Olig2, Nkx2.2, MyT1 and Sox2 (Watanabe et al. 2004; Fancy et al. 2004; Shen et al. 2008). The activation of OPCs is likely to be in response to acute injury-induced changes in microglia and astrocytes, two cell types exquisitely sensitive to disturbance in tissue homeostasis (Glezer et al. 2006; Rhodes et al. 2006). These two cell types, themselves activated by injury, are the major source of factors that induce the rapid proliferative response of OPCs to demyelinating injury. This response is modulated by the levels of the cell cycle regulatory proteins p27Kip-1 and Cdk2 (Crockett et al. 2005; Caillava et al. 2011) and is promoted by the growth factors PDGF and FGF (Murtie et al. 2005; Woodruff et al. 2004) and doubtless other factors associated with acute inflammatory lesions and demonstrated to have OPC mitogenic effects in tissue culture (Vela et al. 2002). Semaphorins have emerged as important regulators of OPC migration following demyelination. Developmental studies first identified semaphorins 3A and 3F as repulsive and attractive guidance cues for OPCs, respectively. It has subsequently been shown that adult OPCs express class 3 semaphorin receptors, neuropilins and plexins and that neuropilin expression increases after demyelination. Gain and loss of function experiments have shown that semaphorin 3A impairs OPC recruitment to the demyelinated area, while semaphorin 3F overexpression accelerates not only OPC recruitment but also remyelination rate (Piaton et al. 2011). The population of areas of demyelination by OPCs is referred to as the *recruitment phase* of remyelination and involves by OPC migration in addition to the ongoing proliferation.

For remyelination to be complete, the recruited OPC must next differentiate into remyelinating oligodendrocytes—the *differentiation phase*. This phase encompasses three distinct steps—establishing contact with the axon to be remyelinated, expression of myelin genes and generation of myelin membrane and finally wrapping and compacting to form the sheath. Despite these being fundamental properties of oligodendrocytes, we still have an incomplete understanding about how axoglial contact is established and how this interaction then regulates, within each individual cell process, the morphological changes that constitute myelination. That said, some molecules have been shown to contribute to the regulation of differentiation, and it is clear that the differentiation of OPCs into myelinating oligodendrocytes in development and during the regenerative process shares many similarities (Fancy et al. 2011a). FGF plays a key role in inhibiting differentiation as well as promoting recruitment and thereby regulates the correct transition from the recruitment to the differentiation phases (Armstrong et al. 2002), and IGF-I is another factor that plays major roles in both processes (Mason et al. 2003). Recently, it has been shown that semaphorin 3A, in addition to its role in OPC recruitment (Piaton et al. 2011), is also an inhibitor of OPC differentiation (Syed et al. 2011). LINGO-1, a component of the trimolecular Nogo receptor, has been found to be a negative regulator of oligodendrocyte differentiation in development (Mi et al. 2005), while mice deficient in LINGO-1 or treated with an antibody antagonist against LINGO-1 exhibited increased remyelination and functional recovery from experimental autoimmune encephalomyelitis (EAE) (Mi et al. 2007). The canonical Wnt pathway has recently emerged as a very powerful negative regulator of oligodendrocyte differentiation in both development and remyelination (Ye et al. 2009; Fancy et al. 2009). For example, constitutive induction of Wnt signalling in oligodendrocyte lineage cells by using transgenic mice that actively express a dominant active β-catenin gene results in mice displaying severe tremor and ataxia within 1 week after birth due to delayed oligodendrocyte differentiation and hypomyelination (Fancy et al. 2009). This effect is transient as CNS myelination ultimately catches up and appears normal in adult mice. Experimental demyelination performed on these transgenic mice results in a similar delay in oligodendrocyte differentiation and remyelination, without affecting OPC recruitment.

However, differences in the regulation of development and regeneration of myelin do occur; the transcription factor Olig1, although essential for developmental myelination (Xin et al. 2005), has a less redundant role in remyelination, where it plays a pivotal permissive role in OPC differentiation (Arnett et al. 2004). In contrast, notch signalling pathway, a negative (Wang et al. 1998) or positive (Hu et al. 2003) regulator of differentiation in development (depending on the ligand), is redundant during remyelination since conditional knockout of the notch1 gene in OPCs has no or a limited effect on remyelination (Stidworthy et al. 2004; Zhang et al. 2009).

Recently, the nuclear receptor retinoid X receptor-γ (RXRγ) has emerged a key positive regulator of oligodendrocyte differentiation directly from the analysis of remyelinating tissue (Huang et al. 2011). Microarray analysis of the separate stages of

CNS remyelination in rats revealed that RXR is highly expressed in oligodendrocytes lineage cells during the differentiation phase of CNS remyelination. Transfection of cultured OPCs with siRNAs generated against RXRγ resulted in less morphologically differentiated oligodendrocytes. Analysis of RXRγ knockout mice that have received focal CNS demyelination resulted in the accumulation of undifferentiated OPCs and less mature oligodendrocytes in lesions. These results indicate that RXRγ regulates oligodendrocyte differentiation.

4.5.3 Inflammation and Remyelination

Several studies have provided compelling evidence for a key role of the inflammatory response to demyelination in creating an environment conducive to remyelination. The relationship between inflammation and regeneration is well recognised in many other tissues. However, its involvement in myelin regeneration has been obscured in a field dominated by the immune-mediated pathology of MS and its various animal models such as EAE, where it is unquestionably true that the adaptive immune response mediates tissue damage. Nevertheless, several descriptive studies using experimental models (Ludwin 1980) and MS tissue (Wolswijk 2002) have pointed to an association between inflammation and remyelination. The role of the innate immune response to demyelination in remyelination has become apparent in part through the use of non-immune-mediated, toxin-induced models of demyelination. Depletion or pharmacological inhibition of macrophages following toxin-induced demyelination leads to an impairment of remyelination (Kotter et al. 2005; Li et al. 2005), as does the absence of T cells (Bieber et al. 2003). The pro-inflammatory cytokines Il-1β and TNF-α, lymphotoxin-β receptor or MHCII have also been implicated as mediators of remyelination following cuprizone-induced demyelination (Mason et al. 2001; Plant et al. 2007; Arnett et al. 2001, 2003). A critical role played by phagocytic macrophages is the removal of myelin debris generated during demyelination since CNS myelin contains proteins inhibitory to OPC differentiation both in vitro and during remyelination (Baer et al. 2009; Kotter et al. 2006). The observation that macrophage activation enhances myelination by transplanted OPCs in the myelin-free retinal nerve fibre layer points to additional and as yet undefined regenerative factors produced by these macrophages (Setzu et al. 2006).

4.6 Demyelinated CNS Axons Can Also Be Remyelinated by Schwann Cells

CNS remyelination can also be mediated by Schwann cells, the myelin-forming cells of the peripheral nervous system; this occurs in several experimental animal models of demyelination as well as in human demyelinating disease (Snyder et al. 1975; Itoyama et al. 1983, 1985; Dusart et al. 1992; Felts et al. 2005). Schwann cell remyelination occurs preferentially where astrocytes are absent—for example, where they have been killed along with oligodendrocytes by the demyelinating agent (Blakemore 1975; Itoyama et al. 1985). Remyelinating Schwann cells within the CNS were generally thought to migrate into the CNS from PNS sources such as spinal and cranial roots, meningeal fibres or autonomic nerves following a breach in the *glia limitans* (Franklin and Blakemore 1993). In support of this idea, CNS Schwann cell remyelination typically occurs in proximity to spinal/cranial nerves or around blood vessels (Snyder et al. 1975; Duncan and Hoffman 1997; Sim et al. 2002a). However, recent genetic fate-mapping studies have revealed that very few CNS remyelinating Schwann cells are derived from PNS Schwann cells but instead the majority derive from OPCs (Zawadzka et al. 2010), revealing a remarkable capacity of these cells to differentiate into cells of neural crest lineage as well as all three neuroepithelial lineages (neurons, astrocytes and oligodendrocytes).

The implications of Schwann cell remyelination of CNS axons are unclear. While both Schwann cell and oligodendrocyte remyelination are associated with a return of saltatory conduction (Smith et al. 1979), their relative abilities to promote axon survival, a major function of myelin (Nave and Trapp 2008), have yet to be established. Thus, from a clinical perspective, we do not yet know whether OLP differentiation into Schwann cells has a beneficial or deleterious effect compared to oligodendrocyte remyelination.

4.7 Causes of Remyelination Failure

The efficiency of remyelination is affected by the non-disease-related factors of age and sex (Li et al. 2006; Sim et al. 2002b). These generic factors will have a bearing on the efficiency of remyelination regardless of the disease process involved and will be discussed first.

Like all other regenerative processes, the efficiency of remyelination decreases with age. This manifests as a decrease in the rate at which it occurs and is likely to have a profound bearing on the outcome of a disease process that in the case of MS can occur over many decades. The age-associated effects on remyelination are due to a decrease in the efficiency of both OPC recruitment and differentiation (Sim et al. 2002b). Of these two events, the impairment of differentiation is rate determining since increasing the provision of OPCs by the overexpression of the OPC mitogen and recruitment factor PDGF following demyelination in old mice does not accelerate remyelination (Woodruff et al. 2004). The impairment of OPC differentiation in ageing mirrors the failure of oligodendrocyte lineage differentiation associated with many chronically demyelinated MS plaques (Wolswijk 1998; Kuhlmann et al. 2008).

The basis of the ageing effect is likely to lie in age-associated changes in both the extrinsic environmental signals to which OPCs are exposed in remyelinating lesions and to intrinsic determinants of OPC behaviour. An impaired macrophage response in ageing, associated with a delay in expression of inflammatory cytokines and chemokines (Zhao et al. 2006), leads to poor clearance of myelin debris and the persistence of myelin-associated differentiation-inhibitory proteins. Changes also occur in the expression of remyelination-associated growth factors following toxin-induced demyelination that are commensurate with delays in OPC activation, recruitment and differentiation and are illustrative of age-associated environmental changes in the remyelination (Hinks and Franklin 2000). Both in vitro studies revealing age-associated changes in growth factor responsiveness of adult OPCs of different age (Tang et al. 2000) and in vivo studies demonstrating slower recruitment of transplanted old adult precursor cells compared to young adult-derived cells into precursor-depleted white matter (Chari et al. 2003) are indicative of intrinsic changes occurring with OPC during adult ageing. A recent study confirms these observations, revealing a critical age-associated change in the epigenetic regulation of OPC differentiation during remyelination (Shen et al. 2008). Differentiation of OPCs is associated with the recruitment of histone deacetylases (HDACs) to promoter regions of differentiation inhibitors (Marin-Husstege et al. 2002). In old animals, HDAC recruitment is impaired, resulting in prolonged expression of these inhibitors, delayed OPC differentiation and hence slower remyelination. This effect can be replicated following induction of demyelination in young animals with the use of the HDAC antagonist valproic acid. A key question relating to the development of remyelination therapies is the extent to which age-associated changes can be reversed. Intriguing experiments on skeletal muscle regeneration using the technique of heterochronic parabiosis provide clear proof of principle that poor regeneration in old animals can be rejuvenated (Conboy et al. 2005).

In addition to these generic factors, remyelination could also be incomplete or fail for disease-specific reasons. The strongest evidence for remyelination failure is provided by MS, and the subsequent discussion will specifically relate to this disease, although the issues discussed will be relevant to other diseases with a demyelinating component. Theoretically, remyelination could fail because of (1) a primary deficiency in precursor cells, (2) a failure of precursor cell recruitment or (3) a failure of precursor cell differentiation and maturation.

Early speculation on remyelination failure focussed on the first of these mechanisms, that is, the process of remyelination itself would deplete an area of CNS of its precursor cells so that subsequent episodes of demyelination occurring at or around the same site would fail to remyelinate due to a lack of OPCs. However, data from experimental studies indicate that OPCs are remarkably efficient at repopulating regions from which they have been depleted (Chari and Blakemore 2002) and that repeat episodes of focal demyelination in the same area (Penderis et al. 2003a) neither deplete OPCs nor prevent subsequent remyelination. The situation may be different, however, when the same area of tissue is exposed to a sustained demyelinating insult, where remyelination impairment seems to be due, at least in part, to a deficiency in OPC availability (Armstrong et al. 2006; Mason et al. 2004).

In the second mechanism, MS lesions fail to remyelinate not because of a shortage of available precursor cells but rather because of a failure of OPC recruitment: proliferation, migration and repopulation of areas of demyelination. Here, descriptions of demyelinated areas from which oligodendrocyte lineage (OL) cells are absent do indicate that this may account for failure of remyelination in at least a proportion of lesions. Why lesions should become deficient in OPCs is not clear, but one possibility is that they are direct targets of the disease process within the lesion. The identification of patients with antibodies recognising OPC-expressed antigens (NG2) supports this possibility (Niehaus et al. 2000). Failure of OPC recruitment into areas of demyelination may arise due to disturbances in the local expression of the OPC migration guidance cues semaphorins 3A and 3F (Williams et al. 2007a). In situations where OPCs need to be recruited into lesions from surrounding intact tissue, the size of the lesion will clearly have a bearing on the efficiency of remyelination, larger lesions requiring a greater OPC recruitment impetus than smaller ones, especially in ageing where older OPCs appear intrinsically less responsive to recruitment signals.

The best evidence at present supports the third mechanism, a failure of differentiation and maturation, as several sets of observations based on the detection of oligodendrocyte lineage cells within areas of demyelination indicate that this stage of remyelination is the most vulnerable to failure in MS. The presence of OPCs apparently unable to differentiate within MS lesions was initially shown with the OL marker O4 (Wolswijk 1998) and subsequently with NG2 (Chang et al. 2000), with PLP (to reveal pre-myelinating oligodendrocytes) (Chang et al. 2002) and with Olig2 and Nkx2.2 (Kuhlmann et al. 2008). Even though the density of OPCs within chronic lesions is on average lower than in normal white matter, the density can be as high as that in normal white matter or remyelinated lesions, showing that OPC availability is not a limiting factor for remyelination.

One possible explanation for this failure of differentiation is that chronically demyelinated lesions contain factors that inhibit precursor differentiation. First implicated was the notch-jagged pathway, a negative regulator of OPC differentiation: notch and its downstream activator Hes5 were detected in OPCs and jagged in astrocytes within chronic demyelinated MS lesions (John et al. 2002). However, the expression of notch by OPCs and jagged by other cells within lesions undergoing remyelination and, more informatively, the limited remyelination phenotype in experimental models following conditional deletion of notch in OL cells suggest that notch-jagged signalling is not a critical non-redundant negative regulator of remyelination (Zhang et al. 2009; Stidworthy et al. 2004). The ability of inhibitors of γ-secretase, an enzyme involved in the notch pathway, to enhance recovery following EAE might be indicative of an inhibitory role for notch signalling in remyelination, but is difficult to interpret given the additional expression of notch in the inflammatory effector cells.

Other potential inhibitory factors have been identified in other experimental and pathological studies. The accumulation of the glycosaminoglycan inhibitor of OPC differentiation hyaluronan within MS lesions may contribute to an environment within chronic lesions that is not conducive to remyelination by inhibiting OPC function via TLR2 signalling (Back et al. 2005; Sloane et al. 2010). The demyelinated axon itself has been implicated since demyelinated axons have been shown to express the adhesion molecule PSA-NCAM (Charles et al. 2002), which inhibits myelination in cell culture (Charles et al. 2000). The possibility that the properties of the OPC within areas of demyelination might be regulated by synaptic input from axons represents an exciting new development in the understanding of the complexity of regulatory factors that govern remyelination and by extension account for remyelination failure (Etxeberria et al. 2010).

While many studies in the last few years have concentrated on putative inhibitory signals to account for the failure of OPC to undergo complete differentiation within demyelinated MS plaques, an alternative explanation is that these lesions fail to remyelinate because of a deficiency of signals that induce differentiation. This hypothesis, based on the absence of factors, is difficult to prove but is consistent with a model of remyelination in which the acute inflammatory events play a key role in precursor activation and creating an environment conducive to remyelination (see above). While MS lesions are rarely devoid of any inflammatory activity, chronic lesions are relatively non-inflammatory compared to the acute lesions and constitute a less active environment in which OPC differentiation might become quiescent. Acute inflammatory lesions are characterised by reactive astrocytes that are the sources of many pro-remyelination-signalling factors (Williams et al. 2007b; Moore et al. 2011). In contrast, chronic quiescent lesions are characterised by scarring astrocytes that are transcriptionally quiet compared to reactive astrocytes. The scarring astrocyte is better viewed as a consequence of remyelination failure and not its cause. Thus, neither the reactive nor the scarring astrocytes (both confusingly contributing to astrogliosis) should be viewed as primary drivers of remyelination failure.

The two possibilities that remyelination failure reflects the presence of negative factors or the absence of positive factors are not, of course, mutually exclusive. Moreover, it has become apparent from many studies in recent years that there are a multitude of interacting factors, both environmental and intrinsic, that guide the behaviour of OL cells through the various stages of remyelination. Efficient remyelination may depend as much on the precise timing of action as on the presence or absence of these factors. In an earlier review, we articulated this in a model called the dysregulation hypothesis in which remyelination failure reflects an inappropriate sequence of events (Franklin and ffrench-Constant 2008; Franklin 2002; Fancy et al. 2011a). While the causes of remyelination failure in such a varied disease as MS are likely to be multiple, we still regard hypothesis as useful for understanding remyelination failure in the majority of cases.

4.8 Enhancing Endogenous Remyelination

Since remyelination can occur completely and that the cells responsible are abundant throughout the adult CNS, even within demyelinated lesions, a conceptually attractive approach to enhancing remyelination is to target the endogenous regenerative process. This approach is predicated on the view that if the mechanisms of remyelination can be understood and non-redundant pathways described, then the causes of remyelination failure and hence plausible therapeutic targets will be identified. From the preceding sections, it will be clear that remyelination failure is associated with either insufficient OPC recruitment or, more commonly, failed

OPC/oligodendrocyte differentiation. However, the underlying biology of these two phases of remyelination is different and sometimes mutually exclusive. The implication is therefore that pro-recruitment therapies may not promote remyelination where the primary problem is OPC differentiation and vice versa.

A further consideration in the development of remyelination enhancement therapies is the use of appropriate animal models. In the chronic demyelinated plaques of MS, remyelination is assumed to have failed, and hence the requirement is for an intervention that will reactivate a dormant process. In contrast, in many of the demyelination models used to test enhancement of remyelination such as the toxin-based models, demyelination does not fail, and so at best, it is possible to achieve acceleration of an already effective ongoing process. That said, definitive knowledge of the dynamics of remyelination in MS tissue, that is, whether it has stopped or slowed, is always difficult to assess from the "snapshots" provided by biopsy or post-mortem tissue. Nevertheless, this problem can in part be overcome in two ways: first, by using aged animals where the slow rate of remyelination is suboptimal presenting an opportunity for its enhancement and second, by modifications of standard lesion models where the endogenous process is compromised, such as in the chronic cuprizone model (Armstrong et al. 2006). The Theiler's virus-induced demyelination model has also proven useful for demonstrating enhanced remyelination (Njenga et al. 1999). However, assessment of remyelination is especially complicated in EAE, where the processes of demyelination and remyelination can occur concurrently. This can make it very difficult to distinguish an effect that renders the environment less hostile to remyelination, allowing it to proceed at its natural rate, from one in which the rate of remyelination to proceed is accelerated. For example, systemic delivery of putative remyelination-enhancing factors can affect the balance of myelin damage and regeneration via effects on cells other than oligodendroglial cells, such as those of the immune system. This may account for the discrepancy in the studies of IGF-I and GGF-2 administered systemically in EAE and delivered locally in non-immune-mediated models of demyelination (O'Leary et al. 2002; Yao et al. 1995; Penderis et al. 2003b; Cannella et al. 1998).

Despite caveats regarding models and methods of analysis, several recent studies provide proof of principle for enhancement of remyelination. An especially intriguing line of investigation has been the identification of polyreactive IgM autoantibodies that react with oligodendrocyte surface antigens and promote remyelination (Warrington et al. 2000), although the mechanisms of this effect remain unclear.

Over the last few years, three pathways have emerged amenable to pharmacological manipulation that holds considerable promise for the development of drug-based remyelination-enhancing medicines (Fancy et al. 2010). First, humanised monoclonal antibodies against LINGO-1 have been developed and are already in phase 1 clinical trials. Second, the growing interest to develop pharmacological inhibitors against the Wnt pathway in cancer therapy means that it might be possible in the near future to use Wnt inhibitors to stimulate OPC differentiation. Particularly relevant here is the recent report that small-molecule inhibitors of tankyrase, an ADP-polyribosylating enzyme that stabilises Axin and thereby releases OPCs from Wnt pathway-mediated differentiation block, can induce precocious OCP differentiation and thus accelerate remyelination (Fancy et al. 2011b). Third, chemical agonists and

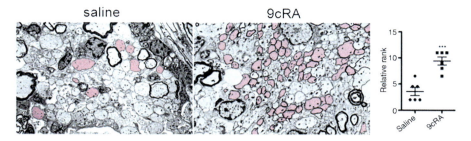

Fig. 4.4 9-*cis*-Retinoic acid enhances CNS remyelination. Ultrastructural microscopy shows many remyelinated axons (shown in *pink*) compared to axons that remained demyelinated following systemic 9cRA administration to ageing rats in which demyelination has been induced in the cerebellar peduncle by injection of ethidium bromide. Ranking analysis of semi-thin sections shows increased remyelination in 9cRA-treated compared to saline-treated animals. Highest rank = highest degree of CNS remyelination (from Huang et al. 2011)

antagonists of RXR signalling, or rexinoids, are widely available and are showing promise in the treatment of certain types of cancers and metabolic disorders (Altucci et al. 2007). When cultured OPCs are exposed to the RXR selective antagonists, HX531 and PA452, oligodendrocyte differentiation is severely impaired (Huang et al. 2011). By contrast, when OPCs are exposed to the RXR agonists, 9-*cis*-retinoic acid (9cRA), HX630, or PA024, oligodendrocytes are stimulated to differentiate and form myelin membrane-like sheets in culture. Further experiments testing the effect of 9cRA on aged rats that received focal demyelination resulted in the significant acceleration of remyelination (Fig. 4.4).

4.9 Conclusions: Translating Remyelination Biology into Remyelination Medicine

Bridging the gap from regenerative biology to regenerative medicine is not unique to remyelination and CNS, but is being explored in many other tissues including liver, skin, pancreas, bone and blood. At least some of the mechanisms for the generation of new tissue-specific cells from endogenous stem and precursor cells will be shared, so the current interest in stem cell medicine will inform future remyelination therapies. The significant advances in remyelination biology in recent years provide cautious optimism that these can be translated into remyelination therapy; it is realistic to think that progress over the next decade will be rapid.

References

Albert M, Antel J, Bruck W, Stadelmann C (2007) Extensive cortical remyelination in patients with chronic multiple sclerosis. Brain Pathol 17:129–138
Altucci L, Leibowitz MD, Ogilvie KM, de Lera AR, Gronemeyer H (2007) RAR and RXR modulation in cancer and metabolic disease. Nat Rev Drug Discov 6:793–810

Armstrong RC, Le TQ, Frost EE, Borke RC, Vana AC (2002) Absence of fibroblast growth factor 2 promotes oligodendroglial repopulation of demyelinated white matter. J Neurosci 22:8574–8585

Armstrong RC, Le TQ, Flint NC, Vana AC, Zhou YX (2006) Endogenous cell repair of chronic demyelination. J Neuropathol Exp Neurol 65:245–256

Arnett HA, Mason J, Marino M, Suzuki K, Matsushima GK, Ting JPY (2001) TNF alpha promotes proliferation of oligodendrocyte progenitors and remyelination. Nat Neurosci 4:1116–1122

Arnett HA, Wang Y, Matsushima GK, Suzuki K, Ting JP (2003) Functional genomic analysis of remyelination reveals importance of inflammation in oligodendrocyte regeneration. J Neurosci 23:9824–9832

Arnett HA, Fancy SPJ, Alberta JA, Zhao C, Plant SR, Raine CS, Rowitch DH, Franklin RJM, Stiles CD (2004) The bHLH transcription factor Olig1 is required for repair of demyelinated lesions in the CNS. Science 306:2111–2115

Back SA, Tuohy TM, Chen H, Wallingford N, Craig A, Struve J, Luo NL, Banine F, Liu Y, Chang A, Trapp BD, Bebo BF, Rao MS, Sherman LS (2005) Hyaluronan accumulates in demyelinated lesions and inhibits oligodendrocyte progenitor maturation. Nat Med 11:966–972

Baer AS, Syed YA, Kang SU, Mitteregger D, Vig R, ffrench-Constant C, Franklin RJM, Altmann F, Lubec G, Kotter MR (2009) Myelin-mediated inhibition of oligodendrocyte precursor differentiation can be overcome by pharmacological modulation of Fyn-RhoA and protein kinase C signalling. Brain 132:465–481

Bieber AJ, Kerr S, Rodriguez M (2003) Efficient central nervous system remyelination requires T cells. Ann Neurol 53:680–684

Blakemore WF (1975) Remyelination by Schwann cells of axons demyelinated by intraspinal injection of 6-aminonicotinamide in the rat. J Neurocytol 4:745–757

Blakemore WF, Franklin RJM (2008) Remyelination in experimental models of toxin-induced demyelination. Curr Top Microbiol Immunol 318:193–212

Brinkmann BG, Agarwal A, Serada MW, Garratt AN, Mueller T, Wende H, Stassart RM, Nawaz S, Humml C, Velanac V, Radyuschkin K, Goebbels S, Fischer TM, Franklin RJM, Lai C, Ehrenreich H, Birchmeier C, Schwab MH, Nave KA (2008) Neuregulin-1/ErbB signaling serves distinct functions in myelination of the peripheral and central nervous system. Neuron 59:581–594

Caillava C, Vandenbosch R, Jablonska B, Deboux C, Spigoni G, Gallo V, Malgrange B, Baron-Van Evercooren A (2011) Cdk2 loss accelerates precursor differentiation and remyelination in the adult central nervous system. J Cell Biol 193:397–407

Cannella B, Hoban CJ, Gao YL, Garcia-Arenas R, Lawson D, Marchionni M, Gwynne D, Raine CS (1998) The neuregulin, glial growth factor 2, diminishes autoimmune demyelination and enhances remyelination in a chronic relapsing model for multiple sclerosis. Proc Natl Acad Sci USA 95:10100–10105

Chang A, Nishiyama A, Peterson J, Prineas J, Trapp BD (2000) NG2-positive oligodendrocyte progenitor cells in adult human brain and multiple sclerosis lesions. J Neurosci 20:6404–6412

Chang A, Tourtellotte WW, Rudick R, Trapp BD (2002) Premyelinating oligodendrocytes in chronic lesions of multiple sclerosis. N Engl J Med 346:165–173

Chari DM, Blakemore WF (2002) Efficient recolonisation of progenitor-depleted areas of the CNS by adult oligodendrocyte progenitor cells. Glia 37:307–313

Chari DM, Crang AJ, Blakemore WF (2003) Decline in rate of colonization of oligodendrocyte progenitor cell (OPC)-depleted tissue by adult OPCs with age. J Neuropathol Exp Neurol 62:908–916

Charles P, Hernandez MP, Stankoff B, Aigrot MS, Colin C, Rougon G, Zalc B, Lubetzki C (2000) Negative regulation of central nervous system myelination by polysialylated-neural cell adhesion molecule. Proc Natl Acad Sci USA 97:7585–7590

Charles P, Reynolds R, Seilhean D, Rougon G, Aigrot MS, Niezgoda A, Zalc B, Lubetzki C (2002) Re-expression of PSA-NCAM by demyelinated axons: an inhibitor or remyelination in multiple sclerosis? Brain 125:1972–1979

Conboy IM, Conboy MJ, Wagers AJ, Girma ER, Weissman IL, Rando TA (2005) Rejuvenation of aged progenitor cells by exposure to a young systemic environment. Nature 433:760–764

Crockett DP, Burshteyn M, Garcia C, Muggironi M, Casaccia-Bonnefil P (2005) Number of oligodendrocyte progenitors recruited to the lesioned spinal cord is modulated by the levels of the cell cycle regulatory protein p27Kip-1. Glia 49:301–308

Dawson MRL, Polito A, Levine JM, Reynolds R (2003) NG2-expressing glial progenitor cells: an abundant and widespread population of cycling cells in the adult rat CNS. Mol Cell Neurosci 24:476–488

Duncan ID, Hoffman RL (1997) Schwann cell invasion of the central nervous system of the myelin mutants. J Anat 190:35–49

Duncan ID, Brower A, Kondo Y, Curlee JF Jr, Schultz RD (2009) Extensive remyelination of the CNS leads to functional recovery. Proc Natl Acad Sci USA 106:6832–6836

Dusart I, Marty S, Peschanski M (1992) Demyelination and remyelination by Schwann cells and oligodendrocytes after kainate-induced neuronal depletion in the central nervous system. Neuroscience 5:137–148

Edgar JM, McLaughlin M, Yool D, Zhang SC, Fowler JH, Montague P, Barrie JA, McCulloch MC, Duncan ID, Garbern J, Nave KA, Griffiths IR (2004) Oligodendroglial modulation of fast axonal transport in a mouse model of hereditary spastic paraplegia. J Cell Biol 166:121–131

Etxeberria A, Mangin JM, Aguirre A, Gallo V (2010) Adult-born SVZ progenitors receive transient synapses during remyelination in corpus callosum. Nat Neurosci 13:287–289

Fancy SPJ, Zhao C, Franklin RJM (2004) Increased expression of Nkx2.2 and Olig2 identifies reactive oligodendrocyte progenitor cells responding to demyelination in the adult CNS. Mol Cell Neurosci 27:247–254

Fancy SPJ, Baranzini SE, Zhao C, Yuk DI, Irvine KA, Kaing S, Sanai N, Franklin RJM, Rowitch DH (2009) Dysregulation of the Wnt pathway inhibits timely myelination and remyelination in the mammalian CNS. Genes Dev 23:1571–1585

Fancy SPJ, Kotter MR, Harrington EP, Huang JK, Zhao C, Rowitch DH, Franklin RJM (2010) Overcoming remyelination failure in multiple sclerosis and other myelin disorders. Exp Neurol 225:18–23

Fancy SPJ, Chan JR, Baranzini SE, Franklin RJM, Rowitch DH (2011a) Myelin regeneration: a recapitulation of development? Annu Rev Neurosci 34:19–41

Fancy SPJ, Harrington EP, Yuen TJ, Silbereis JC, Zhao C, Baranzini SE, Bruce CC, Otero JJ, Huang EJ, Nusse R, Franklin RJM, Rowitch DH (2011b) Axin2 as regulatory and therapeutic target in newborn brain injury and remyelination. Nat Neurosci 14(8):1009–1016

Felts PA, Woolston AM, Fernando HB, Asquith S, Gregson NA, Mizzi OJ, Smith KJ (2005) Inflammation and primary demyelination induced by the intraspinal injection of lipopolysaccharide. Brain 128:1649–1666

Flores AI, Narayanan SP, Morse EN, Shick HE, Yin X, Kidd G, Avila RL, Kirschner DA, Macklin WB (2008) Constitutively active Akt induces enhanced myelination in the CNS. J Neurosci 28:7174–7183

Franklin RJM (2002) Why does remyelination fail in multiple sclerosis? Nat Rev Neurosci 3:705–714

Franklin RJM, Blakemore WF (1993) Requirements for Schwann cell migration within CNS environments: a viewpoint. Int J Dev Neurosci 11:641–649

Franklin RJM, ffrench-Constant C (2008) Remyelination in the CNS: from biology to therapy. Nat Rev Neurosci 9:839–855

Franklin RJM, Hinks GL (1999) Understanding CNS remyelination – clues from developmental and regeneration biology. J Neurosci Res 58:207–213

Garbern JY, Yool DA, Moore GJ, Wilds IB, Faulk MW, Klugmann M, Nave KA, Sistermans EA, van der Knaap MS, Bird TD, Shy ME, Kamholz JA, Griffiths IR (2002) Patients lacking the major CNS myelin protein, proteolipid protein 1, develop length-dependent axonal degeneration in the absence of demyelination and inflammation. Brain 125:551–561

Glezer I, Lapointe A, Rivest S (2006) Innate immunity triggers oligodendrocyte progenitor reactivity and confines damages to brain injuries. FASEB J 20:750–752

Goldschmidt T, Antel J, Konig FB, Bruck W, Kuhlmann T (2009) Remyelination capacity of the MS brain decreases with disease chronicity. Neurology 72:1914–1921

Griffiths I, Klugmann M, Anderson T, Yool D, Thomson C, Schwab MH, Schneider A, Zimmermann F, McCulloch M, Nadon N, Nave KA (1998) Axonal swellings and degeneration in mice lacking the major proteolipid of myelin. Science 280:1610–1613

Hampton DW, Anderson J, Pryce G, Irvine KA, Giovannoni G, Fawcett JW, Compston A, Franklin RJM, Baker D, Chandran S (2008) An experimental model of secondary progressive multiple sclerosis that shows regional variation in gliosis, remyelination, axonal and neuronal loss. J Neuroimmunol 201–202:200–211

Harrington EP, Zhao C, Fancy SPJ, Kaing S, Franklin RJM, Rowitch DH (2010) Oligodendrocyte PTEN required for myelin and axonal integrity not remyelination. Ann Neurol 68:703–726

Hinks GL, Franklin RJM (1999) Distinctive patterns of PDGF-A, FGF-2, IGF-I and TGF-beta1 gene expression during remyelination of experimentally-induced spinal cord demyelination. Mol Cell Neurosci 14:153–168

Hinks GL, Franklin RJM (2000) Delayed changes in growth factor gene expression during slow remyelination in the CNS of aged rats. Mol Cell Neurosci 16:542–556

Horner PJ, Power AE, Kempermann G, Kuhn HG, Palmer TD, Winkler J, Thal LJ, Gage FH (2000) Proliferation and differentiation of progenitor cells throughout the intact adult rat spinal cord. J Neurosci 20:2218–2228

Hu QD, Ang BT, Karsak M, Hu WP, Cui XY, Duka T, Takeda Y, Chia W, Sankar N, Ng YK, Ling EA, Maciag T, Small D, Trifonova R, Kopan R, Okano H, Nakafuku M, Chiba S, Hirai H, Aster JC, Schachner M, Pallen CJ, Watanabe K, Xiao ZC (2003) F3/Contactin acts as a functional ligand for Notch during oligodendrocyte maturation. Cell 115:163–175

Huang JK, Jarjour AA, Nait Oumesmar B, Kerninon C, Williams A, Krezel W, Kagechika H, Bauer J, Zhao C, Baron van Evercooren A, Chambon P, ffrench-Constant C, Franklin RJM (2011) Retinoid X receptor gamma signaling accelerates CNS remyelination. Nat Neurosci 14:45–53

Irvine KA, Blakemore WF (2008) Remyelination protects axons from demyelination-associated axon degeneration. Brain 131:1464–1477

Itoyama Y, Webster HD, Richardson EP Jr, Trapp BD (1983) Schwann cell remyelination of demyelinated axons in spinal cord multiple sclerosis lesions. Ann Neurol 14:339–346

Itoyama Y, Ohnishi A, Tateishi J, Kuroiwa Y, Webster HD (1985) Spinal cord multiple sclerosis lesions in Japanese patients: Schwann cell remyelination occurs in areas that lack glial fibrillary acidic protein (GFAP). Acta Neuropathol (Berl) 65:217–223

Jeffery ND, Crang AJ, O'Leary MT, Hodge SJ, Blakemore WF (1999) Behavioural consequences of oligodendrocyte progenitor cell transplantation into demyelinating lesions in rat spinal cord. Eur J Neurosci 11:1508–1514

John GR, Shankar SL, Shafit-Zagardo B, Massimi A, Lee SC, Raine CS, Brosnan CF (2002) Multiple sclerosis: re-expression of a developmental pathway that restricts oligodendrocyte maturation. Nat Med 8:1115–1121

Kornek B, Storch MK, Weissert R, Wallstroem E, Stefferl A, Olsson T, Linington C, Schmidbauer M, Lassmann H (2000) Multiple sclerosis and chronic autoimmune encephalomyelitis: a comparative quantitative study of axonal injury in active, inactive, and remyelinated lesions. Am J Pathol 157:267–276

Kotter MR, Zhao C, van Rooijen N, Franklin RJM (2005) Macrophage-depletion induced impairment of experimental CNS remyelination is associated with a reduced oligodendrocyte progenitor cell response and altered growth factor expression. Neurobiol Dis 18:166–175

Kotter MR, Li WW, Zhao C, Franklin RJM (2006) Myelin impairs CNS remyelination by inhibiting oligodendrocyte precursor cell differentiation. J Neurosci 26:328–332

Kuhlmann T, Miron V, Cuo Q, Wegner C, Antel J, Bruck W (2008) Differentiation block of oligodendroglial progenitor cells as a cause for remyelination failure in chronic multiple sclerosis. Brain 131(Pt 7):1749–1758

Lappe-Siefke C, Goebbels S, Gravel M, Nicksch E, Lee J, Braun PE, Griffiths IR, Nave KA (2003) Disruption of Cnp1 uncouples oligodendroglial functions in axonal support and myelination. Nat Genet 33:366–374

Lasiene J, Shupe L, Perlmutter S, Horner P (2008) No evidence for chronic demyelination in spared axons after spinal cord injury in a mouse. J Neurosci 28:3887–3896

Li WW, Setzu A, Zhao C, Franklin RJM (2005) Minocycline-mediated inhibition of microglia activation impairs oligodendrocyte progenitor cell responses and remyelination in a non-immune model of demyelination. J Neuroimmunol 158:58–66

Li WW, Penderis J, Zhao C, Schumacher M, Franklin RJM (2006) Females remyelinate more efficiently than males following demyelination in the aged but not young adult CNS. Exp Neurol 202:250–254

Liebetanz D, Merkler D (2006) Effects of commissural de- and remyelination on motor skill behaviour in the cuprizone mouse model of multiple sclerosis. Exp Neurol 202:217–224

Linington C, Engelhardt B, Kapocs G, Lassman H (1992) Induction of persistently demyelinated lesions in the rat following the repeated adoptive transfer of encephalitogenic T cells and demyelinating antibody. J Neuroimmunol 40:219–224

Ludwin SK (1980) Chronic demyelination inhibits remyelination in the central nervous system. Lab Invest 43:382–387

Marin-Husstege M, Muggironi M, Liu A, Casaccia-Bonnefil P (2002) Histone deacetylase activity is necessary for oligodendrocyte lineage progression. J Neurosci 22:10333–10345

Mason JL, Suzuki K, Chaplin DD, Matsushima GK (2001) Interleukin-1beta promotes repair of the CNS. J Neurosci 21:7046–7052

Mason JL, Xuan S, Dragatsis I, Efstratiadis A, Goldman JE (2003) Insulin-like growth factor (IGF) signaling through type 1 IGF receptor plays an important role in remyelination. J Neurosci 23:7710–7718

Mason JL, Toews A, Hostettler JD, Morell P, Suzuki K, Goldman JE, Matsushima GK (2004) Oligodendrocytes and progenitors become progressively depleted within chronically demyelinated lesions. Am J Pathol 164:1673–1682

Menn B, Garcia-Verdugo JM, Yaschine C, Gonzalez-Perez O, Rowitch D, Alvarez-Buylla A (2006) Origin of oligodendrocytes in the subventricular zone of the adult brain. J Neurosci 26:7907–7918

Merkler D, Ernsting T, Kerschensteiner M, Bruck W, Stadelmann C (2006) A new focal EAE model of cortical demyelination: multiple sclerosis-like lesions with rapid resolution of inflammation and extensive remyelination. Brain 129:1972–1983

Mi S, Miller RH, Lee X, Scott ML, Shulag-Morskaya S, Shao Z, Chang J, Thill G, Levesque M, Zhang M, Hession C, Sah D, Trapp B, He Z, Jung V, McCoy JM, Pepinsky RB (2005) LINGO-1 negatively regulates myelination by oligodendrocytes. Nat Neurosci 8:745–751

Mi S, Hu B, Hahm K, Luo Y, Kam Hui ES, Yuan Q, Wong WM, Wang L, Su H, Chu TH, Guo J, Zhang W, So KF, Pepinsky B, Shao Z, Graff C, Garber E, Jung V, Wu EX, Wu W (2007) LINGO-1 antagonist promotes spinal cord remyelination and axonal integrity in MOG-induced experimental autoimmune encephalomyelitis. Nat Med 13:1228–1233

Michailov GV, Sereda MW, Brinkmann BG, Fischer TM, Haug B, Birchmeier C, Role L, Lai C, Schwab MH, Nave KA (2004) Axonal neuregulin-1 regulates myelin sheath thickness. Science 304:700–703

Moore CS, Milner R, Nishiyama A, Frausto RF, Serwanski DR, Pagarigan RR, Whitton JL, Miller RH, Crocker SJ (2011) Astrocytic tissue inhibitor of metalloproteinase-1 (TIMP-1) promotes oligodendrocyte differentiation and enhances CNS myelination. J Neurosci 31:6247–6254

Murtie JC, Zhou YX, Le TQ, Vana AC, Armstrong RC (2005) PDGF and FGF2 pathways regulate distinct oligodendrocyte lineage responses in experimental demyelination with spontaneous remyelination. Neurobiol Dis 19:171–182

Nait-Oumesmar B, Decker L, Lachapelle F, Avellana-Adalid V, Bachelin C, Van Evercooren AB (1999) Progenitor cells of the adult mouse subventricular zone proliferate, migrate and differentiate into oligodendrocytes after demyelination. Eur J Neurosci 11:4357–4366

Nave KA, Trapp BD (2008) Axon-glial signaling and the glial support of axon function. Annu Rev Neurosci 31:535–561

Niehaus A, Shi J, Grzenkowski M, Diers-Fenger M, Archelos J, Hartung HP, Toyka K, Bruck W, Trotter J (2000) Patients with active relapsing-remitting multiple sclerosis synthesize antibodies recognizing oligodendrocyte progenitor cell surface protein: implications for remyelination. Ann Neurol 48:362–371

Njenga MK, Murray PD, McGavern D, Lin X, Drescher KM, Rodriguez M (1999) Absence of spontaneous central nervous system remyelination in class II-deficient mice infected with Theiler's virus. J Neuropathol Exp Neurol 58:78–91

O'Leary MT, Hinks GL, Charlton HM, Franklin RJM (2002) Increasing local levels of IGF-I mRNA expression using adenoviral vectors does not alter oligodendrocyte remyelination in the CNS of aged rats. Mol Cell Neurosci 19:32–42

Patani R, Balaratnam M, Vora A, Reynolds R (2007) Remyelination can be extensive in multiple sclerosis despite a long disease course. Neuropathol Appl Neurobiol 33:277–287

Patrikios P, Stadelmann C, Kutzelnigg A, Rauschka H, Schmidbauer M, Laursen H, Sorensen PS, Bruck W, Lucchinetti C, Lassmann H (2006) Remyelination is extensive in a subset of multiple sclerosis patients. Brain 129:3165–3172

Penderis J, Shields SA, Franklin RJM (2003a) Impaired remyelination and depletion of oligodendrocyte progenitors does not occur following repeated episodes of focal demyelination in the rat CNS. Brain 126:1382–1391

Penderis J, Woodruff RH, Lakatos A, Li WW, Dunning MD, Zhao C, Marchionni M, Franklin RJM (2003b) Increasing local levels of neuregulin (glial growth factor-2) by direct infusion into areas of demyelination does not alter remyelination in the rat CNS. Eur J Neurosci 18:2253–2264

Piaton G, Williams A, Seilhean D, Lubetzki C (2009) Remyelination in multiple sclerosis. Prog Brain Res 175:453–464

Piaton G, Aigrot MS, Williams A, Moyon S, Tepavcevic V, Moutkine I, Gras J, Matho KS, Schmitt A, Soellner H, Huber AB, Ravassard P, Lubetzki C (2011) Class 3 semaphorins influence oligodendrocyte precursor recruitment and remyelination in adult central nervous system. Brain 134:1156–1167

Plant SR, Iocca HA, Wang Y, Thrash JC, O'Connor BP, Arnett HA, Fu YX, Carson MJ, Ting JP (2007) Lymphotoxin beta receptor (Lt betaR): dual roles in demyelination and remyelination and successful therapeutic intervention using Lt betaR-Ig protein. J Neurosci 27:7429–7437

Pohl HB, Porcheri C, Mueggler T, Bachmann LC, Martino G, Riethmacher D, Franklin RJ, Rudin M, Suter U (2011) Genetically induced adult oligodendrocyte cell death is associated with poor myelin clearance, reduced remyelination, and axonal damage. J Neurosci 31:1069–1080

Prayoonwiwat N, Rodriguez M (1993) The potential for oligodendrocyte proliferation during demyelinating disease. J Neuropathol Exp Neurol 52:55–63

Rhodes KE, Raivich G, Fawcett JW (2006) The injury response of oligodendrocyte precursor cells is induced by platelets, macrophages and inflammation-associated cytokines. Neuroscience 140:87–100

Richardson WD, Young KM, Tripathi RB, McKenzie I (2011) NG2-glia as multipotent neural stem cells: fact or fantasy? Neuron 70:661–673

Setzu A, Lathia JD, Zhao C, Wells KA, Rao M, ffrench-Constant C, Franklin RJM (2006) Inflammation stimulates myelination by transplanted oligodendrocyte precursor cells. Glia 54:297–303

Shen S, Sandoval J, Swiss V, Li J, Dupree J, Franklin RJM, Casaccia-Bonnefil P (2008) Age-dependent epigenetic control of differentiation inhibitors: a critical determinant of remyelination efficiency. Nat Neurosci 11:1024–1034

Sim FJ, Zhao C, Li WW, Lakatos A, Franklin RJM (2002a) Expression of the POU domain transcription factors SCIP/Oct-6 and Brn-2 is associated with Schwann cell but not oligodendrocyte remyelination of the CNS. Mol Cell Neurosci 20:669–682

Sim FJ, Zhao C, Penderis J, Franklin RJM (2002b) The age-related decrease in CNS remyelination efficiency is attributable to an impairment of both oligodendrocyte progenitor recruitment and differentiation. J Neurosci 22:2451–2459

Sloane JA, Batt C, Ma Y, Harris ZM, Trapp B, Vartanian T (2010) Hyaluronan blocks oligodendrocyte progenitor maturation and remyelination through TLR2. Proc Natl Acad Sci USA 107:11555–11560

Smith KJ, Blakemore WF, McDonald WI (1979) Central remyelination restores secure conduction. Nature 280:395–396

Snyder DH, Valsamis MP, Stone SH, Raine CS (1975) Progressive demyelination and reparative phenomena in chronic experimental allergic encephalomyelitis. J Neuropathol Exp Neurol 34:209–221

Stidworthy MF, Genoud S, Suter U, Mantei N, Franklin RJM (2003) Quantifying the early stages of remyelination following cuprizone-induced demyelination. Brain Pathol 13:329–339

Stidworthy MF, Genoud S, Li WW, Leone DP, Mantei N, Suter U, Franklin RJM (2004) Notch1 and Jagged1 are expressed after CNS demyelination but are not a major rate-determining factor during remyelination. Brain 127:1928–1941

Syed YA, Hand E, Mobius W, Zhao C, Hofer M, Nave KA, Kotter MR (2011) Inhibition of CNS remyelination by the presence of semaphorin 3A. J Neurosci 31:3719–3728

Tang DG, Tokumoto YM, Raff MC (2000) Long-term culture of purified postnatal oligodendrocyte precursor cells. Evidence for an intrinsic maturation program that plays out over months. J Cell Biol 148:971–984

Trapp BD, Nave KA (2008) Multiple sclerosis: an immune or neurodegenerative disorder? Annu Rev Neurosci 31:247–269

Tripathi RB, Rivers LE, Young KM, Jamen F, Richardson WD (2010) NG2 glia generate new oligodendrocytes but few astrocytes in a murine experimental autoimmune encephalomyelitis model of demyelinating disease. J Neurosci 30:16383–16390

Vela JM, Molina-Holgado E, Arevalo-Martin A, Almazan G, Guaza C (2002) Interleukin-1 regulates proliferation and differentiation of oligodendrocyte progenitor cells. Mol Cell Neurosci 20:489–502

Wang S, Sdrulla AD, diSibio G, Bush G, Nofziger D, Hicks C, Weinmaster G, Barres BA (1998) Notch receptor activation inhibits oligodendrocyte differentiation. Neuron 21:63–75

Warrington AE, Asakura K, Bieber AJ, Ciric B, Van KV, Kaveri SV, Kyle RA, Pease LR, Rodriguez M (2000) Human monoclonal antibodies reactive to oligodendrocytes promote remyelination in a model of multiple sclerosis. Proc Natl Acad Sci USA 97:6820–6825

Watanabe M, Hadzic T, Nishiyama A (2004) Transient upregulation of Nkx2.2 expression in oligodendrocyte lineage cells during remyelination. Glia 46:311–322

Werner HB, Kuhlmann K, Shen S, Uecker M, Schardt A, Dimova K, Orfaniotou F, Dhaunchak A, Brinkmann BG, Mobius W, Guarente L, Casaccia-Bonnefil P, Jahn O, Nave KA (2007) Proteolipid protein is required for transport of sirtuin 2 into CNS myelin. J Neurosci 27:7717–7730

Williams A, Piaton G, Aigrot MS, Belhadi A, Theaudin M, Petermann F, Thomas JL, Zalc B, Lubetzki C (2007a) Semaphorin 3A and 3F: key players in myelin repair in multiple sclerosis? Brain 130:2554–2565

Williams A, Piaton G, Lubetzki C (2007b) Astrocytes – friends or foes in multiple sclerosis? Glia 55:1300–1312

Wolswijk G (1998) Chronic stage multiple sclerosis lesions contain a relatively quiescent population of oligodendrocyte precursor cells. J Neurosci 18:601–609

Wolswijk G (2002) Oligodendrocyte precursor cells in the demyelinated multiple sclerosis spinal cord. Brain 125:338–349

Wolswijk G, Noble M (1989) Identification of an adult-specific glial progenitor cell. Development 105:387–400

Wolswijk G, Noble M (1992) Cooperation between PDGF and FGF converts slowly dividing O-2Aadult progenitors to rapidly dividing cells with characteristics of O-2Aperinatal progenitor cells. J Cell Biol 118:889–900

Woodruff RH, Fruttiger M, Richardson WD, Franklin RJM (2004) Platelet-derived growth factor regulates oligodendrocyte progenitor numbers in adult CNS and their response following CNS demyelination. Mol Cell Neurosci 25:252–262

Xin M, Yue T, Ma Z, Wu FF, Gow A, Lu QR (2005) Myelinogenesis and axonal recognition by oligodendrocytes in brain are uncoupled in Olig1-null mice. J Neurosci 25:1354–1365

Yao DL, Liu X, Hudson LD, Webster HD (1995) Insulin-like growth factor I treatment reduces demyelination and up-regulates gene expression of myelin-related proteins in experimental autoimmune encephalomyelitis. Proc Natl Acad Sci USA 92:6190–6194

Ye F, Chen Y, Hoang T, Montgomery RL, Zhao XH, Bu H, Hu T, Taketo MM, van Es JH, Clevers H, Hsieh J, Bassel-Duby R, Olson EN, Lu QR (2009) HDAC1 and HDAC2 regulate oligodendrocyte differentiation by disrupting the beta-catenin-TCF interaction. Nat Neurosci 12:829–838

Zawadzka M, Rivers LE, Fancy SPJ, Zhao C, Tripathi R, Jamen F, Young K, Goncharevich A, Pohl H, Rizzi M, Rowitch DH, Kessaris N, Suter U, Richardson WD, Franklin RJM (2010) CNS-resident glial progenitor/stem cells produce Schwann cells as well as oligodendrocytes during repair of CNS demyelination. Cell Stem Cell 6:578–590

Zhang Y, Argaw AT, Gurfein BT, Zameer A, Snyder BJ, Ge C, Lu QR, Rowitch DH, Raine CS, Brosnan CF, John GR (2009) Notch1 signaling plays a role in regulating precursor differentiation during CNS remyelination. Proc Natl Acad Sci USA 106(45):19162–19167

Zhao C, Li WW, Franklin RJM (2006) Differences in the early inflammatory responses to toxin-induced demyelination are associated with the age-related decline in CNS remyelination. Neurobiol Aging 27:1298–1307

Chapter 5
Exogenous Cell Myelin Repair and Neuroprotection in Multiple Sclerosis

Ian D. Duncan and Yoichi Kondo

5.1 Introduction

The loss of myelin has serious functional consequences in multiple sclerosis (MS) and other demyelinating disorders and may place demyelinated axons at risk of subsequent degeneration. Hence, remyelination has two major consequences: it will restore and speed impulse conduction (Smith et al. 1979) and it may protect axons against degeneration, thus acting as a form of neuroprotection (Irvine and Blakemore 2008; Kornek et al. 2000). The current available treatments of MS do not promote remyelination as far as is known; hence there is a critical need for such a restorative therapy. It is well known that in experimental demyelinating disease, the CNS has remarkable ability to be remyelinated by an endogenous response (Blakemore 1973; Franklin and Ffrench-Constant 2008; Ludwin 1978), and recently, it was clearly demonstrated that widespread endogenous remyelination can lead to restoration of function (Duncan et al. 2009). Likewise in MS, extensive remyelination occurs early in the disease (Kornek et al. 2000; Prineas and Connell 1979; Raine and Wu 1993) and can also be seen at later stages (Patani et al. 2007; Patrikios et al. 2006) although it is not clear how long the human CNS can sustain an endogenous response. However it is likely that the aging CNS remyelinates less efficiently (Goldschmidt et al. 2009; Shen et al. 2008; Shields et al. 1999). In later stages of the disease, endogenous remyelination will only occur if the remaining cells of the oligodendrocyte lineage, either progenitors or mature cells, in or very close to lesions, can be mobilized and differentiate into myelinating oligodendrocytes. As there are no proven strategies available that will promote such a response in the human CNS, the transplantation of cells into focal areas of demyelination or multiple sites using a more disseminated delivery approach may be important therapeutically.

I.D. Duncan (✉) • Y. Kondo
Department of Medical Sciences, School of Veterinary Medicine,
University of Wisconsin-Madison, 2015 Linden Drive, Madison, WI 53706, USA
e-mail: duncani@svm.vetmed.wisc.edu

In this chapter, we will discuss the case for exogenous cell-based remyelination, while the means of potentially promoting endogenous repair will be discussed elsewhere in this book (Franklin and Ffrench-Constant 2008). Exogenous and endogenous remyelination may not be mutually exclusive however as, for example, the transplantation of growth factor-expressing cells may be used to promote endogenous remyelination or in some way modify the milieu to enhance this process (Milward et al. 2000). The possible therapeutic application of exogenous remyelination strategies in MS has received growing attention over the years, based primarily on a wealth of data that has demonstrated the extensive myelinating capability of numerous cell types transplanted into the brain or spinal cord in many different myelin-deficient model systems (Duncan et al. 1997). However, the idea that similar approaches may be used in MS is not without criticism (Keirstead 2005; Lassmann 2005). Three major concerns have been expressed. Firstly, the lesions to be transplanted may contain endogenous inhibitory factors that prevent persistent oligodendrocyte progenitor cells (OPCs) from differentiation (Back et al. 2005; Chang et al. 2002; Charles et al. 2002; Kremer et al. 2011). Hence transplanted cells may face the same barrier. Secondly, as MS usually affects many levels of the CNS, the delivery systems currently available are inadequate to disseminate cells throughout the parenchyma of the CNS, and repairing single lesions may not be clinically useful (Comi 2008). Thirdly, the optimal cell to be delivered to remyelinate the target area has not yet been identified although this is a subject of intense investigation (Duncan et al. 2008; Martino et al. 2010). These and other concerns will be addressed at different points in this review. However, the current inability to remyelinate chronically demyelinated axons in the CNS of MS patients resulting in conduction failure and putting axons at risk for degeneration demands that careful scrutiny is paid to the opportunities identified in the literature on experimental remyelination and how such strategies might be applied to MS. In addition, application of exogenous cell therapies in MS with delivery of cells into the CNS parenchyma hinges on whether focal or multifocal repair is likely to produce clinical benefit.

In this chapter, therefore, we will present and discuss the scientific basis of the application of exogenous cell therapy of lesions in MS and issues that remain unresolved. A detailed discussion of what we have learned from the extensive animal experiments will set the scene for the discussion on the choice of cell to be used. A point of debate is the choice between myelinating cells of CNS or PNS origin, that is, cells of the oligodendrocyte or Schwann cell lineage. Within the lineage known to give rise to oligodendrocytes, multiple choices exist based upon their stage of differentiation, for example, neural stem cells (NSCs) or OPCs. Likewise in the Schwann cell lineage, the choice between Schwann cell progenitors and mature cells is to be considered (Chap. 6 by Zujovic and Baron). The stage of the disease at which the transplantation of cells would be performed is a key question. A number of other issues such as the monitoring of repair, behavioral outcomes, and long-term survival of grafted cells and myelin generated are also very important and will be discussed in the light of experimental data that addresses each of these points.

5.2 How Are Myelination/Remyelination Strategies Tested?

Evidence for the successful myelination by transplanted cells has come from a wide variety of experimental models. Perhaps a critical distinction among these models is whether the transplanted cells are required to make myelin on axons that have never been myelinated, such as in the genetic models of myelin disease or whether they are grafted into areas of demyelination, more akin to repairing demyelinated axons in MS. This issue may be settled if it is proven that myelination and remyelination are fundamentally the same process (Fancy et al. 2011), albeit with some dissimilarities which nonetheless are redundant and do not prevent transplanted cells from making myelin, no matter in what type of lesion.

Testing the myelinating capacity of exogenous cells was first attempted in myelin-deficient animals, the so-called myelin mutants, in the early 1980s (Duncan 1995; Griffiths 1996). These have been extensively used to explore myelination by a wide variety of cell types (see Sect. 5.4). The second major model system that has been used to test the myelinating capacity of transplanted cells is the model of focal demyelination created by the injection of myelin-toxic chemicals, lysolecithin or ethidium bromide, into the brain or spinal cord (Blakemore et al. 1995a; Blakemore and Franklin 1991). Following injection of either of these compounds, endogenous remyelination occurs; hence transplanted cells must compete with the host response, perhaps similar to what may occur in transplanting cells into MS lesions where host OPCs persist. To generate a more "amenable" lesion, Blakemore devised an X-irradiation protocol prior to cell transplantation that resulted in focal, persistent plaques of demyelination following the killing of endogenous OPCs at the site (Blakemore and Patterson 1978). This technique has not been widely used by others, however, perhaps because of the technical difficulty in delivery of the appropriate dose of irradiation and also a concern that irradiation may have additional effects on the milieu that might influence the transplant results. In addition to lysolecithin and ethidium bromide, the ingestion of cuprizone with its resultant focal demyelination of the corpus callosum and cerebellar peduncles has also been used in a limited fashion to provide a model to test cell therapies (Czepiel et al. 2011; Einstein et al. 2009; Mason et al. 2004). Finally, injection of antibodies to galactocerebroside with complement can also produce areas of focal demyelination in the spinal cord that persist if the area is irradiated prior to injection (Keirstead et al. 1998).

While these models have been extremely instructive in determining the myelination capabilities of cells on transplantation, they lack certain key features of the pathology of MS that could provide additional challenges. Thus cell transplantation has also been performed in experimental autoimmune encephalomyelitis (EAE), the bona fide model of MS in both mice and rats (Einstein et al. 2003; Tourbah et al. 1997) and in nonhuman primates (Pluchino et al. 2009). The difficulty with EAE has been targeting demyelinated lesions as these can be scattered throughout the CNS, and their localization by imaging may not always be practical. Cells therefore have been transplanted into the lateral ventricles to test their ability to migrate into the brain, or directly into the parenchyma of the brain and spinal cord. The difficulty in transplanting cells directly into focal inflammatory lesions may be circumvented using an alternative model of focal EAE (Kerschensteiner et al. 2004). Lewis rats are immunized with myelin oligodendrocyte glycoprotein (MOG) to create subclinical EAE followed by focal injection of tumor necrosis factor-α (TNF-α) and interferon-γ (IFN-γ) into the spinal cord (Kerschensteiner et al. 2004), resulting in focal inflammation, demyelination, and axonal degeneration. Difficulty in reproducing these lesions, however, led to a modification of this protocol and injection of vascular endothelial growth factor into the spinal cord instead of cytokines, resulting in more consistent areas of demyelination associated with an inflammatory infiltrate (Sasaki et al. 2010). We have used the focal injection of zymosan, a yeast cell wall protein carbohydrate complex that creates intense inflammation on injection into the CNS (Popovich et al. 2002; Schonberg et al. 2007) leading to axonal degeneration and demyelination, as a model to test cell therapy (Wang et al. 2011b). Finally, cell transplantation into the spinal cord of mice demyelinated by infection with mouse hepatitis virus (MHV), and where there is notable inflammation, has been used in a similar fashion (Hardison et al. 2006; Hatch et al. 2009; Totoiu et al. 2004).

In summary, it is likely that none of these models provide an identical background to all stages of the pathological course of MS, in which to test exogenous cell therapy. Nonetheless, the adoption of a reductionist approach (Dubois-Dalcq et al. 2005) where the model being used displays individual aspects of the pathology of MS such as acute or chronic demyelination, inflammation, or gliosis is useful to study the effects of these pathologies on exogenous remyelination. The ideal scenario would then be to combine all of these separate changes in a pathologic milieu that would match the stage or stages of MS that may require therapeutic intervention. One of the greatest needs in MS is the ability to remyelinate axons in chronic, non-repairing MS plaques in which endogenous, non-differentiating OPC persists, perhaps associated with low-grade inflammation. At present, no models are available that accurately represent this stage of MS.

5.3 What Is the Extent of Myelination Following Transplantation?

In a general sense, making functional myelin in a myelin-deficient or demyelinated CNS by transplantation of a wide variety of myelinating cell types is straightforward. This is certainly true compared to exogenous neuronal replacement where the required neurons need to make appropriate, often long-distance connections and synthesize cell-specific transmitters at the correct level (Björklund and Lindvall 2000; Goldman 2005; Lindvall and Kokaia 2006). Myelin repair is much simpler; the transplanted cell must differentiate (assuming the use of immature cells), associate with and ensheath axons, and synthesize the required myelin lipids and proteins.

From the first demonstrations of myelination by transplanted cells in the labs of Blakemore, Gumpel, and Duncan and their colleagues, in the 1970s–1980s grafting of putative myelinating cells has been widely reported. In their original studies, oligodendrocytes (Lachapelle et al. 1983) or Schwann cells (Blakemore 1977; Duncan et al. 1981) derived from either tissue chunks or dissociated tissue, or cultured cells attached to collagen, were transplanted into the brain or spinal cord (Duncan et al. 1988; Gumpel et al. 1985, 1987). In all of these early reports, myelination at the site of grafting was seen. Since that time, however, transplantation studies have used direct injection of cells and not tissue. Cells have been derived from fetal, neonatal, and adult CNS, and in general, the former two have been a better source of myelinating oligodendrocytes. Cells can be injected as mixed glial cell preparations or sorted by fluorescence-activated cell sorting (FACS), immunopanning, or magnetic bead sorting with antibodies recognizing OPCs or immature oligodendrocytes (see Sect. 5.4.3). More recently, cells grown as free-floating spheres in the presence of selected growth factors (Sect. 5.4.3) to allow their culture and expansion have been used for transplantation. In all of these instances, successful myelination has been seen at the site of engraftment. In the case of grafting cells into the myelin mutants, transplant-derived myelin can be grossly recognized by "white" areas in a myelin-deficient background (Fig. 5.1) and confirmed in the myelin-deficient (*md*) rat and shiverer (*shi*) mouse by the presence of the missing protein, proteolipid protein (PLP) (Duncan et al. 1988) or myelin basic protein (MBP) (Gumpel et al. 1985, 1987), respectively (Fig. 5.1). In toluidine blue-stained 1-μm sections, this myelin can be clearly differentiated from adjacent, non-myelinated areas (Duncan et al. 1988; Rosenbluth et al. 1990, 1993; Tontsch et al. 1994) and confirmed on electron microscopy (EM) (Duncan et al. 1988; Gansmuller et al. 1986; Gout et al. 1990; Rosenbluth et al. 1990, 1993) (Fig. 5.2). Myelin sheaths made by transplanted cells are frequently thin as is seen in endogenous remyelination (Fig. 5.2), but this may relate to the length of time post-grafting. We have found evidence of more normal thickness myelin sheaths in the *md* rat (Duncan, unpublished observation) and shaking (*sh*) pup (Archer et al. 1997) when transplanted as neonates and studied after longer periods of engraftment. This is also the case in long-term grafting experiments in the *shi* mouse (Windrem et al. 2008). EM studies of myelin in the mutants also show corrections of ultrastructural defects, notably the presence of a normal major dense (*shi* mouse) or intraperiod lines (*md* rat). Finally, myelin made by the grafted cells is associated with the formation of normal nodes of Ranvier which is critical for restoration of impulse conduction (see Sect. 5.7).

While focal myelination by transplanted cells is easy to achieve, more global repair has also been documented in the myelin mutants. Studies of transplantation into the spinal cord of the *sh* pup showed more extensive myelination across the spinal cord at the grafting site and more rostral and caudal myelination to it (up to 20 mm) than in prior rodent studies (Archer et al. 1997) (Fig. 5.3). This was calculated to be ten times the volume of myelin achieved in the *md* rat. In the *shi* mouse, Mitome et al. (2001) reported extensive myelination in the brain and upper cervical cord by injecting NSCs into the lateral ventricles and cisterna magna. Kondo et al. (2005) demonstrated similar myelination in the brain of *shi* using OPCs but importantly showed that transplantation into the brain and spinal cord of the same mouse resulted in spatially separated areas of repair. This is critical to the potential application

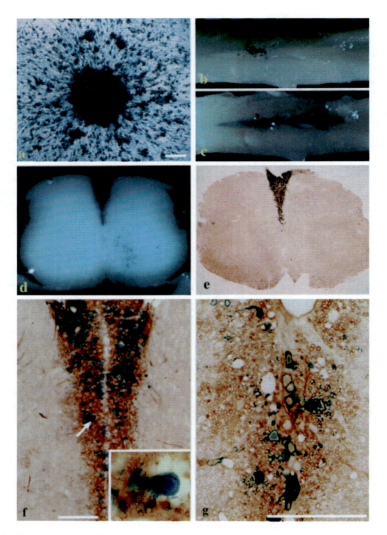

Fig. 5.1 Transplantation of oligosphere cells into *md* rats. Before transplantation, the oligosphere cells were transduced with retrovirus to express β-gal (**a**). Between 10 and 14 days after transplantation, a white streak of myelin was seen along the dorsal surface of the cord (**b**). The *black dots* are sterile charcoal marking the injection site. When the cord was stained with X-gal, the white streak turned to blue (**c**). Cross section of the cord indicated the location of the blue staining mainly in the dorsal funiculus but also in gray matter (**d**). Immunostaining of the transplanted cord showed PLP+ myelin in the dorsal funiculus with blue oligodendrocytes interspersing them (**e, f**). The *inset* in the *lower right* of (**f**) is the magnification of a "blue" oligodendrocyte with processes attaching to PLP+ myelin sheaths (*arrow*). Semithin sections demonstrated that a large number of myelin sheaths and blue cells were in the dorsal funiculus (**g**). *Bar*: 100 μm (from Zhang et al. 1998a)

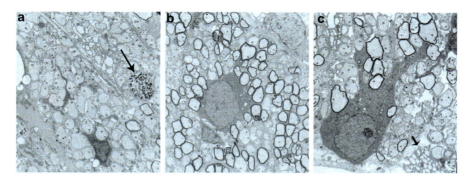

Fig. 5.2 Electron micrographs from representative myelinated areas in an *md* rat produced by rat neurospheres. (**a**) Adjacent to the transplant site, most of the axons are nonmyelinated. As in the uninjected mutant, no normal, native oligodendrocytes are present. On the far right in this field, a single degenerating axon is seen (*arrow*). (**b**) In a transplanted area, the majority of the axons are myelinated with a normal oligodendrocyte present. (**c**) Underneath the pia, a normal appearing oligodendrocyte has extensive processes leading to many myelinated fibers. Adjacent to this cell, the cytoplasm of an abnormal *md* oligodendrocyte is present (*small arrow*) with distended rough endoplasmic reticulum, a characteristic feature in *md* oligodendrocytes

of transplantation to MS, as it proves that different strategic sites in brain and spinal cord could be remyelinated by more than a single transplant. Windrem et al. (2008) extended these results by transplanting OPCs into four sites, in the corpus callosum and cerebellar peduncles. For the first time, they showed that the brain could be almost completely myelinated by transplanted human glia and that this resulted in an improvement in the phenotype and extended survival (1 year or more) of 23 % of the grafted *shi* mice.

Many of the transplant studies in the myelin mutants have been performed in neonatal animals. As MS is a disease that affects the mature CNS, a question arises regarding the feasibility of the successful transplantation cells into the brain or spinal cord of adults, where the cues that drive myelination during development may be missing. Fortunately a number of experiments in different model systems have clearly shown that grafting cells into the mature CNS can result in myelination by the transplanted cells, including transplantation *shi* (Buchet et al. 2011; Mothe and Tator 2008) and the *sh* pup (Archer et al. 1997). In addition, this has been tested using both NSCs and OPCs. Indeed, transplanting cells into focally demyelinated lesions in the spinal cord or brain is usually performed in mature rodents, confirming observations in the older myelin mutant studies. Finally, confirmation of the ability to myelinate CNS axons throughout life comes from studies where NSCs or OPCs were grafted into the non-myelinated portion of the retina in both neonates (Ader et al. 2000; Setzu et al. 2004) and mature rats (Setzu et al. 2004).

Extending the spatial extent of remyelination by transplanted cells may be achieved by promoting the migration and division of transplanted cells. In all neural

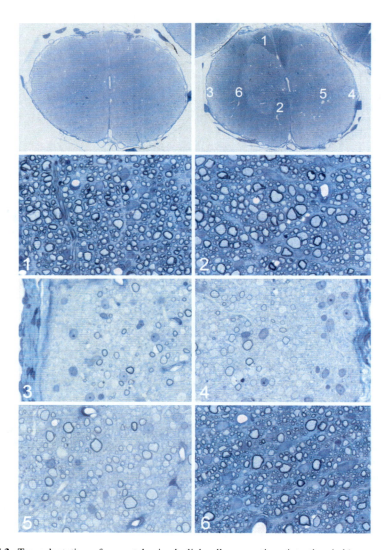

Fig. 5.3 Transplantation of neonatal mixed glial cell preparations into the shaking pup spinal cord. The montage illustrates the result of transplantations of such cells into a 7-day-old *sh* pup thoracic spinal cord, 8 weeks after transplantation. The *top left figure* shows a spinal cord segment 5 cm away from the transplant site and the *top right figure*, at the site of transplantation. At the site of transplantation, there is clearly evidence of myelin in the dorsal, lateral (*left*), and ventral column (*right*). Areas from the spinal cord (1–6) are shown in the lower subsequent figures. In 1, 2, and 6, there is complete myelination by the transplanted cells. In the lateral columns, the superficial left lateral column (3) and the entire right column (5) compared to the deep left lateral column (4) indicating extensive but incomplete migration and myelination by the transplanted cells

transplants, significant death of cells occurs acutely, and the division of surviving cells will be critical to success. It has been shown that both NSCs and OPCs may divide on engraftment (Milward et al. 2000; Windrem et al. 2008). It has also been shown that the division of OPCs can be increased by their co-transplantation with growth factor-producing cells. We have shown that an OPC cell line, CG-4, co-grafted in the *md* rat with neuroblastoma cells (B104), divided more than OPCs alone and hence occupied a greater area of neuropil (Milward et al. 2000) (Fig. 5.4). In addition, these studies demonstrated that the direction of migration can be influenced; when the two cells noted above were grafted at different sites, the OPCs migrated preferentially toward the B104 cells (Milward et al. 2000) (Fig. 5.4). More recently, we have shown that similar chemotactic attraction occurs when OPCs are transplanted adjacent to focal areas of inflammation (Wang et al. 2011b). The migration of transplanted cells through a non-myelinated neuropil, such as in the myelin mutants, may be relatively straightforward, but the ability of these cells to migrate through normal areas of white matter, for example, from one MS plaque to another, may present difficulties. Evidence suggesting that normal white matter does not support such migration has been reported (O'Leary and Blakemore 1997), yet a more recent study of transplanting human NSCs into demyelinated lesions of adult nude mouse spinal cord showed extensive migration, rostral and caudal to the transplant site and through areas of normal myelin (Buchet et al. 2011).

Migration of OPCs in vitro has been studied extensively, and the development and normal myelination of the CNS are dependent on their expression of key molecules and their interaction with the extracellular matrix (De Castro and Bribián 2005; Jarjour and Kennedy 2004). Remyelination may call for the extension of these interactions in vivo following transplantation of cells. In MS lesions, molecules such as semaphorin 3A and F (Piaton et al. 2011; Syed et al. 2011) and chemokines and their receptors (Kerstetter et al. 2009) may influence migration and in some cases the differentiation of OPCs in a positive or negative way. Transduction of NSCs with two molecules known to be involved in migration of OPCs in vitro, FGF2 and PSA-NCAM, has been studied to determine their effects on migration and differentiation both in vitro and in vivo. FGF2, a known mitogen of OPCs, was transduced into the OPC cell line, CG-4 (Magy et al. 2003), and shown to enhance proliferation and migration in vitro though not inhibiting differentiation. Transduction of NSCs with PSA-NCAM (Franceschini et al. 2004) or oligospheres (Decker et al. 2000; Vitry et al. 2001) and transplantation of these cells into *shi* brain extended the migration of the latter but not the former and, with its downregulation, the differentiation and myelination by other cells (Magy et al. 2003). It is not entirely clear however whether this cell manipulation significantly extended migration of the cells compared to controls. While these approaches are experimentally interesting and important, it seems likely that inducing expression of transgenes in human cells that are to be used in clinical trials would increase the difficulty in achieving Institutional Review Board and the Food and Drug Administration approval. A more detailed analysis of small molecules, cytokines, etc. that may promote endogenous remyelination is presented in Chap. 4 by Franklin et al.

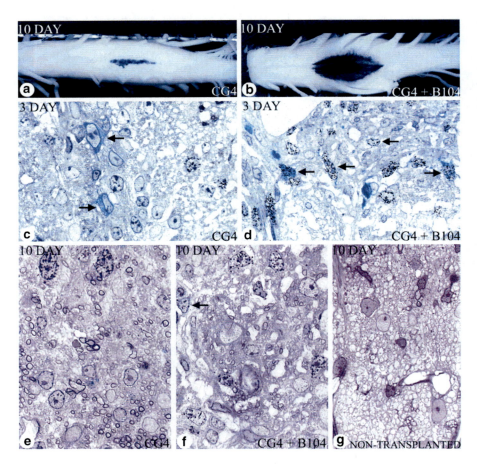

Fig. 5.4 Effects of cografting growth factor secreting cells on oligodendroglial progenitor cells in vivo. (**a**) When 50,000 CG-4 cells were injected alone, they were localized to the center of the dorsal column. (**b**) In contrast to (**a**), when an identical number of CG-4 cells were transplanted with 5,000 B104 cells, there was much greater lateral spread with more dense X-gal labeling of the spinal cord 10 days after transplantation. (**c, d**) At 3 days after transplantation in rats that received CG-4 cells alone (**c**) or CG-4 and B104 cells (**d**), more blue cells (CG-4) (*arrows*) are seen in animals that received the cograft. More of the blue cells in the cograft are labeled with silver grains, indicating greater division of CG-4 cells. At 10 days after transplantation, myelin formed by the transplanted cells was seen in both grafts (**e, f**) compared with the nontransplanted *md* rat, in which no myelin is present (**g**). However, more myelin was seen in those rats that received CG-4 cells alone (**e**) than in the cograft (**f**). In addition, more dividing cells were still seen in the cografted rats (**f**; seven in this field); many of those cells that were labeled with silver grains (e.g., *arrow*) showed thin blue perinuclear X-gal staining, (**a–g**) after X-gal histochemical stain; (**c–g**) 1-μm sections from rats labeled with [3H]-thymidine. Sections from X-gal-stained tissue and counterstained with toluidine blue (from Milward et al. 2000)

As with the established data of cell transplantation into the myelin mutants, much is known and has been documented on the transplantation of cells into focal, chemically demyelinated lesions. This work has been pioneered by Blakemore and Franklin, and through many detailed, wide ranging studies, they have examined the glial cell–axonal interactions that occur in this model system (Blakemore et al. 1995a; Blakemore and Franklin 2000; Franklin and Blakemore 1997). An instant advantage over the myelin mutants is that these studies can be performed in adult animals which is not always possible in the myelin mutants, yet there are some disadvantages. For example, in animals where the spinal cord or brain is irradiated to prevent endogenous repair, it is only possible to study test animals up to 1 month after irradiation, as necrosis at the site will eventually occur. From the perspective of transplanting cells into MS lesions, a number of important issues have been highlighted by these studies and these include (1) as with endogenous remyelination, transplant-induced repair results in thin myelin sheaths. It remains to be shown however whether this is deleterious to long-term conduction in such axons and whether thin myelin remains neuroprotective (Franklin 1993); (2) OPCs are a better source of remyelination than adult oligodendrocytes (Crang et al. 1998), and their division is required for remyelination to occur; (3) astrocytes can be both helpful and inhibitory to exogenous cell remyelination (Blakemore et al. 2003; Franklin et al. 1991). Astrocytes are known to produce growth factors that play a key role in myelination in development (Moore et al. 2011). In demyelinated lesions where astrocytes are present, the success of remyelination by transplanted OPCs is dependent upon the "activation" status of astrocytes which will differ in acute and chronic demyelination (Blakemore et al. 2003). In regard to chronic MS lesions, the degree of gliosis is likely to be very important to the success of exogenous cells, if only through the inhibition of migration of cells in highly gliotic lesions; (4) OPCs transplanted into a normal neuropil do not migrate or survive well, but if the area is irradiated or in any way pathologic, then cells will survive (O'Leary and Blakemore 1997). This is an important issue as it will determine whether cells can migrate from one MS plaque to another through normal white matter. As they acknowledge, however, this issue is debated (Franklin and Blakemore 1997), and as discussed earlier here, other studies including a recently published report (Buchet et al. 2011) would appear to suggest that in certain experimental situations, this can occur. It was also noted that transplanted glial cells migrated over greater distances and remyelinated axons faster than endogenous cells (Blakemore et al. 2000). This implies that inhibitory signals for endogenous OPCs in MS plaques may not interfere with transplanted exogenous cells; hence one of the proposed major hurdles in MS may be unfounded; (5) chronically demyelinated tissue may not be permissive for exogenous cell repair. Thus, chronic plaques may need to be "primed" by promotion of inflammation at the site, for transplanted OPCs to differentiate and myelinate (Foote and Blakemore 2005); (6) remyelination protects demyelinated axons from degeneration. In studies in cuprizone-treated mice, irradiation of the brain to create persistent demyelination resulted in significant axon death which was prevented by transplantation of labeled NSCs into the lateral ventricles (Irvine and Blakemore 2008). Protection of axons was attributed to their remyelination by the transplanted cells. This is an important observation as it confirms the neuroprotective properties of remyelination therapy; (7) tissue matching of allografts may not be essential for graft survival. Adult oligodendrocytes can be induced to express major histocompatibility (MHC) class I or II except if induced with IFN-γ in vitro. However, OPCs do not express either in vitro or in vivo after transplantation (Tepavcevic and Blakemore 2006). Mixed glial cell transplants (including astrocytes and microglia) are rejected after 1 month but pure OPCs are not (Tepavcevic and Blakemore 2005). This highlights the importance of purifying cell preparations prior to grafting and using primarily OPCs in allograft transplants. However, it should be noted that subsequent rejection of allografted cells may not be entirely negative as the resultant inflammation may provoke endogenous remyelination (Blakemore et al. 1995b). Thus in chronic MS plaques where OPCs are present but non-differentiating, the rejection of transplanted cells may trigger endogenous remyelination by the persisting OPCs and underscore the view that exogenous and endogenous repair are not mutually exclusive.

5.4 What Cells Can Be Used as an Exogenous Source of Remyelination?

Without doubt, one of the most important issues on the transplantation of cells into MS lesions is the choice and source of cells to be used (Duncan 2008; Duncan et al. 2008). This section will deal with cells of the oligodendrocyte lineage that may be candidates for therapeutic consideration, while this will be followed by a chapter on the role that Schwann cells may have in remyelination in MS. The lineage of the oligodendrocyte is one of the most studied of all the cells of the CNS (see Chap. 1 by Miller). In dissecting apart the lineage from the perspective of identifying stages that could be used for exogenous repair, it is worthwhile considering the ideal characteristics of cells for use in remyelinating MS lesions. These cells should be (1) able to migrate long distances, (2) capable of many divisions but do not become transformed, (3) able to myelinate many axons, (4) unlikely to provoke an immune response or this can be prevented, and (5) available in large numbers and can be cryopreserved.

Based on these characteristics, what is known about the development and differentiation of oligodendrocytes and their precursors that will fulfill the criteria required? While it is certain that the cells to be used in the first clinical trials in MS will be of human origin, much of what is known about oligodendrocytes comes from studies on the rodent CNS (see Chap. 1 by Miller). As can be seen in Fig. 5.5, the oligodendrocyte lineage is complex with multiple, defined, and likely, yet to be defined stages, originating as all cells from the embryonic stem cell (ESC) and developing through the NSC to the mature myelinating oligodendrocyte. Much of

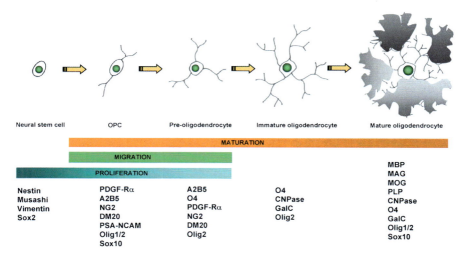

Fig. 5.5 Schematic diagram of the four-step maturation of neural stem cells toward the oligodendroglial lineage. Oligodendrocyte progenitor cells (OPCs), preoligodendrocytes, immature oligodendrocytes, and mature (myelinating) oligodendrocytes can be distinguished by their increasingly complex morphology; by their proliferating, migrating, and myelinating potentials; as well as by their expression of a wide range of well-defined markers. These markers may not be identical between rodents and humans (modified from Buchet and Baron-Van Evercooren 2009 with permission)

this information has been derived from in vitro studies and correlated and confirmed where possible in vivo using both wild type and transgenic mice and rats. In summary, the development of oligodendrocytes from the first cells of the embryo, the ESCs, requires the critical temporal expression of a wide range of transcription factors (Nicolay et al. 2007), surface markers, and myelin genes (Fig. 5.5) (Buchet and Baron-Van Evercooren 2009). While many of the transcription factors may not be essential for development, Olig1/2, Sox 10, and Nkx2.2 are the most important and require signals such as sonic hedgehog (SH) for their correct spatial–temporal expression. The membrane proteins, NG-2, and the platelet-derived growth factor receptor-α (PDGF-Rα) are markers of the OPC as well as the sulfatide O4. Premyelinating then myelinating oligodendrocytes will express O4 then O1 (Galc), also surface markers, followed by the expression of myelin genes in mature cells, such as MBP, PLP, myelin-associated glycoprotein (MAG), and MOG. As the cell differentiates, it progresses from the classic bipolar OPC (at neonatal stages) to the early multipolar, immature oligodendrocyte to the mature oligodendrocyte with multiple processes and membrane sheets (Fig. 5.6). This in vitro scheme is also likely similar or identical to oligodendrocyte development in vivo. The identification of cells that will myelinate or remyelinate axons on transplantation into the CNS is most assured if they are shown to follow such a pattern in tissue culture prior to grafting.

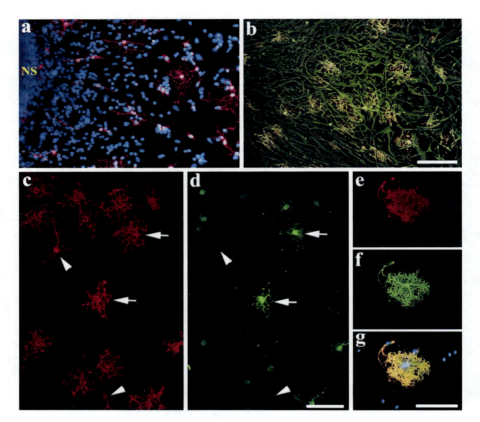

Fig. 5.6 Differentiation of human oligodendroglia from neural precursor cells. (**a**) A neurosphere (NS; passage 3) was cultured in the differentiation condition for 7 days in vitro (DIV) and stained with O4, showing that O4 oligodendroglia emerged from the neurosphere. Note the morphological transition from bipolar cells adjacent to the sphere to more branched oligodendroglia in the periphery. (**b**) A similar culture (from passage 7) for 15 DIV was stained with O1 (in *red*) and GFAP (in *green*). A 23-day-old differentiating culture was stained with O4 (**c**) and MBP (**d**) showing that the well-branched (*arrows*) but not the morphologically simple (*arrowheads*) oligodendroglia expressed MBP. (**e–g**) An oligodendrocyte that grew outside of the astrocytic layer and exhibited membrane-like structure. Cell nuclei in (**a**) and (**g**) were stained with DAPI (in *blue*). Bar = 100 μm (from Zhang et al. 1999)

5.4.1 Embryonic Stem Cells

As the pluripotent cell that gives rise to all cells of the body and is capable of providing an almost infinite number of cells, ESCs have been explored as a source of oligodendrocytes. In the original study of Brüstle et al. (1999), three mouse ESC lines were grown in the presence of leukemia inhibiting factor (LIF) which was then withdrawn with resultant formation of embryoid bodies (EBs). Using conditions that favor differentiation toward NSCs with media containing growth factors, then with addition of FGF2, epidermal growth factor (EGF), and PDGF, a population of cells that were positive for the surface marker, A2B5, was isolated. Upon removal of the growth factors, around 35 % of the cells differentiated into either oligodendrocytes or astrocytes (Brüstle et al. 1999). Transplantation of the cells prior to growth factor withdrawal demonstrated their oligodendrocyte differentiation and myelination in the *md* rat as well as their differentiation in vivo into astrocytes (Brüstle et al. 1999). A second, similar study published 1 year later showed evidence of oligodendrocyte differentiation of cells derived from mouse ESCs, in areas of chemically induced demyelination in the spinal cord and also following transplantation into the *shi* spinal cord (Liu et al. 2000). An increase in the differentiation of NSCs derived from mouse ESCs to oligodendrocytes has been reported when they were transduced to express an interleukin-6/soluble interleukin-6-receptor fusion protein (Zhang et al. 2006). This also increased myelination when these cells were transplanted into *shi* mice. Induction of Olig2 in mouse ESCs promoted the differentiation to oligodendrocytes during gliogenesis in the absence of sonic hedgehog, previously shown essential for oligodendrocyte differentiation from NSCs (Du et al. 2006).

Despite being isolated many years after mouse ESCs, human ESCs (hESCs) have taken center stage since 1998 (Thomson et al. 1998) as the potential source of cells for much of regenerative medicine and in particular here, as a source of NSCs that could be differentiated into neurons or glia and used in the therapy of neurodegenerative disease (Lindvall and Kokaia 2006). In 2001, Zhang et al. provided proof that NSCs subsequently OPCs then oligodendrocytes could be derived from hESCs (Zhang et al. 2001). Like mouse ESCs, the hESCs were cultured with the development of embryoid bodies, though in the presence of FGF2. Within these structures, neural rosettes develop that can be manipulated to high purity, NSC-containing cultures, with the production finally of neurons, astrocytes, and a minor proportion of cells of the oligodendrocyte lineage (5 % or less (Zhang et al. 2001)). In essence, this culture system is long (up to 3 months) and complex involving the induction of NSCs, patterning of Olig2 progenitors, differentiation of Olig2/Nkx2.2 expressing pre-OPCs, then the generation of OPCs. This calls for sequential culturing in chemically defined media, followed by exposure to retinoic acid and SH then FGF2. The differentiation of OPCs is then dependent on more standard protocols for these cells (Hu et al. 2009). This culture protocol can lead to up to 80 % of cells becoming OPCs, with 40 % becoming immature oligodendrocytes yet few differentiating to pure MBP-expressing mature oligodendrocytes (Gumpel et al. 1983). This culture system underscores the difficulty of isolating significant numbers of mature, human oligodendrocytes yet large numbers of OPCs. Other protocols have been reported that documents NSC (Joannides et al. 2007; Reubinoff et al. 2001) derivation from hESCs.

In contrast to the reports described above, the method of isolating mature oligodendrocytes from hESCs of Keirstead and colleagues (Nistor et al. 2005) resulted in a purity of up to 95 % of cells expressing the mature oligodendrocyte marker Galc. Perhaps the major difference between this and other protocols is the selection of cell aggregates (spheres) growing in the presence of retinoic acid selected for their yellow color (perhaps indicating viability), followed by subsequent growth in retinoic acid-free media, leading to the purified cultures noted above. More recent studies on oligodendrocyte differentiation from NSCs derived from hESCs have adopted certain differences in the culture protocol to maximize the generation of OPCs (Izrael et al. 2007; Kang et al. 2007). Both produced enriched populations of OPCs and in the first study, the addition of noggin, the bone morphogenetic protein antagonist, increased the number of mature oligodendrocytes (Izrael et al. 2007).

The clinical future for pluripotent stem cells may rest however with induced pluripotent cells (iPSCs) as in principle, they can be generated on an individual patient basis and cells derived from them are used as autologous grafts; hence immunosuppression would not be necessary (Kiskinis and Eggan 2010; Koch et al. 2009). Recent results from generating mouse iPSCs from skin fibroblasts are promising as the differentiation to NSCs and then OPCs resulted in around 18 % of cells becoming mature oligodendrocytes expressing MBP (Czepiel et al. 2011). Transplantation of OPCs derived in this fashion into chronic cuprizone lesions in the corpus callosum resulted in scattered MBP-positive areas (Czepiel et al. 2011). As with hESCs, the differentiation of human iPSCs toward OPCs and oligodendrocytes is not straightforward, and using different reprogramming techniques resulted in variable differentiation efficiencies (Hu et al. 2010). The current intense search and refinement in reprogramming methods of adult somatic cells (Amabile and Meissner 2009; Yamanaka and Blau 2010) and improvement of differentiation protocols suggest that iPSCs may become the eventual source of human OPCs and oligodendrocytes for transplantation, though perhaps not always on an individual patient basis (Kiskinis and Eggan 2010). A caveat however is that the epigenetic events of reprogramming can cause DNA damage, and the significance of these events may have to the translational use of the cells remains to be decided. Alternately, the direct conversion from adult somatic cells to OPCs, skipping the conversion to pluripotent cells as has been shown for the conversion of mouse and human fibroblasts to neurons (Pang et al. 2011), will also be a target.

5.4.2 Neural Stem Cells

Since the report of Reynolds and Weiss in 1992 of the isolation and culture of multipotent NSCs in the presence of EGF and FGF2, as free-floating spheres called neurospheres, from the adult striatum of mice (Reynolds and Weiss 1992) and subsequently embryonic brain (Reynolds and Weiss 1996), this protocol has become the optimal method of studying NSCs and their isolation for therapeutic purposes (Pastrana et al. 2011). Removal of the growth factors leads to the cells differentiating into neurons, astrocytes, and oligodendrocytes, though a minority of cells become oligodendrocytes in vitro (Reynolds and Weiss 1992). This pattern of differentiation however can be influenced by the in vivo environment into which they are grafted if cues driving differentiation toward a gliogenic fate and particularly toward an oligodendrocyte need are present. In a study on the transplantation of rat

NSCs into the *md* rat, Hammang et al. (1997) showed that these cells preferentially differentiated into OPCs then oligodendrocytes which myelinated significantly large areas of the spinal cord at the site of injection. Since then, numerous other studies have confirmed that NSCs derived from the fetal and adult CNS can give rise to myelinating oligodendrocytes (Mothe et al. 2008; Mothe and Tator 2008). NSCs have been used quite extensively in spinal cord injury models where functional recovery has been associated with their development into both oligodendrocytes and Schwann cells. In addition, mouse NSCs (Karimi-Abdolrezaee et al. 2006) or those derived from ESCs (Kumagai et al. 2009) or iPSC (Tsuji et al. 2010) have also been used in spinal cord injury models where they took part in remyelination. Human NSCs have also been shown to give rise to oligodendrocytes in vitro and in vivo following lysolecithin demyelination of the corpus callosum (Neri et al. 2010). In addition, mouse iPSC-derived NSCs have also been shown to remyelinate axons in a spinal cord injury model (Tsuji et al. 2010). However, NSCs may have pleiotropic effects on transplantation into the CNS and not only be capable of remyelination. Thus when transplanted into EAE models where they ameliorate disease, this is thought to be principally through immunomodulation (Pluchino et al. 2003, 2005), or promotion of endogenous remyelination as shown in chronic cuprizone intoxication (Einstein et al. 2009). If their primary use is to repair myelin however, it remains undecided whether NSCs are a poorer source of OPCs than using OPCs directly (Smith and Blakemore 2000). We would argue that the study of Hammang et al. (1997) and subsequent reports (Mothe et al. 2008; Mothe and Tator 2008) have shown that NSCs are a viable source of cells for remyelination, though a head-to-head comparison of NSCs vs. OPCs in an experimental model would be required to determine if one or other leads to greater myelination. A concern that NSCs, which by definition are multipotent, will give rise to ectopic neurons or astrocytes in white matter has not been found to be excessive or inhibit myelin repair.

5.4.3 *Oligodendrocyte Progenitor Cells and Oligodendrocytes*

Conventional wisdom would suggest that OPCs are the best cell for myelin repair as they fit many of the criteria defined earlier. They are migratory, mitotic, and known to differentiate into oligodendrocytes. While this is self-evident, they may also be the source of remyelinating Schwann cells (Zawadzka et al. 2010) in astrocyte-free areas. Many studies however have confirmed their differentiation into myelinating oligodendrocytes in vivo following transplantation.

The OPC or as it was originally known, the O-2A cell, is one of the most studied cells of the CNS. Its acquisition from the CNS, and as used by Raff, Noble, Barres, Miller, and Richardson and others who studied it in vitro, was from the dissociation and culture of mixed glial cell preparations from the optic nerve or brain. Pure populations of OPCs, neonatal, and adult OPCs have been isolated by FACS, immunopanning, or magnetic bead sorting using antibodies to appropriate stage markers such as A2B5 (e.g., Shi et al. 1998), PSA-NCAM (e.g., Windrem et al. 2008), O1 (e.g., Duncan et al. 1992), and more recently CD140a (Sim et al. 2011). The limitation of these methods of isolating OPCs for transplantation has been in the numbers of cells generated from some of the protocols in terms of what would be required for human therapeutic application and in the complexity of the isolation procedure. An alternative strategy has been to generate OPC cell lines, though the downside of using, for example, SV40 transduction and immortalization is in the potential for transformation and tumor growth (Barnett et al. 1993; Trotter et al. 1993). However this approach resulted in the demonstration of extensive myelination by a retrovirally immortalized O-2A cell line (Groves et al. 1993). Perhaps the most useful OPC cell line has been CG-4, which was produced by growth factor expansion of OPCs (Louis et al. 1992). In culture, CG-4 cells differentiate into almost pure oligodendrocytes with few astrocytes though increasing serum in the media results in pure astrocyte populations. When transplanted into either the *md* rat or focal, glial-deficient demyelinated areas, CG-4 cells will give rise to myelinating oligodendrocytes (Franklin et al. 1996; Tontsch et al. 1994) and some astrocytes (Duncan 1996; Franklin et al. 1995).

In order to generate a greater supply of OPCs, Avellana-Adalid et al. (1996) developed a technique to isolate OPCs from neonatal rat brain by producing clusters of free-floating cells called oligospheres, purified by removal of astrocytes and oligodendrocytes by selective adhesion. Exposure to conditioned media from the B-104 neuroblastoma cell line resulted in a pure populations of OPCs which in low serum became pure oligodendrocytes while in high serum, astrocytes (Zhang et al. 1998a, b). These OPCs made myelin on transplantation into the *shi* mouse. A similar strategy was performed with mouse brain (Vitry et al. 1999). We chose a different method of purifying expandable collections of OPCs, isolating cells from the neonatal striatum of rats and canines, and culturing them initially as neurospheres in the presence of EGF and FGF2. By the gradual substitution of these growth factors with B-104 conditioned media, the cells switched to the oligodendrocyte lineage with over 95 % of cells differentiating into immature oligodendrocytes (Zhang et al. 1998a, b, 1999, 2000).

As noted at the beginning of this section, the cells that will be used in the first trials will be human. Thus, does what we know of rodent and higher animal oligodendrocyte development apply to humans? In short, similar principles apply to the differentiation of NSCs and OPCs in the human lineage as illustrated in Fig. 5.6. Indeed the maturation of human oligodendrocytes from neurospheres illustrates the classic features of OPC development toward the mature oligodendrocyte (Fig. 5.6). However there are significant differences between animal and human cells that lead to continued challenges in isolating sufficient OPCs for clinical use. A major gap in knowledge is a lack of known mitogens of the human OPC. Likewise, seemingly arcane yet potentially important differences, such as human O4-positive cells being non-mitotic, unlike rodent cells (Zhang et al. 2000) may be important. At present therefore, the generation of large numbers of human OPCs (or NSCs) for human clinical trial would seem to be most likely achievable using pluripotent cells as the source.

5.5 How Can Cells Be Tracked After Transplantation and Myelination Monitored?

It will be important to be able to track cells after grafting to determine whether they remain within the targeted lesion or migrate away to other sites. This question has been explored experimentally by labeling either NSCs, OPCs, Schwann cells, or OECs with superparamagnetic iron particles and following them by MRI (Bulte et al. 1999, 2001). Successful monitoring of cells has been achieved on transplantation into the *md* and Long Evans shaker (*les*) rats (Bulte et al. 1999, 2001), in focal demyelinated lesions (Dunning et al. 2004) and in EAE (Ben-Hur et al. 2007; Muja et al. 2011; Politi et al. 2007). It has been shown that labeling of cells does not interfere with their myelinating function (Bulte et al. 1999). It is also possible to follow cells in the mouse brain that express the luciferase gene using bioluminescence (Sher et al. 2009), but this is less likely to be used in patients because of the genetic manipulation of the cells and the technical limitations in detecting labeled cells in larger brains.

Of equal importance to following the transplanted cells will be monitoring remyelination in vivo. The primary method to achieve this will be with MRI, but the definitive MRI parameters of remyelination are still debated and require definitive data from an experimental system. Nonetheless, pre- and posttransplant MRIs, using magnetic transfer ratios, diffusion tensor imaging, and the newer method imaging "myelin water fraction" (Laule et al. 2007) will still provide the baseline information. Beyond MRI imaging, PET scanning using myelin-specific radiotracers (Stankoff et al. 2006; Wang et al. 2011a) has been used to follow myelination in development and in myelin mutants. A further advance using near-infrared imaging of myelination using a fluorescent probe that binds to myelin has been reported and studied in cuprizone-induced demyelination and remyelination (Wang et al. 2011a). However with both this and PET imaging, it remains to be proven that the techniques have the resolution to accurately image remyelination.

5.6 Do Exogenous Cells Remyelinate Axons in Inflammatory Disease?

There is growing interest in the role of inflammation in remyelination (Arnett et al. 2003; Hohlfeld 2007; Kondo et al. 2011; Popovich and Longbrake 2008; Setzu et al. 2006). Certainly endogenous remyelination occurs at the time of inflammatory cell infiltrate into the CNS, and a positive role of certain cytokines, once thought harmful to repair, has been postulated (Arnett et al. 2003). Likewise in experimental disorders, it has been shown that inflammation can promote both myelination (Setzu et al. 2006) and remyelination (Foote and Blakemore 2005). Myelination by oligodendrocytes derived from mouse ESCs has also been demonstrated in a focal, inflammatory model in the adult rat spinal cord (Perez-Bouza et al. 2005). The function of transplanted OPCs or other cell types has been most studied in EAE with

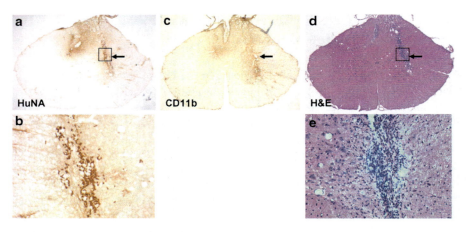

Fig. 5.7 Human cells survive after transplantation into areas of inflammation. Four weeks after transplantation of a glial-restricted human precursor cells (264) into Dark Agouti rats with EAE, human cells are indentified by immunolabeling for human nuclear antigen (HuNA) (**a**, **b**). They are located in areas of inflammation as noted in adjacent H + E stained sections (**d**, **e**) and in areas of microglial activation (OX-42) (**c**) (prepared by Dr. E. Larsen)

some surprising results, most of them positive. While the original goal may have been to use transplanted cells to remyelinate axons, it has been shown that they may have pleiotropic effects. An important finding on transplanting cells into the CNS of animals with EAE is that both animal and human cells survive and that inflammation promotes their migration (Einstein et al. 2003; Tourbah et al. 1997) (Fig. 5.7). Confirmation of this has been reported following transplantation of OPCs adjacent to focal areas of inflammation created by injection of zymosan in the spinal cord (Wang et al. 2011b). Indeed, OPCs grafted adjacent to the lesions preferentially migrated into the lesion where they survived suggesting chemotactic signals in the lesion (Wang et al. 2011b) and some differentiated into myelinating oligodendrocytes.

In a series of detailed experiments from Ben-Hur and colleagues, transplantation of NSCs into the lateral ventricles of mice with EAE demonstrated that encephalitis promoted the migration of the cells into the brain parenchyma (Einstein et al. 2003, 2006a–c). The cells differentiated into glial cells where they decreased inflammation and lessened axon loss (Aharonowiz et al. 2008; Einstein et al. 2006b). More recently, the same group showed that when NSCs were transplanted into the ventricles of mice treated with cuprizone, where there is a marked microglial response but no T-cell infiltration, the cells promoted endogenous OPCs to remyelinate axons in the corpus callosum. The transplantation of NSCs by a combined intravenous and intrathecal approach has also been shown to ameliorate EAE in a mouse model (Pluchino et al. 2003, 2005). These important experiments concluded that improvement was primarily through peripheral immunomodulation by the NSCs and not through remyelination. These data are discussed in greater detail by Martino and Ben-Hur (Chap. 7).

Transplantation of NSCs or glial-committed progenitors in mice with inflammatory demyelination resulting from MHV also resulted in lessening of disease, remyelination, and axonal protection. Firstly it was shown that transplanted cells survived and migrated within the demyelinated, inflamed spinal cord (Totoiu et al. 2004). These mice showed behavioral improvement associated with extensive remyelination and sparing of axons (Totoiu et al. 2004). In a follow-up study, they demonstrated that the remyelination was not associated with a decrease in inflammation as measured by T-cell and macrophage infiltration and expression of proinflammatory cytokines such as TNF-α (Hardison et al. 2006). In neither of these studies were they able to determine whether the remyelination resulted from the transplanted cells or from endogenous repair. The same group transplanted OPCs derived from hESCs into MHV-induced demyelinated spinal cord, but despite treatment with high-dose cyclosporine A, the cells were rejected after 2 weeks (Hatch et al. 2009). However extensive remyelination was seen at the site, and it was suggested to have resulted from an endogenous OPC response, stimulated by the xenograft rejection (Hatch et al. 2009). Finally, myelination by transplanted OPCs into the non-myelinated retina was enhanced by concurrent injection of zymosan which produced focal inflammation (Setzu et al. 2006), thus adding to the observations summarized above that highlight the importance of inflammation in repair.

5.7 How Is Functional Recovery Evaluated After Transplantation?

Evaluation of functional recovery after the transplantation of myelinating cells in animal models is documented by multiple levels of evidence (Zhang and Duncan 2000). The end goal of course is for MS patients receiving such a therapy to show clinical improvement. Evidence of functional improvement from the animal models will therefore provide assurance of potential success in human trials and is based on (1) the restoration of the myelin sheath (perhaps still thin) with normal nodes of Ranvier, (2) evidence that these structures have restored impulse conduction, and (3) evidence of behavioral improvement, for example, in muscle strength/movement and vision.

5.7.1 Myelin Sheath Thickness and Nodes of Ranvier

Remyelination both by endogenous and exogenous cells usually results in the formation of thinner-than-normal myelin sheaths around demyelinated/dysmyelinated axons. However, these myelin sheaths are capable of almost normal impulse conduction (Smith et al. 1981; Utzschneider et al. 1994). The reason that myelin sheaths remain thin on remyelination is not known though it may require axons to be growing for restoration of normal sheath thickness to occur. Certainly, transplantation into neonates may result in normal myelin sheaths after a prolonged period (Windrem et al. 2008). Nerve conduction is dependent on the formation of normal nodes of

Ranvier. The node of Ranvier is defined by ultrastructural specializations at the node and paranode and a normal molecular architecture at these sites (Poliak and Peles 2003). A transplant study in the *md* rat showed normal axoglial junctions at the node, both by routine ultrastructural evaluation and freeze fracture (Rosenbluth et al. 1993) in areas of transplant myelin. Two reports of transplantation of rodent NSCs (Eftekharpour et al. 2007) or human OPCs (Windrem et al. 2008) into the *shi* examined the distribution of voltage-gated sodium and potassium channels, as well as another important paranodal molecule, contactin-associated protein (Caspr), in repaired areas, at nodes and paranodal areas. These studies confirmed that transplanted glia and the myelin made by these cells resulted in normal formation of nodes of Ranvier, thus providing the structural basis of normal impulse transmission.

5.7.2 Restoration of Impulse Conduction

The first proof that myelin produced by transplanted cells came from ex vivo studies on the spinal cords of *md* rats (Utzschneider et al. 1994). These studies clearly showed that the nerve conduction velocity as well as other physiological parameters were restored through the transplant site, in contrast to non-myelinated areas rostral and caudal to it (Utzschneider et al. 1994). The transplant site was grossly visible in the dorsal column of the isolated spinal cord allowing placement of stimulatory and recording electrodes at desired sites (Fig. 2 in Duncan and Milward 1995). In two studies of conduction in grafted areas of the CNS in the *shi* mouse following transplantation, conduction velocity was also determined to be increased, though not normal in the brain (Windrem et al. 2008) and spinal cord (Eftekharpour et al. 2007). Two studies on the physiological effect of transplanted cells in inflammatory disease have been reported. In the first, human NSCs were injected into either the cisterna magna or the tail vein of marmosets that had been immunized with MOG peptide resulting in development of EAE (Pluchino et al. 2009). Although there were improvements in central conduction following transcranial stimulation in transplanted marmosets, it is not certain that this resulted from remyelination as was shown histologically, and the authors suggested that clinical improvement related to the immunomodulatory effects of the NSCs. In the second study, transplantation of glial-restricted precursors (Walczak et al. 2011) was carried out into focal inflammatory lesions in the spinal cord. Evaluation of somatosensory-evoked potentials in the transplanted group showed a decrease in latency. Whether this was related to remyelination by the transplanted cells is also not entirely clear.

5.7.3 Improvement in the Phenotype After Transplantation

Reports of transplantation of cells in the myelin mutants have rarely reported improvement in function as the myelination is predominately focal. However, two studies on the *shi* mouse have provided evidence of functional improvement. In the first, a mouse NSC cell line was transplanted into the newborn *shi* brain, and evidence of its widespread migration as seen by expression of β-galactosidase was seen (Yandava et al. 1999). A loss of tail tremor was purported to represent improvement in function though definitive proof of widespread myelination in the brain was lacking. A later study in *shi* in which human OPCs were injected into neonatal mice at five sites produced remarkable chimerism of the brain, with generalized spread of the cells and production of MBP-positive myelin (Windrem et al. 2008). Remarkably, many but not all of the mice lived for 1 year or more with reduced seizures and tremor (Windrem et al. 2008). To date, this is the most convincing evidence that exogenous cell therapy resulting in widespread myelination can restore function. Amelioration of disease following transplantation of human NSCs has also been reported in EAE in marmosets (Pluchino et al. 2009) as well as preventing their early death.

In models of focal demyelination resulting from injection of myelinotoxic chemicals, one study where lesions were created with ethidium bromide in the cervical spinal cord demonstrated improvement in function (judged by a beam-walking test), after transplantation and remyelination (Jeffery et al. 1999). However, this improvement only occurred in animals where axon loss was minimal; the loss of axons was thought to be associated with prior X-irradiation. It is worth noting that several models of focal spinal cord injury in which transplantation of OPCs or glial precursors has been performed showed evidence of functional recovery, potentially through remyelination and/or other mechanisms (Cao et al. 2005; Keirstead et al. 2005; Kumagai et al. 2009). Importantly, acute spinal cord lesions in the rat at the C5 level that were transplanted with OPCs generated from hESCs showed less severe pathology, remyelination, and behavioral improvement (Erceg et al. 2010; Sharp et al. 2010), supporting the idea that cervical cord lesions in MS may be amenable to therapy.

5.8 What Stage of Disease Would Be Targeted and How Will It Be Performed?

The choice of stage or stages in the course of MS where exogenous cell therapy would be useful is highly debated. Given the heterogeneous clinical course of MS and its underlying pathology, the timing of cell transplantation remains a critical question. A key point however is that any intervention should not interfere with endogenous remyelination. For that reason, transplantation during acute disease carries this risk; thus early relapsing-remitting disease would not seem to be a primary target. Perhaps the most useful stage to instigate cellular therapy is in lesions where there is persistent demyelination, a significant population of surviving axons (Fig. 5.8) and perhaps some persistent but low-grade inflammation that would help stimulate exogenous NSCs or OPCs to differentiate and myelinate axons. The selection of such lesions would begin by the careful clinical documentation of the time course of neurologic dysfunction related to specific lesions, for example, in the spinal cord. At the same time, state-of-the-art MRI will follow these lesions,

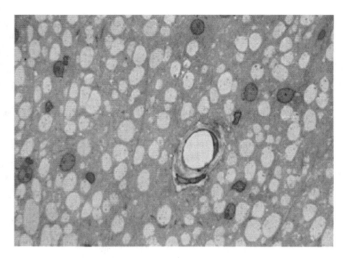

Fig. 5.8 The target for remyelination. Part of a chronic MS plaque that shows many preserved large-diameter axons embedded in a dense gliotic matrix. Although it is not possible to state the length of time they have been demyelinated, remyelination of these axons could lead to restored function in this area. Toluidine blue and safranin (modified from Prineas and McDonald 1997, with permission)

specifically to monitor the persistence of axons and perhaps spinal cord atrophy that could not be accounted for only by demyelination (indicating a severe loss of axons or the development of a "black hole") (Barkhof et al. 2009). Finally, PET scanning to provide evidence of microglial activation may be useful, at least in determining the activity of macrophages/microglia in the target site (Banati 2002; Vowinckel et al. 1997).

5.9 Selection of a Lesion Site and Surgical Approach

While lesions in MS are disseminated throughout the neuraxis, many are clinically silent; hence the targeting of focal repair should focus on strategically placed lesions. Three sites that have been discussed are the spinal cord, cerebellar peduncles, and optic nerve (Compston 1997). The latter has certain attractions as it is isolated from the CNS; hence other areas would not be negatively affected by the procedure. However it has been difficult to successfully transplant cells experimentally into the optic nerve (Baron personal communication). The nerve has a tough dura which is difficult to penetrate, and it is compartmentalized by large astrocyte processes that may inhibit the migration of transplanted cells. The cerebellar peduncle could be targeted stereotactically, and if successful, the lessening of intention tremor which results from lesions here could be monitored. Focal lesions of the spinal cord are important as 74–90 % of MS patients have plaques in the cervical cord on MRI

(Kidd et al. 1993; Vaithianathar et al. 2003), and at postmortem 87 % of randomly selected MS patients had cord lesions (Ikuta and Zimmerman 1976). Indeed, myelopathy can dominate the clinical presentation in MS though the patient may have MRI evidence of lesions in the brain; these are clinically silent. However, transplantation into the spinal cord may be risky if adjacent intact tracts are damaged. Despite this caveat, the clinical importance of lesions in the spinal cord, and their prevalence in MS, and the fact that there is a wealth of experimental data on spinal cord transplantation, makes it the optimal site.

Many surgical options exist for transplanting cells into the spinal cord. Depending on the site of the targeted lesion, a simple burr hole in the dorsal lamina or injection through the intervertebral space could be performed. This is currently being carried out in spinal cord injury patients using a custom-built stereotactic frame transplanting on oligodendrocyte cell line generated by Geron Corporation (Menlo Park, CA). A less invasive procedure would be to use a fiber-optic probe inserted into the lumbar space and followed by fluoroscopy. This could potentially target any level of spinal cord and certain sites in the brain. The ultimate goal will be to deliver cells in an entirely noninvasive way through vascular or intrathecal approaches, though the former does not appear to enable widespread intra-parenchymal access. While there is experimental evidence that cells injected into the lateral ventricles of animals with EAE can migrate into CNS parenchyma, direct injection into the lesion remains the best way of ensuring cells get access to the demyelinated axons.

5.10 Immunosuppression Following Transplantation

There is considerable clinical experience of immunosuppression in patients with Parkinson's disease (PD) who have received fetal dopaminergic cells as allografts (Björklund and Lindvall 2000; Lindvall and Kokaia 2006). While the CNS is not an immunologically privileged site, transplanted cells are not surveilled immunologically as sensitively as cells grafted to other organs. Nonetheless, MS patients receiving grafts will undoubtedly be immunosuppressed, but with the experience of PD, drugs such as cyclosporin A can be given at low levels and potentially discontinued at some stage (Winkler et al. 2005). Cyclosporin A has been used in glial cell xenografts and has been shown to prevent rejection but not inhibit the function of the transplanted cells with successful myelination (Archer et al. 1994; Perez-Bouza et al. 2005; Rosenbluth et al. 1993). Interestingly, it has been shown in vitro and in vivo to increase the number of NSCs, suggesting that it may also increase the cells available for differentiation and remyelination when used to prevent rejection in patients receiving transplanted cells (Hunt et al. 2010). Most recently, it was demonstrated that immunosuppression with cyclosporin A promoted the endogenous NSC response in ischemic injury (Erlandsson et al. 2011), thus supporting its possible role in tissue repair in MS through endogenous or potentially exogenous cells recruitment and enhancement. Cyclosporin A also been used as a therapy in MS (The Multiple Sclerosis Study Group 1990) but needs to be given at low doses and potentially discontinued to prevent long-term side effects such as renal damage. Tacrolimus (FK506) may be more likely to be used in future trials. The choice of immunosuppressive drugs and regimens supports the position that rejection of transplanted cells is not the greatest hurdle in glial cell transplantation in MS.

5.11 Conclusions

The first clinical trial of cell transplantation into focal or more disseminated lesions in MS seems close. The roadblock to such a trial has primarily been the cell to be used, but progress here would suggest that this may be resolved in the near future. The other key issues are what stage of disease to transplant, will there be sufficient axons to remyelinate, and will the milieu be supportive? Only the latter point, that is, whether the environment will support and promote repair, is perhaps impossible to predict or model, and it will then truly be a human experiment. The goal is to do no harm, restore function, and protect intact tissue. To quote loosely from the Chinese proverb "we live in interesting times" and pioneering trials in MS would seem to be on the horizon.

Acknowledgments The work supported from the I. D. Duncan's lab cited here has been supported by the NMSS Translational Research Partnership on Nervous System Repair and Protection in MS (TR-3761), the Myelin Project, the Elizabeth Elser Doolittle Charitable Trust, and the Oscar Rennebohm Foundation. We are grateful to many past members of the lab for their scientific and technical contributions. This manuscript was skillfully prepared by Abigail Radcliff and Naomi Dahnert.

References

Ader M, Meng J, Schachner M, Bartsch U (2000) Formation of myelin after transplantation of neural precursor cells into the retina of young postnatal mice. Glia 30(3):301–310

Aharonowiz M, Einstein O, Fainstein N, Lassmann H, Reubinoff B, Ben-Hur T (2008) Neuroprotective effect of transplanted human embryonic stem cell-derived neural precursors in an animal model of multiple sclerosis. PLoS One 3:e3145

Amabile G, Meissner A (2009) Induced pluripotent stem cells: current progress and potential for regenerative medicine. Trends Mol Med 15:59–68

Archer DR, Leven S, Duncan ID (1994) Myelination by cryopreserved xenografts and allografts in the myelin-deficient rat. Exp Neurol 125:268–277

Archer DR, Cuddon PA, Lipsitz D, Duncan ID (1997) Myelination of the canine central nervous system by glial cell transplantation: a model for repair of human myelin disease. Nat Med 3:54–59

Arnett HA, Wang Y, Matsushima GK, Suzuki K, Ting JPY (2003) Functional genomic analysis of remyelination reveals importance of inflammation in oligodendrocyte regeneration. J Neurosci 23:9824–9832

Avellana-Adalid V, Nait-Oumesmar B, Lachapelle F, Baron-Van Evercooren A (1996) Expansion of rat oligodendrocyte progenitors into proliferative "oligospheres" that retain differentiation potential. J Neurosci Res 45:558–570

Back SA, Tuohy TM, Chen H, Wallingford N, Craig A, Struve J, Luo NL, Banine F, Liu Y, Chang A, Trapp BD, Bebo BF Jr, Rao MS, Sherman LS (2005) Hyaluronan accumulates in demyelinated lesions and inhibits oligodendrocyte progenitor maturation. Nat Med 11:966–972

Banati RB (2002) Visualising microglial activation in vivo. Glia 40:206–217

Barkhof F, Calabresi PA, Miller DH, Reingold SC (2009) Imaging outcomes for neuroprotection and repair in multiple sclerosis trials. Nat Rev Neurol 5:256–266

Barnett SC, Franklin RJM, Blakemore WF (1993) In vitro and in vivo analysis of a rat bipotential O-2A progenitor cell line containing the temperature-sensitive mutant gene of the SV40 large T antigen. Eur J Neurosci 5:1247–1260

Ben Hur T, van Heeswijk RB, Einstein O, Aharonowiz M, Xue R, Frost EE, Mori S, Reubinoff BE, Bulte JW (2007) Serial in vivo MR tracking of magnetically labeled neural spheres transplanted in chronic EAE mice. Magn Reson Med 57:164–171

Björklund A, Lindvall O (2000) Cell replacement therapies for central nervous system disorders. Nat Neurosci 3:537–544

Blakemore WF (1973) Remyelination of the superior cerebellar peduncle in the mouse following demyelination induced by feeding cuprizone. J Neurol Sci 20:73–83

Blakemore WF (1977) Remyelination of CNS axons by Schwann cells transplanted from the sciatic nerve. Nature 266:68–69

Blakemore WF, Franklin RJM (1991) Transplantation of glial cells into the CNS. Trends Neurosci 14:323–327

Blakemore WF, Franklin RJ (2000) Transplantation options for therapeutic central nervous system remyelination. Cell Transplant 9:289–294

Blakemore WF, Patterson RC (1978) Suppression of remyelination in the CNS by X-irradiation. Acta Neuropathol 42:105–113

Blakemore WF, Crang AJ, Franklin RJM (1995a) Transplantation of glial cells. In: Ransom BR, Kettenmann H (eds) Neuroglial cells. Oxford University Press, Cambridge, pp 869–882

Blakemore WF, Crang AJ, Franklin RJM, Tang K, Ryder S (1995b) Glial cell transplants that are subsequently rejected can be used to influence regeneration of glial cell environments in the CNS. Glia 13:79–91

Blakemore WF, Gilson JM, Crang AJ (2000) Transplanted glial cells migrate over a greater distance and remyelinate demyelinated lesions more rapidly than endogenous remyelinating cells. J Neurosci Res 61:288–294

Blakemore WF, Gilson JM, Crang AJ (2003) The presence of astrocytes in areas of demyelination influences remyelination following transplantation of oligodendrocyte progenitors. Exp Neurol 184:955–963

Brüstle O, Jones KN, Learish RD, Karram K, Choudhary K, Wiestler OD, Duncan ID, McKay RDG (1999) Embryonic stem cell-derived glial precursors: a source of myelinating transplants. Science 285:754–756

Buchet D, Baron-Van Evercooren A (2009) In search of human oligodendroglia for myelin repair. Neurosci Lett 456:112–119

Buchet D, Garcia C, Deboux C, Nait-Oumesmar B, Baron-Van Evercooren A (2011) Human neural progenitors from different foetal forebrain regions remyelinate the adult mouse spinal cord. Brain 134:1168–1183

Bulte JWM, Zhang SC, van Gelderen P, Herynek V, Jordan EK, Duncan ID, Frank JA (1999) Neurotransplantation of magnetically labeled oligodendrocyte progenitors: magnetic resonance tracking of cell migration and myelination. Proc Natl Acad Sci USA 96:15256–15261

Bulte JWM, Douglas T, Witwer B, Zhang SC, Strable E, Lewis BK, Zywicke H, Miller B, van Gelderen P, Moskowitz BM, Duncan ID, Frank JA (2001) Magnetodendrimers allow endosomal magnetic labeling and in vivo tracking of stem cells. Nat Biotechnol 19:1141–1147

Cao Q, Xu XM, DeVries WH, Enzmann GU, Ping P, Tsoulfas P, Wood PM, Bunge MB, Whittemore SR (2005) Functional recovery in traumatic spinal cord injury after transplantation of multi-neurotrophin-expressing glial-restricted precursor cells. J Neurosci 25:6947–6957

Chang A, Tourtellotte WW, Rudick R, Trapp BD (2002) Premyelinating oligodendrocytes in chronic lesions of multiple sclerosis. N Engl J Med 346:165–173

Charles P, Reynolds R, Seilhean D, Rougon G, Aigrot MS, Niezgoda A, Zalc B, Lubetzki C (2002) Re-expression of PSA-NCAM by demyelinated axons: an inhibitor of remyelination in multiple sclerosis? Brain 125:1972–1979

Comi G (2008) Is it clinically relevant to repair focal multiple sclerosis lesions? J Neurol Sci 265:17–20

Compston A (1997) Remyelination in multiple sclerosis: a challenge for therapy. The 1996 European Charcot Foundation Lecture. Mult Scler 3:51–70

Crang AT, Gilson J, Blakemore WF (1998) The demonstration by transplantation of the very restricted remyelinating potential of post-mitotic oligodendrocytes. J Neurocytol 27:541–553

Czepiel M, Balasubramaniyan V, Schaafsma W, Stancic M, Mikkers H, Huisman C, Boddeke E, Copray S (2011) Differentiation of induced pluripotent stem cells into functional oligodendrocytes. Glia 59:882–892

De Castro F, Bribián A (2005) The molecular orchestra of the migration of oligodendrocyte precursors during development. Brain Res Rev 49:227–241

Decker L, Avellana-Adalid V, Nait-Oumesmar B, Durbec P, Baron-Van Evercooren A (2000) Oligodendrocyte precursor migration and differentiation: combined effects of PSA residues, growth factors, and substrates. Mol Cell Neurosci 16:422–439

Du ZW, Li XJ, Nguyen GD, Zhang SC (2006) Induced expression of Olig2 is sufficient for oligodendrocyte specification but not for motoneuron specification and astrocyte repression. Mol Cell Neurosci 33:371–380

Dubois-Dalcq M, Ffrench-Constant C, Franklin RJ (2005) Enhancing central nervous system remyelination in multiple sclerosis. Neuron 48:9–12

Duncan ID (1995) Inherited disorders of myelination of the central nervous system. In: Ransom BR, Kettenmann HR (eds) Neuroglial cells. Oxford University Press, Cambridge, pp 990–1009

Duncan ID (1996) Glial cell transplantation and remyelination of the CNS. Neuropathol Appl Neurobiol 22:87–100

Duncan ID (2008) Replacing cells in multiple sclerosis. J Neurol Sci 265:89–92

Duncan ID, Milward EA (1995) Glial cell transplants: experimental therapies of myelin diseases. Brain Pathol 5:301–310

Duncan ID, Aguayo AJ, Bunge RP, Wood PM (1981) Transplantation of rat Schwann cells grown in tissue culture into the mouse spinal cord. J Neurol Sci 49:241–252

Duncan ID, Hammang JP, Jackson KF, Wood PM, Bunge RP, Langford LA (1988) Transplantation of oligodendrocytes and Schwann cells into the spinal cord of the myelin-deficient rat. J Neurocytol 17:351–360

Duncan ID, Paino C, Archer DR, Wood PM (1992) Functional capacities of transplanted cell-sorted adult oligodendrocytes. Dev Neurosci 14:114–122

Duncan ID, Grever WE, Zhang SC (1997) Repair of myelin disease: strategies and progress in animal models. Mol Med Today 3:554–561

Duncan ID, Goldman S, Macklin WB, Rao M, Weiner LP, Reingold SC (2008) Stem cell therapy in multiple sclerosis: promise and controversy. Mult Scler 14:541–546

Duncan ID, Brower A, Kondo Y, Curlee JF Jr, Schultz RD (2009) Extensive remyelination of the CNS leads to functional recovery. Proc Natl Acad Sci USA 106:6832–6836

Dunning MD, Lakatos A, Loizou L, Kettunen M, Ffrench-Constant C, Brindle KM, Franklin RJ (2004) Superparamagnetic iron oxide-labeled Schwann cells and olfactory ensheathing cells can be traced in vivo by magnetic resonance imaging and retain functional properties after transplantation into the CNS. J Neurosci 24:9799–9810

Eftekharpour E, Karimi-Abdolrezaee S, Wang J, El Beheiry H, Morshead C, Fehlings MG (2007) Myelination of congenitally dysmyelinated spinal cord axons by adult neural precursor cells results in formation of nodes of Ranvier and improved axonal conduction. J Neurosci 27:3416–3428

Einstein O, Karussis D, Grigoriadis N, Mizrachi-Kol R, Reinhartz E, Abramsky O, Ben-Hur T (2003) Intraventricular transplantation of neural precursor cell spheres attenuates acute experimental allergic encephalomyelitis. Mol Cell Neurosci 27:1074–1082

Einstein O, Fainstein N, Vaknin I, Mizrachi-Kol R, Reihartz E, Grigoriadis N, Lavon I, Baniyash M, Lassmann H, Ben Hur T (2006a) Neural precursors attenuate autoimmune encephalomyelitis by peripheral immunosuppression. Ann Neurol 61:209–218

Einstein O, Grigoriadis N, Mizrachi-Kol R, Reinhartz E, Polyzoidou E, Lavon I, Milonas I, Karussis D, Abramsky O, Ben-Hur T (2006b) Transplanted neural precursor cells reduce brain inflammation to attenuate chronic experimental autoimmune encephalomyelitis. Exp Neurol 198:275–284

Einstein O, Menachem-Tzidon O, Mizrachi-Kol R, Reinhartz E, Grigoriadis N, Ben Hur T (2006c) Survival of neural precursor cells in growth factor-poor environment: implications for transplantation in chronic disease. Glia 53:449–455

Einstein O, Friedman-Levi Y, Grigoriadis N, Ben-Hur T (2009) Transplanted neural precursors enhance host brain-derived myelin regeneration. J Neurosci 29:15694–15702

Erceg S, Ronaghi M, Oria M, Rosello MG, Arago MA, Lopez MG, Radojevic I, Moreno-Manzano V, Rodriguez-Jimenez FJ, Bhattacharya SS, Cordoba J, Stojkovic M (2010) Transplanted oligodendrocytes and motoneuron progenitors generated from human embryonic stem cells promote locomotor recovery after spinal cord transection. Stem Cells 28:1541–1549

Erlandsson A, Lin CH, Yu F, Morshead CM (2011) Immunosuppression promotes endogenous neural stem and progenitor cell migration and tissue regeneration after ischemic injury. Exp Neurol 230:48–57

Fancy SP, Chan JR, Baranzini SE, Franklin RJ, Rowitch DH (2011) Myelin regeneration: a recapitulation of development? Annu Rev Neurosci 34:21–43

Foote AK, Blakemore WF (2005) Inflammation stimulates remyelination in areas of chronic demyelination. Brain 128:528–539

Franceschini I, Vitry S, Padilla F, Casanova P, Tham TN, Fukuda M, Rougon G, Durbec P, Dubois-Dalcq M (2004) Migrating and myelinating potential of neural precursors engineered to over-express PSA-NCAM. Mol Cell Neurosci 27:151–162

Franklin RJM (1993) Reconstructing myelin-deficient environments in the CNS by glial cell transplantation. Neurosciences 5:443–452

Franklin RJM, Blakemore WF (1997) Transplanting oligodendrocyte progenitors into the adult CNS. J Anat 190:23–33

Franklin RJ, Ffrench-Constant C (2008) Remyelination in the CNS: from biology to therapy. Nat Rev Neurosci 9:839–855

Franklin RJM, Crang AJ, Blakemore WF (1991) Transplanted type-1 astrocytes facilitate repair of demyelinating lesions by host oligodendrocytes in adult rat spinal cord. J Neurocytol 20:420–430

Franklin RJM, Bayley SA, Milner R, Ffrench-Constant C, Blakemore WF (1995) Differentiation of the O-2A progenitor cell line CG-4 into oligodendrocytes and astrocytes following transplantation into glia-deficient areas of CNS white matter. Glia 13:39–44

Franklin RJM, Bayley SA, Blakemore WF (1996) Transplanted CG4 cells (an oligodendrocyte progenitor cell line) survive, migrate, and contribute to repair of areas of demyelination in X-irradiated and damaged spinal cord but not in normal spinal cord. Exp Neurol 137:263–276

Gansmuller A, Lachapelle F, Baron-Van Evercooren A, Hauw JJ, Baumann N, Gumpel M (1986) Transplantations of newborn CNS fragments into the brain of shiverer mutant mice: extensive myelination by transplanted oligodendrocytes. Dev Neurosci 8:197–207

Goldman S (2005) Stem and progenitor cell-based therapy of the human central nervous system. Nat Biotechnol 23:862–871

Goldschmidt T, Antel J, Konig FB, Brück W, Kuhlmann T (2009) Remyelination capacity of the MS brain decreases with disease chronicity. Neurology 72:1914–1921

Gout O, Gansmüller A, Gumpel M (1990) Remyelination of a chemically induced demyelinated lesion in the spinal cord of the adult shiverer mouse by transplanted oligodendrocytes. In: Jeserich G, Althaus HH, Waehneldt TV (eds) Cellular and molecular biology of myelination, vol 43. Springer, Berlin, pp 185–198

Griffiths IR (1996) Myelin mutants: model systems for the study of normal and abnormal myelination. Bioessays 18:789–797

Groves AK, Barnett SC, Franklin RJM, Crang AJ, Mayer M, Blakemore WF, Noble M (1993) Repair of demyelinated lesions by transplantation of purified O-2A progenitor cells. Nature 362:453–455

Gumpel M, Baumann N, Raoul M, Jacque C (1983) Survival and differentiation of oligodendrocytes from neural tissue transplanted into new-born mouse brain. Neurosci Lett 37:307–311

Gumpel M, Lachapelle F, Baumann N (1985) Central nervous tissue transplantation into mouse brain: differentiation of myelin from transplanted oligodendrocytes. In: Björklund A, Stenevi U (eds) Neural grafting in the mammalian CNS. Elsevier, Amsterdam, pp 151–158

Gumpel M, Lachapelle F, Gansmüller A, Baulac M, Baron-Van Evercooren A, Baumann N (1987) Transplantation of human embryonic oligodendrocytes into shiverer brain. Ann NY Acad Sci 495:71–85

Hammang JP, Archer DR, Duncan ID (1997) Myelination following transplantation of EGF-responsive neural stem cells into a myelin-deficient environment. Exp Neurol 147:84–95

Hardison JL, Nistor G, Gonzalez R, Keirstead HS, Lane TE (2006) Transplantation of glial-committed progenitor cells into a viral model of multiple sclerosis induces remyelination in the absence of an attenuated inflammatory response. Exp Neurol 197:420–429

Hatch MN, Schaumburg CS, Lane TE, Keirstead HS (2009) Endogenous remyelination is induced by transplant rejection in a viral model of multiple sclerosis. J Neuroimmunol 212:74–81

Hohlfeld R (2007) Does inflammation stimulate remyelination? J Neurol 254(Suppl 1):I-47–I-54

Hu BY, Du ZW, Zhang SC (2009) Differentiation of human oligodendrocytes from pluripotent stem cells. Nat Protoc 4:1614–1622

Hu BY, Weick JP, Yu J, Ma LX, Zhang XQ, Thomson JA, Zhang SC (2010) Neural differentiation of human induced pluripotent stem cells follows developmental principles but with variable potency. Proc Natl Acad Sci USA 107:4335–4340

Hunt J, Cheng A, Hoyles A, Jervis E, Morshead CM (2010) Cyclosporin A has direct effects on adult neural precursor cells. J Neurosci 30:2888–2896

Ikuta F, Zimmerman HM (1976) Distribution of plaques in seventy autopsy cases of multiple sclerosis in the United States. Neurology 26:26–28

Irvine KA, Blakemore WF (2008) Remyelination protects axons from demyelination-associated axon degeneration. Brain 131:1464–1477

Izrael M, Zhang PL, Kaufman R, Shinder V, Ella R, Amit M, Itskovitz-Eldor J, Chebath J, Revel M (2007) Human oligodendrocytes derived from embryonic stem cells: effect of noggin on phenotypic differentiation in vitro and on myelination in vivo. Mol Cell Neurosci 34:310–323

Jarjour AA, Kennedy TE (2004) Oligodendrocyte precursors on the move: mechanisms directing migration. Neuroscientist 10:99–105

Jeffery ND, Crang AJ, O'Leary MT, Hodge SJ, Blakemore WF (1999) Behavioural consequences of oligodendrocyte progenitor cell transplantation into experimental demyelinating lesions in the rat spinal cord. Eur J Neurosci 11:1508–1514

Joannides AJ, Fiore-Heriche C, Battersby AA, Athauda-Arachchi P, Bouhon IA, Williams L, Westmore K, Kemp PJ, Compston A, Allen ND, Chandran S (2007) A scaleable and defined system for generating neural stem cells from human embryonic stem cells. Stem Cells 25:731–737

Kang SM, Cho MS, Seo H, Yoon CJ, Oh SK, Choi YM, Kim DW (2007) Efficient induction of oligodendrocytes from human embryonic stem cells. Stem Cells 25:419–424

Karimi-Abdolrezaee S, Eftekharpour E, Wang J, Morshead CM, Fehlings MG (2006) Delayed transplantation of adult neural precursor cells promotes remyelination and functional neurological recovery after spinal cord injury. J Neurosci 26:3377–3389

Keirstead HS (2005) Stem cells for the treatment of myelin loss. Trends Neurosci 28:677–683

Keirstead HS, Levine JM, Blakemore WF (1998) Response of the oligodendrocyte progenitor cell population (defined by NG2 labelling) to demyelination of the adult spinal cord. Glia 22:161–170

Keirstead HS, Nistor G, Bernal G, Totoiu M, Cloutier F, Sharp K, Steward O (2005) Human embryonic stem cell-derived oligodendrocyte progenitor cell transplants remyelinate and restore locomotion after spinal cord injury. J Neurosci 25:4694–4705

Kerschensteiner M, Stadelmann C, Buddeberg BS, Merkler D, Bareyre FM, Anthony DC, Linington C, Brück W, Schwab ME (2004) Targeting experimental autoimmune encephalomyelitis lesions to a predetermined axonal tract system allows for refined behavioral testing in an animal model of multiple sclerosis. Am J Pathol 164:1455–1469

Kerstetter AE, Padovani-Claudio DA, Bai L, Miller RH (2009) Inhibition of CXCR2 signaling promotes recovery in models of multiple sclerosis. Exp Neurol 220:44–56

Kidd D, Thorpe JW, Thompson AJ, Kendall BE, Moseley IF, MacManus DG, McDonald WI, Miller DH (1993) Spinal cord MRI using multi-array coils and fast spin echo II: findings in multiple sclerosis. Neurology 43:2632–2637

Kiskinis E, Eggan K (2010) Progress toward the clinical application of patient-specific pluripotent stem cells. J Clin Invest 120:51–59

Koch P, Kokaia Z, Lindvall O, Brustle O (2009) Emerging concepts in neural stem cell research: autologous repair and cell-based disease modelling. Lancet Neurol 8:819–829

Kondo Y, Wenger DA, Gallo V, Duncan ID (2005) Galactocerebrosidase-deficient oligodendrocytes maintain stable central myelin by exogenous replacement of the missing enzyme in mice. Proc Natl Acad Sci USA 102:18670–18675

Kondo Y, Adams JM, Vanier MT, Duncan ID (2011) Macrophages counteract demyelination in a mouse model of globoid cell leukodystrophy. J Neurosci 31:3610–3624

Kornek B, Storch MK, Weissert R, Wallstroem E, Stefferl A, Olsson T, Linington C, Schmidbauer M, Lassmann H (2000) Multiple sclerosis and chronic autoimmune encephalomyelitis: a comparative quantitative study of axonal injury in active, inactive, and remyelinated lesions. Am J Pathol 157:267–276

Kremer D, Aktas O, Hartung HP, Kury P (2011) The complex world of oligodendroglial differentiation inhibitors. Ann Neurol 69:602–618

Kumagai G, Okada Y, Yamane J, Nagoshi N, Kitamura K, Mukaino M, Tsuji O, Fujiyoshi K, Katoh H, Okada S, Shibata S, Matsuzaki Y, Toh S, Toyama Y, Nakamura M, Okano H (2009) Roles of ES cell-derived gliogenic neural stem/progenitor cells in functional recovery after spinal cord injury. PLoS One 4:e7706

Lachapelle F, Gumpel M, Baulac M, Jacque C, Duc P, Baumann N (1983) Transplantation of CNS fragments into the brain of shiverer mutant mice: extensive myelination by implanted oligodendrocytes. I. Immunohistochemical studies. Dev Neurosci 6:325–334

Lassmann H (2005) Stem cell and progenitor cell transplantation in multiple sclerosis: the discrepancy between neurobiological attraction and clinical feasibility. J Neurol Sci 233:83–86

Laule C, Vavasour IM, Kolind SH, Li DK, Traboulsee TL, Moore GR, MacKay AL (2007) Magnetic resonance imaging of myelin. Neurotherapeutics 4:460–484

Lindvall O, Kokaia Z (2006) Stem cells for the treatment of neurological disorders. Nature 441:1094–1096

Liu S, Stewart TJ, Howard MJ, Chakrabortty S, Holekamp TF, McDonald JW (2000) Embryonic stem cells differentiate into oligodendrocytes and myelinate in culture and after spinal cord transplantation. Proc Natl Acad Sci USA 97:6126–6131

Louis JC, Magal E, Muir D, Manthorpe M, Varon S (1992) CG-4, a new bipotential glial cell line from rat brain, is capable of differentiating in vitro into either mature oligodendrocytes or type-2 astrocytes. J Neurosci Res 31:193–204

Ludwin SK (1978) Central nervous system demyelination and remyelination in the mouse. Lab Invest 39:597–612

Magy L, Mertens C, Avellana-Adalid V, Keita M, Lachapelle F, Nait-Oumesmar B, Fontaine B, Baron-Van Evercooren A (2003) Inducible expression of FGF2 by a rat oligodendrocyte precursor cell line promotes CNS myelination in vitro. Exp Neurol 184:912–922

Martino G, Franklin RJ, Van Evercooren AB, Kerr DA (2010) Stem cell transplantation in multiple sclerosis: current status and future prospects. Nat Rev Neurol 6:247–255

Mason JL, Toews A, Hostettler JD, Morell P, Suzuki K, Goldman JE, Matsushima GK (2004) Oligodendrocytes and progenitors become progressively depleted within chronically demyelinated lesions. Am J Pathol 164:1673–1682

Milward EA, Zhang SC, Zhao M, Lundberg C, Ge B, Goetz BD, Duncan ID (2000) Enhanced proliferation and directed migration of oligodendroglial progenitors co-grafted with growth factor-secreting cells. Glia 32:264–270

Mitome M, Low HP, van den Pol A, Nunnari JJ, Wolf MK, Billings-Gagliardi S, Schwartz WJ (2001) Towards the reconstruction of central nervous system white matter using neural precursor cells. Brain 124:2147–2161

Moore CS, Abdullah SL, Brown A, Arulpragasam A, Crocker SJ (2011) How factors secreted from astrocytes impact myelin repair. J Neurosci Res 89:13–21

Mothe AJ, Tator CH (2008) Transplanted neural stem/progenitor cells generate myelinating oligodendrocytes and Schwann cells in spinal cord demyelination and dysmyelination. Exp Neurol 213:176–190

Mothe AJ, Kulbatski I, Parr A, Mohareb M, Tator CH (2008) Adult spinal cord stem/progenitor cells transplanted as neurospheres preferentially differentiate into oligodendrocytes in the adult rat spinal cord. Cell Transplant 17:735–751

Muja N, Cohen ME, Zhang J, Kim H, Gilad AA, Walczak P, Ben-Hur T, Bulte JW (2011) Neural precursors exhibit distinctly different patterns of cell migration upon transplantation during either the acute or chronic phase of EAE: a serial MR imaging study. Magn Reson Med 65:1738–1749

Neri M, Maderna C, Ferrari D, Cavazzin C, Vescovi AL, Gritti A (2010) Robust generation of oligodendrocyte progenitors from human neural stem cells and engraftment in experimental demyelination models in mice. PLoS One 5:e10145

Nicolay DJ, Doucette JR, Nazarali AJ (2007) Transcriptional control of oligodendrogenesis. Glia 55:1287–1299

Nistor GI, Totoiu MO, Haque N, Carpenter MK, Keirstead HS (2005) Human embryonic stem cells differentiate into oligodendrocytes in high purity and myelinate after spinal cord transplantation. Glia 49:385–396

O'Leary MT, Blakemore WF (1997) Oligodendrocyte precursors survive poorly and do not migrate following transplantation into the normal adult central nervous system. J Neurosci Res 48:159–167

Pang ZP, Yang N, Vierbuchen T, Ostermeier A, Fuentes DR, Yang TQ, Citri A, Sebastiano V, Marro S, Sudhof TC, Wernig M (2011) Induction of human neuronal cells by defined transcription factors. Nature 476(7359):220–223

Pastrana E, Silva-Vargas V, Doetsch F (2011) Eyes wide open: a critical review of sphere-formation as an assay for stem cells. Cell Stem Cell 8:486–498

Patani R, Balaratnam M, Vora A, Reynolds R (2007) Remyelination can be extensive in multiple sclerosis despite a long disease course. Neuropathol Appl Neurobiol 33:277–287

Patrikios P, Stadelmann C, Kutzelnigg A, Rauschka H, Schmidbauer M, Laursen H, Sorensen PS, Brück W, Lucchinetti C, Lassmann H (2006) Remyelination is extensive in a subset of multiple sclerosis patients. Brain 129:3165–3172

Perez-Bouza A, Glaser T, Brüstle O (2005) ES cell-derived glial precursors contribute to remyelination in acutely demyelinated spinal cord lesions. Brain Pathol 15:208–216

Piaton G, Aigrot MS, Williams A, Moyon S, Tepavcevic V, Moutkine I, Gras J, Matho KS, Schmitt A, Soellner H, Huber AB, Ravassard P, Lubetzki C (2011) Class 3 semaphorins influence oligodendrocyte precursor recruitment and remyelination in adult central nervous system. Brain 134:1156–1167

Pluchino S, Quattrini A, Brambilla E, Gritti A, Salani G, Dina G, Galli R, Del Carro U, Amadio S, Bergami A, Furlan R, Comi G, Vescovi AL, Martino G (2003) Injection of adult neurospheres induces recovery in a chronic model of multiple sclerosis. Nature 422:688–694

Pluchino S, Zanotti L, Rossi B, Brambilla E, Ottoboni L, Salani G, Martinello M, Cattalini A, Bergami A, Furlan R, Comi G, Constantin G, Martino G (2005) Neurosphere-derived multipotent precursors promote neuroprotection by an immunomodulatory mechanism. Nature 436:266–271

Pluchino S, Gritti A, Blezer E, Amadio S, Brambilla E, Borsellino G, Cossetti C, Del Carro U, Comi G, t Hart B, Vescovi A, Martino G (2009) Human neural stem cells ameliorate autoimmune encephalomyelitis in non-human primates. Ann Neurol 66:343–354

Poliak S, Peles E (2003) The local differentiation of myelinated axons at nodes of Ranvier. Nat Rev Neurosci 4:968–980

Politi LS, Bacigaluppi M, Brambilla E, Cadioli M, Falini A, Comi G, Scotti G, Martino G, Pluchino S (2007) Magnetic-resonance-based tracking and quantification of intravenously injected neural stem cell accumulation in the brains of mice with experimental multiple sclerosis. Stem Cells 25:2583–2592

Popovich PG, Longbrake EE (2008) Can the immune system be harnessed to repair the CNS? Nat Rev Neurosci 9:481–493

Popovich PG, Guan Z, McGaughy V, Fisher L, Hickey WF, Basso DM (2002) The neuropathological and behavioral consequences of intraspinal microglial/macrophage activation. J Neuropathol Exp Neurol 61:623–633

Prineas J, Connell F (1979) Remyelination in multiple sclerosis. Ann Neurol 5:22–31

Prineas JW, McDonald WI (1997) Demyelinating diseases. In: Graham DI, Lantos PL (eds) Greenfield's neurophathology, vol 1, 6 edn. Edward Arnonld, London, pp 813–896

Raine CS, Wu E (1993) Multiple sclerosis: remyelination in acute lesions. J Neuropathol Exp Neurol 52:199–204

Reubinoff BE, Itsykson P, Turetsky T, Pera MF, Reinhartz E, Itzik A, Ben-Hur T (2001) Neural progenitors from human embryonic stem cells. Nat Biotechnol 19:1134–1140

Reynolds BA, Weiss S (1992) Generation of neurons and astrocytes from isolated cells of the adult mammalian central nervous system. Science 255:1707–1710

Reynolds BA, Weiss S (1996) Clonal and population analyses demonstrate that an EGF-responsive mammalian embryonic CNS precursor is a stem cell. Dev Biol 175:1–13

Rosenbluth J, Hasegawa M, Shirasaki N, Rosen CL, Liu Z (1990) Myelin formation following transplantation of normal fetal glia into myelin-deficient rat spinal cord. J Neurocytol 19:718–730

Rosenbluth J, Liu Z, Guo D, Schiff R (1993) Myelin formation by mouse glia in myelin-deficient rats treated with cyclosporine. J Neurocytol 22:967–977

Sasaki M, Lankford KL, Brown RJ, Ruddle NH, Kocsis JD (2010) Focal experimental autoimmune encephalomyelitis in the Lewis rat induced by immunization with myelin oligodendrocyte glycoprotein and intraspinal injection of vascular endothelial growth factor. Glia 58:1523–1531

Schonberg DL, Popovich PG, McTigue DM (2007) Oligodendrocyte generation is differentially influenced by toll-like receptor (TLR) 2 and TLR4-mediated intraspinal macrophage activation. J Neuropathol Exp Neurol 66:1124–1135

Setzu A, Ffrench-Constant C, Franklin RJM (2004) CNS axons retain their competence for myelination throughout life. Glia 45:307–311

Setzu A, Lathia JD, Zhao C, Wells K, Rao MS, Ffrench-Constant C, Franklin RJ (2006) Inflammation stimulates myelination by transplanted oligodendrocyte precursor cells. Glia 54:297–303

Sharp J, Frame J, Siegenthaler M, Nistor G, Keirstead HS (2010) Human embryonic stem cell-derived oligodendrocyte progenitor cell transplants improve recovery after cervical spinal cord injury. Stem Cells 28:152–163

Shen S, Sandoval J, Swiss VA, Li J, Dupree J, Franklin RJ, Casaccia-Bonnefil P (2008) Age-dependent epigenetic control of differentiation inhibitors is critical for remyelination efficiency. Nat Neurosci 11:1024–1034

Sher F, van Dam G, Boddeke E, Copray S (2009) Bioluminescence imaging of Olig2-neural stem cells reveals improved engraftment in a demyelination mouse model. Stem Cells 27:1582–1591

Shi JY, Marinovich A, Barres BA (1998) Purification and characterization of adult oligodendrocyte precursor cells from the rat optic nerve. J Neurosci 18:4627–4636

Shields SA, Gilson JM, Blakemore WF, Franklin RJM (1999) Remyelination occurs as extensively but more slowly in old rats compared to young rats following gliotoxin-induced CNS demyelination. Glia 28:77–83

Sim FJ, McClain C, Schanz S, Protack TL, Windrem MS, Goldman SA (2011) CD140a identifies a population of highly myelinogenic, migration-competent, and efficiently engrafting human oligodendrocyte progenitor cells. Nat Biotechnol 29(10):934–941

Smith PM, Blakemore WF (2000) Porcine neural progenitors require commitment to the oligodendrocyte lineage prior to transplantation in order to achieve significant remyelination of demyelinated lesions in the adult CNS. Eur J Neurosci 12:2414–2424

Smith KJ, Blakemore WF, McDonald WI (1979) Central remyelination restores secure conduction. Nature 280:395–396

Smith KJ, Blakemore WF, McDonald WI (1981) The restoration of conduction by central remyelination. Brain 104:383–404

Stankoff B, Wang Y, Bottlaender M, Aigrot MS, Dolle F, Wu C, Feinstein D, Huang GF, Semah F, Mathis CA, Klunk W, Gould RM, Lubetzki C, Zalc B (2006) Imaging of CNS myelin by positron-emission tomography. Proc Natl Acad Sci USA 103:9304–9309

Syed YA, Hand E, Mobius W, Zhao C, Hofer M, Nave KA, Kotter MR (2011) Inhibition of CNS remyelination by the presence of semaphorin 3A. J Neurosci 31:3719–3728

Tepavcevic V, Blakemore WF (2005) Glial grafting for demyelinating disease. Philos Trans R Soc Lond B Biol Sci 360:1775–1795

Tepavcevic V, Blakemore WF (2006) Haplotype matching is not an essential requirement to achieve remyelination of demyelinating CNS lesions. Glia 54:880–890

The Multiple Sclerosis Study Group (1990) Efficacy and toxicity of cyclosporine in chronic progressive multiple sclerosis: a randomized, double-blinded, placebo-controlled clinical trial. Ann Neurol 27:591–605

Thomson JA, Itskovitz-Eldor J, Shapiro SS, Waknitz MA, Swiergiel JJ, Marshall VS, Jones JM (1998) Embryonic stem cell lines derived from human blastocysts. Science 282:1145–1147

Tontsch U, Archer DR, Dubois-Dalcq M, Duncan ID (1994) Transplantation of an oligodendrocyte cell line leading to extensive myelination. Proc Natl Acad Sci USA 91:11616–11620

Totoiu MO, Nistor GI, Lane TE, Keirstead HS (2004) Remyelination, axonal sparing, and locomotor recovery following transplantation of glial-committed progenitor cells into the MHV model of multiple sclerosis. Exp Neurol 187:254–265

Tourbah A, Linnington C, Bachelin C, Avellana-Adalid V, Wekerle H, Baron-Van Evercooren A (1997) Inflammation promotes survival and migration of the CG4 oligodendrocyte progenitors transplanted in the spinal cord of both inflammatory and demyelinated EAE rats. J Neurosci Res 50:853–861

Trotter J, Crang AJ, Schachner M, Blakemore WF (1993) Lines of glial precursor cells immortalised with a temperature-sensitive oncogene give rise to astrocytes and oligodendrocytes following transplantation into demyelinated lesions in the central nervous system. Glia 9:25–40

Tsuji O, Miura K, Okada Y, Fujiyoshi K, Mukaino M, Nagoshi N, Kitamura K, Kumagai G, Nishino M, Tomisato S, Higashi H, Nagai T, Katoh H, Kohda K, Matsuzaki Y, Yuzaki M, Ikeda E, Toyama Y, Nakamura M, Yamanaka S, Okano H (2010) Therapeutic potential of appropriately evaluated safe-induced pluripotent stem cells for spinal cord injury. Proc Natl Acad Sci USA 107:12704–12709

Utzschneider DA, Archer DR, Kocsis JD, Waxman SG, Duncan ID (1994) Transplantation of glial cells enhances action potential conduction of amyelinated spinal cord axons in the myelin-deficient rat. Proc Natl Acad Sci USA 91:53–57

Vaithianathar L, Tench CR, Morgan PS, Constantinescu CS (2003) Magnetic resonance imaging of the cervical spinal cord in multiple sclerosis – a quantitative T1 relaxation time mapping approach. J Neurol 250:307–315

Vitry S, Avellana-Adalid V, Hardy R, Lachapelle F, Baron-Van Evercooren A (1999) Mouse oligospheres: from pre-progenitors to functional oligodendrocytes. J Neurosci Res 58(6):735–751

Vitry S, Avellana-Adalid V, Lachapelle F, Van Evercooren AB (2001) Migration and multipotentiality of PSA-NCAM+ neural precursors transplanted in the developing brain. Mol Cell Neurosci 17:983–1000

Vowinckel E, Reutens D, Becher B, Verge G, Evans A, Owens T, Antel JP (1997) PK11195 binding to the peripheral benzodiazepine receptor as a marker of microglia activation in multiple sclerosis and experimental autoimmune encephalomyelitis. J Neurosci Res 50:345–353

Walczak P, All AH, Rumpal N, Gorelik M, Kim H, Maybhate A, Agrawal G, Campanelli JT, Gilad AA, Kerr DA, Bulte JW (2011) Human glial-restricted progenitors survive, proliferate, and preserve electrophysiological function in rats with focal inflammatory spinal cord demyelination. Glia 59:499–510

Wang C, Wu C, Popescu DC, Zhu J, Macklin WB, Miller RH, Wang Y (2011a) Longitudinal near-infrared imaging of myelination. J Neurosci 31:2382–2390

Wang Y, Piao JH, Larsen EC, Kondo Y, Duncan ID (2011b) Migration and remyelination by oligodendrocyte progenitor cells transplanted adjacent to focal areas of spinal cord inflammation. J Neurosci Res 89(11):1737–1746

Windrem MS, Schanz SJ, Guo M, Tian GF, Washco V, Stanwood N, Rasband M, Roy NS, Nedergaard M, Havton LA, Wang S, Goldman SA (2008) Neonatal chimerization with human glial progenitor cells can both remyelinate and rescue the otherwise lethally hypomyelinated shiverer mouse. Cell Stem Cell 2:553–565

Winkler C, Kirik D, Bjorklund A (2005) Cell transplantation in Parkinson's disease: how can we make it work? Trends Neurosci 28:86–92

Yamanaka S, Blau HM (2010) Nuclear reprogramming to a pluripotent state by three approaches. Nature 465:704–712

Yandava BD, Billinghurst LL, Snyder EY (1999) "Global" cell replacement is feasible via neural stem cell transplantation: evidence from the dysmyelinated shiverer mouse brain. Proc Natl Acad Sci USA 96:7029–7034

Zawadzka M, Rivers LE, Fancy SP, Zhao C, Tripathi R, Jamen F, Young K, Goncharevich A, Pohl H, Rizzi M, Rowitch DH, Kessaris N, Suter U, Richardson WD, Franklin RJ (2010) CNS-resident glial progenitor/stem cells produce Schwann cells as well as oligodendrocytes during repair of CNS demyelination. Cell Stem Cell 6:578–590

Zhang SC, Duncan ID (2000) Remyelination and restoration of axonal function by glial cell transplantation. In: Dunnett SB, Björklund A (eds) Functional neural transplantation, vol 2. Elsevier, Amsterdam, pp 515–533

Zhang SC, Lipsitz D, Duncan ID (1998a) Self-renewing canine oligodendroglial progenitor expanded as oligospheres. J Neurosci Res 54:181–190

Zhang SC, Lundberg C, Lipsitz D, O'Connor LT, Duncan ID (1998b) Generation of oligodendroglial progenitors from neural stem cells. J Neurocytol 27:475–489

Zhang SC, Ge B, Duncan ID (1999) Adult brain retains the potential to generate oligodendroglial progenitors with extensive myelination capacity. Proc Natl Acad Sci USA 96:4089–4094

Zhang SC, Ge B, Duncan ID (2000) Tracing human oligodendroglial development in vitro. J Neurosci Res 59:421–429

Zhang SC, Wernig M, Duncan ID, Brüstle O, Thomson JA (2001) In vitro differentiation of transplantable neural precursors from human embryonic stem cells. Nat Biotechnol 19:1129–1133

Zhang PL, Izrael M, Ainbinder E, Ben Simchon L, Chebath J, Revel M (2006) Increased myelinating capacity of embryonic stem cell derived oligodendrocyte precursors after treatment by interleukin-6/soluble interleukin-6 receptor fusion protein. Mol Cell Neurosci 31:387–398

Chapter 6
A Peripheral Alternative to Central Nervous System Myelin Repair

V. Zujovic and A. Baron Van Evercooren

6.1 Introduction

The use of peripheral nervous system (PNS) Schwann cells (SC) in central nervous system (CNS) repair has been an intensely studied strategy to support and myelinate regenerating axons (reviewed in Lavdas et al. 2008; Oudega and Xu, 2006). For myelin repair, likewise, SC-based strategy presented several advantages over the use of oligodendrocytes precursors (OPC) or CNS stem cells. Indeed, while newly formed oligodendrocytes and the myelin they make remain as targets of inflammatory attacks, new peripheral myelin is preserved. Additionally, the thickness of newly formed PNS myelin is closer to developmental CNS myelin than newly formed central myelin seen on remyelination of CNS axons. Furthermore, internodal lengths generated by exogenous SC-derived myelin resemble intact endogenous myelin internodes more closely from a morphological point of view, than newly formed ones. For these reasons, the use of SC transplantation in demyelinating lesions has been extensively explored during the past decades. More recently, olfactory ensheathing cells (OEC) then multipotent or pluripotent stem/precursor cells emerged as candidates of interest to generate peripheral myelin around CNS axons.

In this chapter, we will review the history of the use of committed SC and OEC for myelin repair, the different strategies to increase their repair potential, the other cellular sources of PNS myelin, and the future perspectives that could be considered in generating PNS precursor cells from pluripotent stem cells.

V. Zujovic • A.B. Van Evercooren (✉)
ICM, 5ème étage, pièce 5.031, Centre de Recherche
de l'Institut du Cerveau et de la Moelle Epinière,
UPMC-Paris6, UMR_S 975, Inserm U 975,
CNRS UMR 7225, 47 bd de l'Hôpital, Paris 75013, France
e-mail: anne.baron@upmc.fr

6.2 Committed PNS Glia

6.2.1 Schwann Cells

The specific regenerative properties of SC in the injured PNS lead to their transplantation in the CNS. The first efforts in the field of SC-based therapy focused on committed mature or immature SC. In the late 1920s, Ramon y Cajal implanted peripheral nerve fragments into the CNS and observed that CNS axons invading the PNS graft were surrounded by PNS myelin. Similar observations were made when fragments of peripheral nerves were grafted in the demyelinated spinal cord (Blakemore 1977). Later, SC were harvested from rodent peripheral nerve biopsies, purified, and expanded in vitro (Wood 1976; Brockes et al. 1979). Several experiments of grafting purified neonate or adult SC in various animal models of CNS demyelination provided further proof of the ability of SC to efficiently remyelinate CNS axons. The discovery of neuregulins (NRG), a family of growth factors, which control SC development and differentiation, combined with the recombinant technology provided means to generate substantial SC populations from adult human (Rutkowski et al. 1995; Levi 1996) and monkey (Avellana-Adalid et al. 1998) nerve biopsies. The latter combined with more reliable viral tracing methods allowed the demonstration that autologous macaque SC transplantation in the demyelinated spinal cord leads to robust remyelination of single acute spinal cord lesions (Bachelin et al. 2005). Successful labeling of SC with iron particles was also developed to follow the fate of grafted cells by magnetic resonance imaging (Dunning et al. 2006). However, due to iron release by dying SC and free iron uptake by macrophages, a major constituent of demyelinating lesions, this tracing methodology requires further improvement.

These technological achievements provided the means to assess the repair potential of purified SC in great detail. SC were grafted into toxin-induced areas of demyelination and proven to remyelinate demyelinated axons, competing with endogenous myelin-forming cells (reviewed in Baron-Van Evercooren and Blakemore 2004). It has also been demonstrated that grafted SC sustained stable myelin for up to 5 months after transplantation in the demyelinated mouse spinal cord (Duncan et al. 1981). Interestingly, the myelin formed by the transplanted SC markedly improved conduction of the demyelinated axons (Honmou et al. 1996) and promoted functional recovery (Girard et al. 2005). In all these experiments, the extent of remyelination correlated with the number of SC introduced (Iwashita and Blakemore 2000), their purity (Brierley et al. 2001), and their age. Indeed, SC from neonate and young adult donors achieved more successful repair than SC derived from older animals (Lankford et al. 2002).

While SC neuroregenerative properties have rarely been studied in models of demyelination where minimal axonal injury occurs, numerous studies performed in models of spinal trauma highlighted the capacity of transplanted SC to promote axonal regeneration across the injured area (Oudega and Xu 2006). This property may be of great value for demyelinating diseases such as MS in which axonal injury or degeneration often occurs as a result of long-standing demyelination. The ability of SC to support axonal regeneration is due to multiple factors. They express multiple neurotrophins including nerve growth factor (NGF), neurotrophin-3 (NT3), brain-derived neurotrophic factor (BDNF), fibroblast growth factor (FGF), glial cell line-derived neurotrophic factor (GDNF), and ciliary neurotrophic factor (CNTF) (Lavdas et al. 2008). They also produce extracellular matrix molecules such as laminin and fibronectin that promote axon growth (Kurkinen and Alitalo 1979; Baron-Van Evercooren et al. 1982). Moreover, unlike CNS myelin, they do not seem to be a target of the autoimmune process occurring in MS. These encouraging observations have made SC attractive candidates for cell therapy. In 2002, three MS patients affected with primary progressive MS were transplanted intracranially into single demyelinating lesion with autologous SC isolated from sural nerve biopsies, in a phase I clinical trial. Although the patients had no clinical nor MRI evidence of worsening, the trial was suspended as biopsies from the transplant site failed to show clear evidence of remyelination by the transplanted cells (Martino et al. 2010).

However, several issues have been raised concerning the use of SC as a therapeutic tool to promote remyelination in MS. The major limitation is the inability of SC to migrate efficiently when grafted in the demyelinated or injured CNS. This restriction is likely due to the poor SC interface with astrocytes (Franklin et al. 1993; Lakatos et al. 2003a) since endogenous or exogenous astrocytes (Blakemore et al. 1986; Franklin et al. 1991) limit SC remyelination of CNS-demyelinated axons (Franklin et al. 1993). Moreover, transplantation of SC away from spinal cord lesions showed that their migration abilities depend upon the environment in which they are placed. SC migrate preferentially along meninges, blood vessels, and the spinal cord midline (Bachelin et al. 2010), but do not migrate through white and gray matter (Langford and Owens 1990; Baron-Van Evercooren et al. 1992; Brook et al. 1993; Iwashita et al. 2000). Since this migration failure in the CNS hampers SC ability to participate successfully in CNS remyelination, overcoming this limitation has become a priority for the field in considering remyelination of widely dispersed lesions in MS. Finally little is known about the resistance of transplanted SC to inflammation as well as their ability to sustain clinical recovery after long-term grafting in the context of autoimmune demyelination. However, preliminary data indicate that SC survive, migrate to some extent, and form myelin when grafted in the MOG-induced EAE model (Zujovic et al., unpublished data).

6.2.2 Improving Committed SC

Poor SC integration/migration in the demyelinated CNS has led researchers to explore alternative approaches to optimize their ability to promote CNS repair. These include modifications of the intrinsic properties of SC or of the environment in which they are placed (Fig. 6.1).

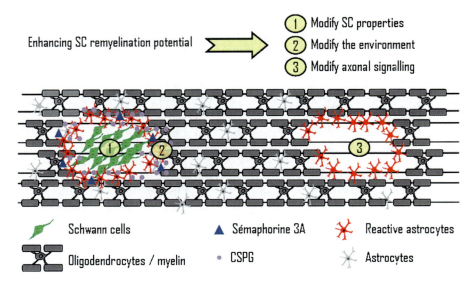

Fig. 6.1 Schematic representation of the different strategies developed to enhance SC migratory potential in the CNS. Different approaches improved SC remyelination capacity whether directly by modifying SC properties, decreasing inhibitory molecules present in the environment, or by adjusting axonal signals

6.2.2.1 Modifying SC Intrinsic Properties

Transduction of SC has been used as a mean to act as gene delivery vehicles. Boosting SC expression of neurotrophins such as BDNF and NT3 has profound effects on neural protection, plasticity, and regeneration after spinal cord injury in the developing and adult CNS (Kromer and Cornbrooks 1985; Martin et al. 1991; Montero-Menei et al. 1992; Levi and Bunge 1994). Since NT3 also promotes oligodendrocyte proliferation, survival and differentiation (McTigue et al. 1998), and SC migration (Yamauchi et al. 2005) in vitro, we explored the benefit of this strategy on remyelination after transplantation into the demyelinated spinal cord. In doing so, we demonstrated that forced expression of BDNF and NT3 in macaque SC promoted functional motor recovery of the demyelinated mouse spinal cord (Girard et al. 2005). This beneficial effect resulted from multiple events including enhanced remyelination by endogenous oligodendrocytes and SC, enhanced neuroprotection, and reduced scar formation, thus highlighting the issue of ex vivo neurotrophin delivery based on SC therapy to enhance global repair of demyelinated lesions. A similar strategy combining overexpression of NT3 and BDNF was also used in spinal cord trauma and found to improve efficiently axonal length and the number of myelinated axons in the lesion (Golden et al. 2007). However this strategy did not distinguish a direct or indirect effect of neurotrophins on myelination, through either creating an increase in axon number and/or acceleration of SC differentiation.

An alternative strategy to enhance their intra-CNS integration/migration has been to modify SC membrane properties. As mentioned above, one major limitation of SC-based therapy is the poor potential of their integration/migration within white and gray matter of the CNS. By contrast, neural precursors and OEC interact more efficiently with astrocytes. Interestingly, both of these cells express the polysialylated form of NCAM (PSA-NCAM), a cell adhesion molecule highly expressed during development and involved in CNS plasticity, remodeling, and repair (reviewed in Rutishauser 2008). Since SC express NCAM but not PSA, it was speculated that forcing NCAM polysialylation in SC would improve their integration/migration in the CNS. Indeed, ectopic expression of PSA by rodent SC enhanced their migratory properties in vitro and functional repair in a trauma model in vivo without altering their myelination potential (Lavdas et al. 2006; Papastefanaki et al. 2007). This strategy was further explored with adult macaque SC grafted in the demyelinated rodent spinal cord. Forced expression of PSA-NCAM by the grafted SC enhanced their recruitment by distant lesions and promoted their ability to remyelinate the demyelinated axons while wild-type SC were unable to do so (Bachelin et al. 2010). This pro-migratory effect resulted in increased SC fluidity and exit from the graft most likely by reduced adhesion among sister SC and SC–astrocytes. Alternatively, it may also have decreased SC sensitivity to inhibitory molecules and/or increased their response to chemotactic cues as demonstrated for PSA-NCAM+ embryonic-stem-cell-derived neural precursors (Glaser et al. 2007).

6.2.2.2 Modifying the Environment

Other strategies have targeted specific molecular cues of the glial scar. Indeed, reducing levels of chondroitin sulfate proteoglycans (CSPG) (Grimpe and Silver 2004; Santos-Silva et al. 2007) and N-cadherin–N-cadherin (Fairless et al. 2005) interaction promoted SC–astrocyte mixing in vitro and in vivo. Interestingly, another inhibitor of axonal regeneration, semaphorin 3A (Sema 3A), was found to be repulsive for SC in vitro and in vivo in a model of spinal trauma. Treatment with a Sema 3A inhibitor was able to reverse this effect and increase SC-mediated myelination and axonal regeneration (Kaneko et al. 2006). Data also indicate that ephrins may play a role in SC–astrocyte non-mixing (Afshari et al. 2010). However, whether neutralizing these axonal growth inhibitors would also improve exogenous SC migration through CNS parenchyma has not been reported so far. Several inhibitors of axonal regeneration are also contained in myelin. However, their role in Schwann cell migration inhibition remains to be elucidated. Interestingly, SC, unlike OEC and OPC, do not secrete metalloendoproteases, enzymes known to degrade inhibitors as extracellular matrix proteoglycans and myelin axonal inhibitors Nogo (Amberger et al. 1997; Gueye et al. 2011).

Another issue with SC-based therapy concerns the control of SC myelination by axon diameter. Injections of SC into areas of small-diameter axons do not result in remyelination (Brook et al. 1993). However, using lentiviral-driven overexpression of NRG 1 type III, Taveggia et al. (2005) induced SC to myelinate small-diameter axons. Furthermore, overexpression of NRG1 type III in transgenic mice increased the thickness of endogenous newly formed PNS but not CNS myelin in response to demyelination (Brinkmann et al. 2008). Interestingly, several members of the ADAM (a disintegrin and metalloprotease) secretase have been implicated in the regulation of NRG 1 type III activity notably by regulating its processing. While some secretases such as BACE (beta secretase) positively regulate the myelination process (Shirakabe et al. 2001; Hu et al. 2006), recent data showed that another secretase TACE (tumor necrosis factor alpha-converting enzyme) is a negative regulator of PNS myelination (La Marca et al. 2011). These last observations underline the therapeutic potential of TACE modulation since several inhibitors of TACE are already under investigation for other disease treatment. The possibility to direct selectively exogenous SC remyelination towards CNS axons irrespectively of their size may thus be useful in enhancing their contribution to CNS remyelination.

6.2.3 OEC

OEC (initially named olfactory SC) are differentiated glial cells very similar to SC that belong to the peripheral olfactory system where they ensheath but do not myelinate the axons belonging to the first cranial nerve (Franklin and Barnett 2000; Vincent et al. 2005).

They share also features with astrocytes expressing "fibrillar" GFAP and forming part of the glial limitans of the olfactory bulb (Doucette 1991). The origin of OEC has long been controversial. During early development, they originate from the nasal placodes and provide essential growth and guidance to sensory axons which they do not myelinate. Although they were believed to be ectodermal derivatives, recent evidences based on the Wnt1Cre fate mapping technology show that at early embryonic stages OEC within the placode derive from neural crest cells (NCC) (Barraud et al. 2010; Forni et al. 2011). These findings could explain the high level of similarities existing between OEC and SC. In view of these recent data, we will consider OEC as an alternative source of neural crest-derived cells rather than cells deriving from another lineage.

In vitro, OEC and SC have remarkable similarities. Both cell types express the low-affinity NGF receptor (p75ntr), S100, GFAP, and cell adhesion molecules such as L1 and NCAM. They also express extracellular matrix proteins such as laminin and fibronectin (Wewetzer et al. 2002) and secrete neurotrophins including NGF, BDNF, FGF, and vascular endothelial growth factor (VEGF) (reviewed in Sasaki et al. 2011). Although OEC do not normally myelinate the very small axons of the first cranial nerve, studies have shown that OEC can (re)myelinate larger axons with myelin that is indistinguishable morphologically and biochemically from that made by SC when transplanted in the CNS. This was first observed in cocultures of OEC with DRG neurons (Devon and Doucette 1992) and further confirmed in vivo with an OEC cell line (Franceschini and Barnett 1996) and primary OEC (Franklin et al. 1996; Smith et al. 2001; Lakatos et al. 2003b). Cell suspensions of acutely dissociated OEC from neonatal rats remyelinated and enhanced axonal conduction, when focally injected into the ethidium-bromide-demyelinated areas of the dorsal columns of the spinal cord (Imaizumi et al. 1998). Moreover, xenotransplanted canine, human, or porcine OEC isolated from the adult olfactory bulb are capable of extensive functional remyelination following transplantation into demyelinated rat CNS (Kato et al. 2000; Smith et al. 2001; Deng et al. 2006). The myelin formed contains the peripheral myelin protein P0, and the cell's association with axons resembles that of SC. Furthermore, when faced with demyelinated axons, transplanted OEC express SC transcription factors Krox-20 and SCIP mRNA, suggesting that similar mechanisms are involved in the manner OEC and SC form myelin (Smith et al. 2001). In addition to these observations, OEC transplantation in trauma models have suggested that OEC have multiple beneficial effects, reducing glial scarring and cavitation, increasing angiogenesis, and providing a cellular scaffold for regenerating axons (reviewed in Toft et al. 2011). Improved regeneration and reduced scarring was also observed after autologous transplantation in a canine model of spinal cord injury (Jeffery et al. 2005). However, the repair capacities of OEC have been challenged given that their ability to support axonal growth was not fully confirmed (Tetzlaff et al. 2011). It was also reported that OEC are less efficient than SC at forming myelin in vitro and in vivo (Plant et al. 2003; Boyd et al. 2004). In addition, it was proposed that a SC contamination could be responsible for all the observed effects (Boyd et al. 2004). However, transplantation of eGFP-OEC in a contusive SCI lesion (Sasaki et al. 2004), in the injured nerve (Dombrowski et al. 2006; Radtke et al. 2009), or in the X-EB lesions (Sasaki et al. 2011) provided proof that OEC have the ability to remyelinate axons with, in some circumstances, recovery of locomotor functions. Moreover, co-transplantation of purified OEC and meningeal cells indicated that this potential can be enhanced by meningeal cells which contaminate OEC preparations (Lakatos et al. 2003a). So far, the behavior of OEC in a context of immune inflammation comparable to MS has not been addressed.

Several studies suggested that OEC migrate better than SC when confronted with CNS components. Indeed, OEC mix better when cocultured with astrocytes (Lakatos et al. 2000), and transplanted OEC elicit a diminished expression of GFAP and CSPG expression in astrocytes compared to SC (Lakatos et al. 2003a; Garcia-Alias et al. 2004; Andrews and Stelzner 2007). This is explained via decreased N-cadherin-mediated adhesion with astrocytes (Fairless et al. 2005) and induction of a weaker astrocyte hypertrophy (Lakatos et al. 2000). Moreover, as mentioned before, OEC secrete matrix metalloproteinases (MMPs), which modulate their migration in vitro (Gueye et al. 2011). Since MMPs also induce changes in astrocyte morphology (Ogier et al. 2006), these enzymes could also play an important role in OEC mixing with astrocytes.

However, the greater migration potential of OEC in the injured spinal cord remains a subject of debate (Deng et al. 2006; Lu et al. 2006; Pearse et al. 2007).

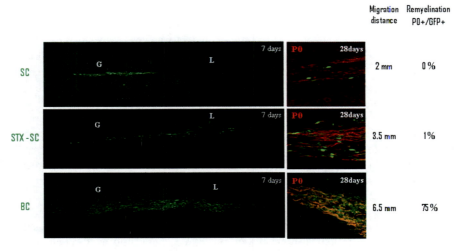

Fig. 6.2 Repair potential of different stages of the SC lineage of mature myelinating SC. The migratory and remyelination properties of mature myelinating SC in the CNS are challenged by their poor interaction with astrocytes and myelin inhibition. In contrast, earlier stages of SC lineages such as BC and SCp present higher repair potential due to their smoother interaction with CNS environment

While several studies show that OEC, SC, and OPC survive and migrate poorly after transplantation in normal white matter (Iwashita et al. 2000; Hinks et al. 2001; Lankford et al. 2008), only one report highlights that transplanted OEC (and OPC), but not SC, migrate extensively in gray and white matter of the X-irradiated spinal cord (Lankford et al. 2008). Although these data stress potential differences between transplanted SC and OEC with respect to their sensitivity to CNS inhibitory cues. However the specific environment of the X-irradiated spinal cord besides NG2+OPC a depletion in NG2+ OPC remains to be identified.

6.2.4 Advantages of OEC over SC

Since OEC myelinate in a manner very similar to SC but mix better with astrocytes in vitro and in vivo, it has been proposed that from a therapeutic point of view, OEC transplantation would have advantages over SC. However, immature stages of the SC lineage migrated profusely on transplantation into the CNS (Fig. 6.2 and 6.3) and elicited similar reduced changes in astrocyte reactivity including CSPG production (Fig. 6.4) (Woodhoo and Sommer 2008; Zujovic et al. 2010). This suggests that OEC represent a stage of the neural crest lineage with plasticity more similar to SC precursors (SCp) and boundary cap cells (BC, see further) than mature SC.

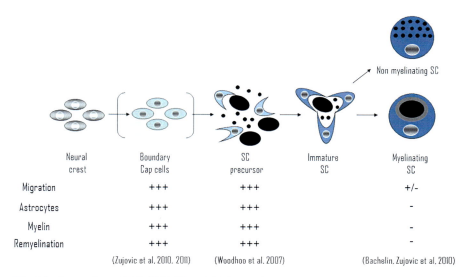

Fig. 6.3 Comparison of BC, STX-SC, and SC interaction with CNS environment. One week post grafting, BC and STX-SC progeny intermingle with astrocyte (*blue, left panel*) while adult SC graft induces the formation of an astrocytic barrier (*blue, left panel*). CSPG (*red, middle panel*) production was lower in BC and STX-SC graft than in SC graft. Only BC were able to migrate in MOG+ areas (*red, right panel*). The table summarizes the behavior of SC, STX-SC, and BC in CNS environment

Whether OEC and SC display species differences in terms of growth factor requirements for expansion has also been investigated. Similar to rodent OEC, OEC from large animal species share many shared features with SC. In fact a recent review by Wewetzer et al. (2011) concluded that rodent, but not canine, human, and monkey, OEC and SC require mitogens for expansion and undergo spontaneous immortalization after prolonged growth factor treatment. These observations highlight that both canine and monkey OEC and SC may serve as valuable translational models for addressing their regenerative capacities at a preclinical stage.

Autologous SC grafts in monkeys and autologous OEC grafts in dogs have been proven feasible and successful in terms of repair. Whether human OEC and SC have equivalent regenerative/remyelinating potentials remain to be defined. Considering autologous grafts studies in humans, lack of adverse effects but no proof of benefit were reported for autologous SC grafts in the injured spinal cord injury 1 year after injury (Saberi et al. 2008). Likewise no conclusions except for safety could be drawn from the suspended MS trial. Autologous olfactory mucosa engraftment (Lima et al. 2006) and autologous olfactory ensheathing cell transplantation (Feron et al. 2005) appeared to be safe and/or provide potential benefit. However, their application to promote remyelination/regeneration in MS patients has not been reported. Clearly more studies on MS-like preclinical models are necessary to evaluate the outcome of grafted cells in an inflammatory demyelinating milieu.

6.3 PNS Progenitors

An alternative to the use of committed SC or OEC for cell therapy would be the use of more immature stages of the neural crest (NC) lineage. While embryonic and adult precursors have been identified, few reports have addressed the repair potential of these PNS stem/precursor cells in the CNS. These include SC precursors (SCp), boundary cap cells (BC), and olfactory epithelial progenitors (OEp).

6.3.1 SCp

Schwann cell precursors are NC derivatives, which undergo various transitional stages before becoming myelin-competent cells (reviewed in Zujovic et al. 2007). Neural crest cells (NCC) differentiate into SC precursors (SCp) before becoming immature SC and mature myelinating or nonmyelinating SC. Woodhoo et al. highlighted the greater capacity of SCp to survive, integrate, and differentiate into myelin-forming cells when transplanted in the normal or demyelinated CNS (Fig. 6.2) (Woodhoo et al. 2007). When grafted into the injured spinal cord, these cells supported axonal growth and sprouting though significant functional improvements were not observed (Agudo et al. 2008).

6.3.2 Boundary Cap Cells

BC are potential stem cells (Hjerling-Leffler et al. 2005) of the embryonic mouse spinal roots (Maro et al. 2004; Aquino et al. 2006) closely related to NCC. When grafted in the demyelinated sciatic nerve, acute BC preparations differentiated into myelin-forming SC only if in vitro priming to committed SC was performed prior engraftment (Aquino et al. 2006). Using cell fate mapping and microdissection, we provided evidence that acutely dissociated BC migrate profusely in the demyelinated CNS and compete aggressively with host myelin-forming cells to remyelinate axons of far distant lesions of the spinal cord while SC are unable to do so (Zujovic et al. 2010) (Fig. 6.4). Although it is not clear why BC and SCp migrate more efficiently through normal white matter, expression of PSA-NCAM may be involved. However, BC, but not SCp, express PSA-NCAM, suggesting that mechanisms other than PSA-NCAM might intervene in efficient intra-CNS migration/integration of the BC/SCp. Moreover while SCp give rise exclusively to peripheral myelin, BC differentiate also into central myelin-forming cells in vitro and in vivo (Zujovic et al. 2011) (Fig. 6.5). While these observations highlight the remarkable therapeutic potential of these new cellular candidates, the use of BC rather than SCp to promote central remyelination might be more promising in view of their self-renewal ability and multipotency (Fig. 6.2). However, their potential for axonal regeneration is unknown.

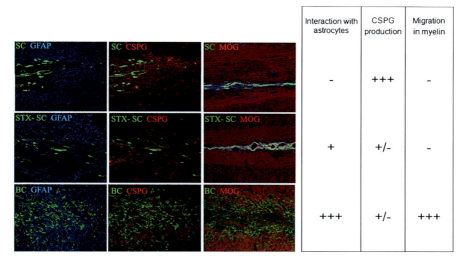

Fig. 6.4 Comparison of the migratory and remyelination potential of mature SC, STX-expressing SC, and BC. In order to increase mature SC migratory potential, two different strategies were developed: by overexpressing STX, a sialyltransferase inducing the polysialylation of NCAM, and by using more immature stages of the SC lineage such as BC. Since mature SC migrate poorly in the CNS (*upper panel*), they do not produce new peripheral myelin (P0, *red*). STX-SC migrate further than mature SC and participate more efficiently in the remyelination process. BC (*lower panel*) demonstrate impressive migratory capacities and participate massively in the remyelination of the lesion

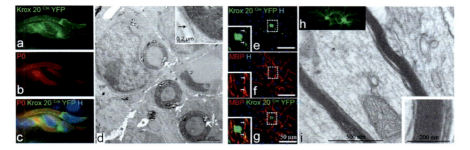

Fig. 6.5 BC directed differentiation towards a peripheral or central myelin-forming cell. When grafted in the demyelinated lesion of the spinal cord, the majority of BC give rise to SC (**a–d**). Four weeks post grafting, there is a clear colocalization of P0 positive myelin (*red* **b**, **c**) with YFP+ cells (*green* **a**, **c**). Bluo-gal precipitates reveal YFP+ BC (*white arrows*, **d**). BC progeny presents a classical morphology of PNS myelin with SC associated in a 1:1 relationship with the axon (**d**). Higher magnification illustrates the compaction of the newly formed PNS myelin and its basal membrane (*black arrow, insert*, **d**). When grafted in the newborn shiverer brain, BC adopt an oligodendroglial fate (**e–i**). YFP+ cells (*green*, **e**, **g**) expressing MBP (*red*, **f**, **g**) and have features of myelin-forming oligodendrocytes. *Insets* (**e**, **g**) are enlarged views of *dotted square* illustrating discrete co-expression of MBP and YFP in processes. Ultrastructural analysis confirms the presence of compact donor-derived central myelin (**i**, *inset*). Illustration of GFP+ oligodendrocytes in corresponding floating sections prior to processing for electron microscopy (**h**)

6.3.3 Adult PNS Stem/Precursor Cells

The persistence of NC-derived stem cells (NCSC) in the adult has been found, most notably in the adult DRG (Li et al. 2007). Li et al. suggested that the NCSC probably originate from satellite cells. Using lineage tracing with P0 and Wnt1-Cre/Floxed EGFP mice, Nagoshi et al. (2008) targeted the NCSC niches in the whisker pad, bone marrow, and the DRG and compared their multipotency and self-renewal ability. Although these various stem cells share the same NC origin and all of them give rise to neurons, glia, and myofibroblasts, the DRG-derived NCSC showed greater multipotency and an enhanced ability to form secondary spheres. Despite the observation that several tissues give rise to stem cells sharing the same origin and similar self-renewing and multipotential capacities, these cells seem to retain tissue-dependent characteristics. Whether they will have similar repair and remyelinating potential has to be addressed.

Recently, Takagi et al. (2011) were able to isolate SCp/immature SC from adult injured peripheral nerve. Based on the capacity of adult SC to differentiate after nerve injury, they collected cell suspensions from injured peripheral nerve at different time point after injury. They noted that while no sphere formation was observed in the intact sciatic nerve, Schwann spheres were obtained 3–10 days after injury. Although spheres differentiated into the SC lineage only, they contained SCp/immature SC with self-renewal capacity and greater neurite outgrowth and myelination potentials when cocultured with neurons, than mature SC.

6.3.4 Olfactory Epithelium Progenitors

Obtaining sufficient quantities of OEC from olfactory bulbs is difficult to achieve and is highly invasive, leading to total loss of olfaction. Another approach consisting in endoscopic biopsy of the olfactory epithelium was found to be less invasive and without adverse effect on olfactory function (Winstead et al. 2005). Interestingly, it is in this structure that OEp were first identified both in rodents and humans. When pieces of olfactory lamina propria containing OEp were placed into injured rat spinal cord, they promoted partial functional recovery in paraplegic rats (Lu et al. 2002). Later OEp were generated from olfactory epithelium from rodent (Tomé et al. 2009), from fresh necropsied tissue, or endoscopic biopsy from patients undergoing sinus surgery. The latter were characterized in vitro and shown to remain equivalent despite the age and sex of donor (Marshall et al. 2005, 2006). Upon transplantation into demyelinated spinal cord lesion, grafted rat OEp provided extensive peripheral remyelination (Markakis et al. 2009). Moreover, human OEp, transplanted in the rodent traumatized spinal cord, rescue axotomized rubrospinal neurons and promote functional recovery (Xiao et al. 2005).Yet, their use in disease models of trauma or demyelination deserves further attention.

6.4 Ectopic Sources

6.4.1 Mesenchymal Stem Cells

Mesenchymal stem cells (MSC) are stromal stem cells that can be isolated from various tissues and differentiate into cells of mesodermal (bone, fat, cartilage) and neuro-ectodermal lineages (neurons, glia). The most commonly studied MSC were isolated from bone marrow (Dezawa et al. 2001; Keilhoff et al. 2006; Krampera et al. 2007) but also from adipose tissue (Krampera et al. 2007; Xu et al. 2008), spleen, and thymus (Krampera et al. 2007). MSC originating from various tissues have the capacity to trans-differentiate into a SC-like phenotype in vitro. Adipocyte- or bone marrow MSC-derived SC were shown to myelinate PC12 neurites in vitro (Keilhoff et al. 2006; Xu et al. 2008) and sustain axonal regeneration when implanted into a biogenic muscle graft to bridge a sciatic nerve gap (Keilhoff et al. 2006). Freshly isolated, purified, or CD133+ subpopulations of bone marrow cells were transplanted in various demyelinating conditions and reported to be associated with either SC or oligodendrocyte remyelination (Sasaki et al. 2001; Akiyama et al. 2002a, b). However, so far, there has not been a clear indication as to whether MSC contribute directly to the genesis of myelin-forming cells, or act simply via a by-standard effect on the host environment-promoting remyelination by host cells.

6.4.2 Skin-Derived Precursors

A number of studies have demonstrated the presence of multipotent cells in the skin that are capable of generating cells of neural, mesenchymal lineage and NC-derived cell types (Toma et al. 2001, 2005; Fernandes et al. 2004; Sieber-Blum et al. 2004; Amoh et al. 2005). The anatomical location and origin of these skin-derived precursors (SKP) was thoroughly investigated, and several "niches" have been identified mainly in the hair follicle. While some studies demonstrated that SKP were present in the hair follicle bulge (Sieber-Blum et al. 2004; Amoh et al. 2005; Wong et al. 2006), others identified the dermal papilla (Fernandes et al. 2004; Biernaskie et al. 2006; Wong et al. 2006) as the main source of SKP. However, the presence of multipotent cells in the human foreskin, a region devoid of hair follicles, highlighted the presence of an extra-follicular niche in the human skin (Toma et al. 2005). Of particular interest is the capacity of SKP to differentiate in vitro into neural crest-derived cell types such as SC (Fernandes et al. 2004, 2006). When rodent and human SKP were cultured with peripheral nerve explants, SKP proliferated and expressed myelin proteins (Fernandes et al. 2004). When grafted into the injured peripheral nerve or into the dysmyelinated neonatal shiverer brain, SKP differentiated into functional SC that associated and remyelinated CNS axons (McKenzie et al. 2006). Biernaskie et al. (2007) demonstrated convincingly that

SC remyelination derived from an SKP graft resulted in functional recovery in a spinal cord contusion model.

Since one of the major issues of autologous cell transplantation therapy is the accessibility of the cell sources, the discovery, isolation, and characterization of SKP offer a large exploratory field of investigation for cell replacement. The recent finding that human bulge stem/precursor cells from adult skin can be isolated reproducibly and can undergo robust ex vivo expansion and directed differentiation raises hope for future cell-based therapies in regenerative medicine (Clewes et al. 2011).

6.4.3 Oligodendrocyte Precursor Cells

Until recently the possibility that SC can be generated from CNS neural precursors was not foreseen. However, this possibility has been raised by both in vitro and in vivo studies. In vitro, cloned CNS neural precursors generated oligodendrocytes and SC (Mujtaba et al. 1998). Moreover transplantation of neural precursors, glial-restricted precursors (Keirstead et al. 1999), and MOG-expressing oligodendrocytes precursors (Crang et al. 2004) in X-irradiated ethidium bromide lesions resulted in both PNS and CNS remyelination. Furthermore, transplantation of cloned human neural precursors, isolated from the adult human brain, into the same type of lesion led also to SC remyelination (Akiyama et al. 2001). These studies underlined an important role for the lesion environment and in particular for astrocytes in regulating the fate of CNS precursors since those that enter areas of demyelination differentiated into oligodendrocytes in the presence of astrocytes but in SC in their absence. More recent data highlighted a role for bone morphogenic protein (BMP) in repressing oligodendrocyte development (Ara et al. 2008) and directing CNS precursor into differentiation SC in vitro (Sailer et al. 2005). Grafted OPC differentiated preferentially into SC in presence of BMP (Talbott et al. 2006), thus suggesting a role for this morphogen in orienting the fate of immature populations when grafted in pathological conditions. While these data indicate that environmental cues can direct the differentiation of CNS neural stem cells into PNS cells, our recent findings suggest that the fate of PNS neural stem cells can also be redirected when forced by transplantation in the CNS. Our fate mapping studies indicate that when BC are grafted in the demyelinated spinal cord, some of these cells differentiate in remyelinating oligodendrocytes (Fig. 6.5) in addition to SC (Zujovic et al. 2010). Furthermore, when transplanted in the developing shiverer brain, BC acquire exclusively a CNS phenotype, giving rise to astrocytes, neurons, and oligodendrocytes (Zujovic et al. 2011). We demonstrated in vitro that this reprogramming was a multistep process depending on a sequential combination of ventralizing morphogens such as noggin, an antagonizer of BMP and Shh, and pro-oligodendroglial growth factors. Taken together, these studies indicate that CNS–PNS boundaries are not as clear-cut and CNS–PNS plasticity prevails in extreme pathological conditions.

6.5 Embryonic Stem Cells and Induced Pluripotent Stem Cells

While the main asset of embryonic stem cells is their capacity to give rise to all the cells of the body, their foremost limitation is the source of these cells. However, one of the major issues of autologous cell transplantation therapy resides in the accessibility of the cell sources. This is the reason why many studies focused on skin as the most accessible organ. Based on the modification of adult fibroblasts transduced into induced pluripotent stem cells (iPS) or directly reprogrammed into differentiated cells, the skin offers great future opportunities for cell replacement and regenerative medicine.

6.5.1 Embryonic Stem Cells

Since embryonic stem cells (ES) are totipotent cells, they can give rise to the myriad of cells that compose the body. While most research on ES differentiation has focused on CNS phenotypes, fewer studies have considered ES as potential source of PNS cells. Using diverse protocols and biological reagents, several research groups have succeeded in differentiating them in a wide range of NC lineage cells. Indeed, the production of SC from ES is a multistep process that begins with the induction of NC-like cells. The first group to successfully obtain NC-like cells used the stromal-derived inducing activity of PA6 fibroblast coculture (Pomp et al. 2005). When Lee et al. (2007) analyzed cells derived from ES neural rosettes more closely, they showed that NC precursors emerged spontaneously in culture of human ES cell-derived neural rosettes and that treatment with extrinsic patterning factors such as FGF2 and BMP 2 increased significantly the hES-NC number. Later, Studer and colleagues established a more defined protocol for neural induction based on the inhibition of SMAD signaling with the dual action of Noggin and SB431542 (Chambers et al. 2009). These protocols were later improved with different strategies using a neurosphere intermediate stage (Pomp et al. 2008), enriching hES-NC by fluorescence-activating cell sorting based on specific markers of the NC lineage cell expression (Lee et al. 2007, 2010; Zhou and Snead 2008; Jiang et al. 2009) treating the cells with small molecule inhibitor of Rho effectors (Hotta et al. 2009). However most of these strategies generated a wide range of NC-derived progeny including sensory and autonomic neurons, myofibroblast, adipocytes, cartilage, and osteogenic cells, but few demonstrated robust and efficient SC differentiation. Many studies report the presence of GFAP+, MBP+, and S100+ cells in the hES-derived culture (Lee et al. 2007; Hotta et al. 2009; Jiang et al. 2009) but only Ziegler et al. (2011) provided evidence of S100, MBP, and PMP 22 positive cells with apparent wrapping of the axons. Nonetheless the process of obtaining SC takes much longer than neuronal differentiation (Lee et al. 2007, 2010; Ziegler et al. 2011), corresponding to the developmental sequence of neurogenesis preceding gliogenesis. Indeed, the capacity of hES-NC to differentiate in SC increases with time in culture, likely uncovering a cell-intrinsic temporal change in their epigenetic state.

6.5.2 Induced Pluripotent Stem Cells

The clue of possible reprogramming of mature differentiated cells into totipotent cells came from the original experiment of Gurdon in 1960 (Gurdon 2006) when he transferred the nucleus of an adult cell into an enucleated egg and obtained totipotent cells capable of giving rise to a whole new frog. More than 40 years later, Takahashi and Yamanaka (2006) published their revolutionary data on transcription factor-based reprogramming of adult mouse fibroblasts. They isolated and identified a cocktail of four transcriptions factors Oct 4, Sox2, Klf4, and Myc representing the minimal set of factors required for reprogramming fibroblasts into pluripotent stem cells.

Using the same culture protocol as that developed from hES cells, iPS were directed towards an NC fate (Chambers et al. 2009) and GFAP+ Sox10+ SC-like cells were generated in vitro (Lee et al. 2010). Wang et al. (2011) exploited the repair potential of these iPS-NC by seeding them in nerve conduits and observed improved regeneration of the sciatic nerve, mostly due to the differentiation of iPS-NC into myelinating SC. However, their exploitation in a CNS lesion model remains to be fully investigated.

Recently, a novel concept of direct cellular reprogramming has emerged since the aim of regenerative medicine is to produce differentiated cell types (Chambers and Studer 2011). Indeed, several studies report the direct reprogramming of fibroblasts into skeletal muscle (Weintraub et al. 1989), functional excitatory neurons (Vierbuchen et al. 2010), cardiomyocytes (Ieda et al. 2010), blood cell progenitors (Szabo et al. 2010), and neural progenitors (Kim et al. 2011), but direct fibroblast-SC reprogramming remains to be documented.

6.6 Conclusion

The repair potential of committed PNS glial cells such as SC and OEC has been widely demonstrated. Although autologous grafting in large animal species has proven to be beneficial, similar successes with human cells remain to be fully established. However, these committed PNS glial cells have some disadvantages: transplanted SC are restricted by astrocytes and OEC purification is highly invasive.

A major advantage has been accomplished over the last decade using immature stages of the NC lineage. Several studies have highlighted the superiority of these immature cells in terms of regenerative potential including migration and remyelination when grafted in the CNS. Although the use of fetal cells raises ethical concerns, stem/precursor cells have been discovered in adult PNS and skin tissue. Proving that these adult NC derivatives are as efficient is essential and has become the subject of active investigation.

New technological developments offer the means to isolate NC derivatives from committed cells such as fibroblasts. Demonstrating that these iPS-derived NC cells have similar therapeutic properties to those that are neurally committed and that they are safe in terms of growth control and differentiation, stability will be priorities in experimental therapy.

Finally experimental peripheral glial cell therapy has been explored in a variety of toxin models of demyelination or spinal trauma established in rodent and primates. Although of great interest, these models do not reflect all aspects of the heterogeneous pathology of MS. It has been speculated that SC are candidates of interest for cell-based therapy of MS. However their resistance to the autoimmune attack has not been proven. It will be therefore crucial that the therapeutic value of the new cellular candidates is validated in MS-like conditions.

Acknowledgements We would like to thank INSERM, ARSEP, and ELA who supported the work performed by the authors and all the past and present members of the Baron's laboratory who made major contributions to the development of this field of research. In particular, we would like to thank Virginia Avellana, François Lachapelle, Corinne Bachelin, and Marie Vidal for their major input in experiments conducted in Paris.

References

Afshari FT, Kwok JC, Fawcett JW (2010) Astrocyte-produced ephrins inhibit Schwann cell migration via VAV2 signaling. J Neurosci 30:4246–4255
Agudo M, Woodhoo A, Webber D, Mirsky R, Jessen KR, McMahon SB (2008) Schwann cell precursors transplanted into the injured spinal cord multiply, integrate and are permissive for axon growth. Glia 56:1263–1270
Akiyama Y, Honmou O, Kato T, Uede T, Hashi K, Kocsis JD (2001) Transplantation of clonal neural precursor cells derived from adult human brain establishes functional peripheral myelin in the rat spinal cord. Exp Neurol 167:27–39
Akiyama Y, Radtke C, Honmou O, Kocsis JD (2002a) Remyelination of the spinal cord following intravenous delivery of bone marrow cells. Glia 39:229–236
Akiyama Y, Radtke C, Kocsis JD (2002b) Remyelination of the rat spinal cord by transplantation of identified bone marrow stromal cells. J Neurosci 22:6623–6630
Amberger VR, Avellana-Adalid V, Hensel T, Baron-van Evercooren A, Schwab ME (1997) Oligodendrocyte-type 2 astrocyte progenitors use a metalloendoprotease to spread and migrate on CNS myelin. Eur J Neurosci 9:151–162
Amoh Y, Li L, Campillo R, Kawahara K, Katsuoka K, Penman S, Hoffman RM (2005) Implanted hair follicle stem cells form Schwann cells that support repair of severed peripheral nerves. Proc Natl Acad Sci USA 102:17734–17738
Andrews MR, Stelzner DJ (2007) Evaluation of olfactory ensheathing and schwann cells after implantation into a dorsal injury of adult rat spinal cord. J Neurotrauma 24:1773–1792
Aquino JB, Hjerling-Leffler J, Koltzenburg M, Edlund T, Villar MJ, Ernfors P (2006) In vitro and in vivo differentiation of boundary cap neural crest stem cells into mature Schwann cells. Exp Neurol 198:438–449
Ara J, See J, Mamontov P, Hahn A, Bannerman P, Pleasure D, Grinspan JB (2008) Bone morphogenetic proteins 4, 6, and 7 are up-regulated in mouse spinal cord during experimental autoimmune encephalomyelitis. J Neurosci Res 86:125–135
Avellana-Adalid V, Bachelin C, Lachapelle F, Escriou C, Ratzkin B, Baron-Van Evercooren A (1998) In vitro and in vivo behaviour of NDF-expanded monkey Schwann cells. Eur J Neurosci 10:291–300
Bachelin C, Lachapelle F, Girard C, Moissonnier P, Serguera-Lagache C, Mallet J, Fontaine D, Chojnowski A, Le Guern E, Nait-Oumesmar B, Baron-Van Evercooren A (2005) Efficient

myelin repair in the macaque spinal cord by autologous grafts of Schwann cells. Brain 128: 540–549

Bachelin C, Zujovic V, Buchet D, Mallet J, Baron-Van Evercooren A (2010) Ectopic expression of polysialylated neural cell adhesion molecule in adult macaque Schwann cells promotes their migration and remyelination potential in the central nervous system. Brain 133:406–420

Baron-Van Evercooren A, Kleinman HK, Ohno S, Marangos P, Schwartz JP, Dubois-Dalcq ME (1982) Nerve growth factor, laminin, and fibronectin promote neurite growth in human fetal sensory ganglia cultures. J Neurosci Res 8:179–193

Baron-Van Evercooren A, Gansmuller A, Duhamel E, Pascal F, Gumpel M (1992) Repair of a myelin lesion by Schwann cells transplanted in the adult mouse spinal cord. J Neuroimmunol 40:235–242

Baron-Van Evercooren A, Blakemore WF. 2004. Remyelination throughengraftment. In: Lazzarini RA, editor. Myelin biology and disorders,Vol. I. San Diego: Elsevier Academic Press. p 143–61.

Barraud P, Seferiadis AA, Tyson LD, Zwart MF, Szabo-Rogers HL, Ruhrberg C, Liu KJ, Baker CV (2010) Neural crest origin of olfactory ensheathing glia. Proc Natl Acad Sci USA 107:21040–21045

Biernaskie JA, McKenzie IA, Toma JG, Miller FD (2006) Isolation of skin-derived precursors (SKPs) and differentiation and enrichment of their Schwann cell progeny. Nat Protoc 1:2803–2812

Biernaskie J, Sparling JS, Liu J, Shannon CP, Plemel JR, Xie Y, Miller FD, Tetzlaff W (2007) Skin-derived precursors generate myelinating Schwann cells that promote remyelination and functional recovery after contusion spinal cord injury. J Neurosci 27:9545–9559

Blakemore WF (1977) Remyelination of CNS axons by Schwann cells transplanted from the sciatic nerve. Nature 266:68–69

Blakemore WF, Crang AJ, Curtis R (1986) The interaction of Schwann cells with CNS axons in regions containing normal astrocytes. Acta Neuropathol 71:295–300

Boyd JG, Lee J, Skihar V, Doucette R, Kawaja MD (2004) LacZ-expressing olfactory ensheathing cells do not associate with myelinated axons after implantation into the compressed spinal cord. Proc Natl Acad Sci USA 101:2162–2166

Brierley CM, Crang AJ, Iwashita Y, Gilson JM, Scolding NJ, Compston DA, Blakemore WF (2001) Remyelination of demyelinated CNS axons by transplanted human schwann cells: the deleterious effect of contaminating fibroblasts. Cell Transplant 10:305–315

Brinkmann BG, Agarwal A, Sereda MW, Garratt AN, Muller T, Wende H, Stassart RM, Nawaz S, Humml C, Velanac V, Radyushkin K, Goebbels S, Fischer TM, Franklin RJ, Lai C, Ehrenreich H, Birchmeier C, Schwab MH, Nave KA (2008) Neuregulin-1/ErbB signaling serves distinct functions in myelination of the peripheral and central nervous system. Neuron 59:581–595

Brockes JP, Fields KL, Raff MC (1979) Studies on cultured rat Schwann cells. I. Establishment of purified populations from cultures of peripheral nerve. Brain Res 165:105–118

Brook GA, Lawrence JM, Raisman G (1993) Morphology and migration of cultured Schwann cells transplanted into the fimbria and hippocampus in adult rats. Glia 9:292–304

Chambers SM, Studer L (2011) Cell fate plug and play: direct reprogramming and induced pluripotency. Cell 145:827–830

Chambers SM, Fasano CA, Papapetrou EP, Tomishima M, Sadelain M, Studer L (2009) Highly efficient neural conversion of human ES and iPS cells by dual inhibition of SMAD signaling. Nat Biotechnol 27:275–280

Clewes O, Narytnyk A, Gillinder KR, Loughney AD, Murdoch AP, Sieber-Blum M (2011) Human epidermal neural crest stem cells (hEPI-NCSC) – characterization and directed differentiation into osteocytes and melanocytes. Stem Cell Rev 7(4):799–814

Crang AJ, Gilson JM, Li WW, Blakemore WF (2004) The remyelinating potential and in vitro differentiation of MOG-expressing oligodendrocyte precursors isolated from the adult rat CNS. Eur J Neurosci 20:1445–1460

Deng C, Gorrie C, Hayward I, Elston B, Venn M, Mackay-Sim A, Waite P (2006) Survival and migration of human and rat olfactory ensheathing cells in intact and injured spinal cord. J Neurosci Res 83:1201–1212

Devon R, Doucette R (1992) Olfactory ensheathing cells myelinate dorsal root ganglion neurites. Brain Res 589:175–179

Dezawa M, Takahashi I, Esaki M, Takano M, Sawada H (2001) Sciatic nerve regeneration in rats induced by transplantation of in vitro differentiated bone-marrow stromal cells. Eur J Neurosci 14:1771–1776

Dombrowski MA, Sasaki M, Lankford KL, Kocsis JD, Radtke C (2006) Myelination and nodal formation of regenerated peripheral nerve fibers following transplantation of acutely prepared olfactory ensheathing cells. Brain Res 1125:1–8

Doucette R (1991) PNS-CNS transitional zone of the first cranial nerve. J Comp Neurol 312: 451–466

Duncan ID, Aguayo AJ, Bunge RP, Wood PM (1981) Transplantation of rat Schwann cells grown in tissue culture into the mouse spinal cord. J Neurol Sci 49:241–252

Dunning MD, Kettunen MI, Ffrench Constant C, Franklin RJ, Brindle KM (2006) Magnetic resonance imaging of functional Schwann cell transplants labelled with magnetic microspheres. Neuroimage 31:172–180

Fairless R, Frame MC, Barnett SC (2005) N-cadherin differentially determines Schwann cell and olfactory ensheathing cell adhesion and migration responses upon contact with astrocytes. Mol Cell Neurosci 28:253–263

Fernandes KJ, McKenzie IA, Mill P, Smith KM, Akhavan M, Barnabe-Heider F, Biernaskie J, Junek A, Kobayashi NR, Toma JG, Kaplan DR, Labosky PA, Rafuse V, Hui CC, Miller FD (2004) A dermal niche for multipotent adult skin-derived precursor cells. Nat Cell Biol 6: 1082–1093

Fernandes KJ, Kobayashi NR, Gallagher CJ, Barnabe-Heider F, Aumont A, Kaplan DR, Miller FD (2006) Analysis of the neurogenic potential of multipotent skin-derived precursors. Exp Neurol 201:32–48

Feron F, Perry C, Cochrane J, Licina P, Nowitzke A, Urquhart S, Geraghty T, Mackay-Sim A (2005) Autologous olfactory ensheathing cell transplantation in human spinal cord injury. Brain 128:2951–2960

Forni PE, Taylor-Burds C, Melvin VS, Williams T, Wray S (2011) Neural crest and ectodermal cells intermix in the nasal placode to give rise to GnRH-1 neurons, sensory neurons, and olfactory ensheathing cells. J Neurosci 31:6915–6927

Franceschini IA, Barnett SC (1996) Low-affinity NGF-receptor and E-N-CAM expression define two types of olfactory nerve ensheathing cells that share a common lineage. Dev Biol 173:327–343

Franklin RJ, Barnett SC (2000) Olfactory ensheathing cells and CNS regeneration: the sweet smell of success? Neuron 28:15–18

Franklin RJ, Crang AJ, Blakemore WF (1991) Transplanted type-1 astrocytes facilitate repair of demyelinating lesions by host oligodendrocytes in adult rat spinal cord. J Neurocytol 20:420–430

Franklin RJ, Crang AJ, Blakemore WF (1993) The reconstruction of an astrocytic environment in glia-deficient areas of white matter. J Neurocytol 22:382–396

Franklin RJ, Gilson JM, Franceschini IA, Barnett SC (1996) Schwann cell-like myelination following transplantation of an olfactory bulb-ensheathing cell line into areas of demyelination in the adult CNS. Glia 17:217–224

Garcia-Alias G, Lopez-Vales R, Fores J, Navarro X, Verdu E (2004) Acute transplantation of olfactory ensheathing cells or Schwann cells promotes recovery after spinal cord injury in the rat. J Neurosci Res 75:632–641

Girard C, Bemelmans AP, Dufour N, Mallet J, Bachelin C, Nait-Oumesmar B, Baron-Van Evercooren A, Lachapelle F (2005) Grafts of brain-derived neurotrophic factor and neurotrophin 3-transduced primate Schwann cells lead to functional recovery of the demyelinated mouse spinal cord. J Neurosci 25:7924–7933

Glaser T, Brose C, Franceschini I, Hamann K, Smorodchenko A, Zipp F, Dubois-Dalcq M, Brustle O (2007) Neural cell adhesion molecule polysialylation enhances the sensitivity of embryonic stem cell-derived neural precursors to migration guidance cues. Stem Cells 25:3016–3025

Golden KL, Pearse DD, Blits B, Garg MS, Oudega M, Wood PM, Bunge MB (2007) Transduced Schwann cells promote axon growth and myelination after spinal cord injury. Exp Neurol 207:203–217

Grimpe B, Silver J (2004) A novel DNA enzyme reduces glycosaminoglycan chains in the glial scar and allows microtransplanted dorsal root ganglia axons to regenerate beyond lesions in the spinal cord. J Neurosci 24:1393–1397

Gueye Y, Ferhat L, Sbai O, Bianco J, Ould-Yahoui A, Bernard A, Charrat E, Chauvin JP, Risso JJ, Feron F, Rivera S, Khrestchatisky M (2011) Trafficking and secretion of matrix metalloproteinase-2 in olfactory ensheathing glial cells: a role in cell migration? Glia 59:750–770

Gurdon JB (2006) From nuclear transfer to nuclear reprogramming: the reversal of cell differentiation. Annu Rev Cell Dev Biol 22:1–22

Hinks GL, Chari DM, O'Leary MT, Zhao C, Keirstead HS, Blakemore WF, Franklin RJ (2001) Depletion of endogenous oligodendrocyte progenitors rather than increased availability of survival factors is a likely explanation for enhanced survival of transplanted oligodendrocyte progenitors in X-irradiated compared to normal CNS. Neuropathol Appl Neurobiol 27:59–67

Hjerling-Leffler J, Marmigere F, Heglind M, Cederberg A, Koltzenburg M, Enerback S, Ernfors P (2005) The boundary cap: a source of neural crest stem cells that generate multiple sensory neuron subtypes. Development 132:2623–2632

Honmou O, Felts PA, Waxman SG, Kocsis JD (1996) Restoration of normal conduction properties in demyelinated spinal cord axons in the adult rat by transplantation of exogenous Schwann cells. J Neurosci 16:3199–3208

Hotta R, Pepdjonovic L, Anderson RB, Zhang D, Bergner AJ, Leung J, Pebay A, Young HM, Newgreen DF, Dottori M (2009) Small-molecule induction of neural crest-like cells derived from human neural progenitors. Stem Cells 27:2896–2905

Hu X, Hicks CW, He W, Wong P, Macklin WB, Trapp BD, Yan R (2006) Bace1 modulates myelination in the central and peripheral nervous system. Nat Neurosci 9:1520–1525

Ieda M, Fu JD, Delgado-Olguin P, Vedantham V, Hayashi Y, Bruneau BG, Srivastava D (2010) Direct reprogramming of fibroblasts into functional cardiomyocytes by defined factors. Cell 142:375–386

Imaizumi T, Lankford KL, Waxman SG, Greer CA, Kocsis JD (1998) Transplanted olfactory ensheathing cells remyelinate and enhance axonal conduction in the demyelinated dorsal columns of the rat spinal cord. J Neurosci 18:6176–6185

Iwashita Y, Blakemore WF (2000) Areas of demyelination do not attract significant numbers of schwann cells transplanted into normal white matter. Glia 31:232–240

Iwashita Y, Fawcett JW, Crang AJ, Franklin RJ, Blakemore WF (2000) Schwann cells transplanted into normal and X-irradiated adult white matter do not migrate extensively and show poor long-term survival. Exp Neurol 164:292–302

Jeffery ND, Lakatos A, Franklin RJ (2005) Autologous olfactory glial cell transplantation is reliable and safe in naturally occurring canine spinal cord injury. J Neurotrauma 22:1282–1293

Jiang X, Gwye Y, McKeown SJ, Bronner-Fraser M, Lutzko C, Lawlor ER (2009) Isolation and characterization of neural crest stem cells derived from in vitro-differentiated human embryonic stem cells. Stem Cells Dev 18:1059–1070

Kaneko S, Iwanami A, Nakamura M, Kishino A, Kikuchi K, Shibata S, Okano HJ, Ikegami T, Moriya A, Konishi O, Nakayama C, Kumagai K, Kimura T, Sato Y, Goshima Y, Taniguchi M, Ito M, He Z, Toyama Y, Okano H (2006) A selective Sema3A inhibitor enhances regenerative responses and functional recovery of the injured spinal cord. Nat Med 12:1380–1389

Kato T, Honmou O, Uede T, Hashi K, Kocsis JD (2000) Transplantation of human olfactory ensheathing cells elicits remyelination of demyelinated rat spinal cord. Glia 30:209–218

Keilhoff G, Stang F, Goihl A, Wolf G, Fansa H (2006) Transdifferentiated mesenchymal stem cells as alternative therapy in supporting nerve regeneration and myelination. Cell Mol Neurobiol 26:1235–1252

Keirstead HS, Ben-Hur T, Rogister B, O'Leary MT, Dubois-Dalcq M, Blakemore WF (1999) Polysialylated neural cell adhesion molecule-positive CNS precursors generate both oligodendrocytes and Schwann cells to remyelinate the CNS after transplantation. J Neurosci 19:7529–7536

Kim J, Efe JA, Zhu S, Talantova M, Yuan X, Wang S, Lipton SA, Zhang K, Ding S (2011) Direct reprogramming of mouse fibroblasts to neural progenitors. Proc Natl Acad Sci USA 108: 7838–7843

Krampera M, Marconi S, Pasini A, Galie M, Rigotti G, Mosna F, Tinelli M, Lovato L, Anghileri E, Andreini A, Pizzolo G, Sbarbati A, Bonetti B (2007) Induction of neural-like differentiation in human mesenchymal stem cells derived from bone marrow, fat, spleen and thymus. Bone 40:382–390

Kromer LF, Cornbrooks CJ (1985) Transplants of Schwann cell cultures promote axonal regeneration in the adult mammalian brain. Proc Natl Acad Sci USA 82:6330–6334

Kurkinen M, Alitalo K (1979) Fibronectin and procollagen produced by a clonal line of Schwann cells. FEBS Lett 102:64–68

La Marca R, Cerri F, Horiuchi K, Bachi A, Feltri ML, Wrabetz L, Blobel CP, Quattrini A, Salzer JL, Taveggia C (2011) TACE (ADAM17) inhibits Schwann cell myelination. Nat Neurosci 14:857–865

Lakatos A, Franklin RJ, Barnett SC (2000) Olfactory ensheathing cells and Schwann cells differ in their in vitro interactions with astrocytes. Glia 32:214–225

Lakatos A, Barnett SC, Franklin RJ (2003a) Olfactory ensheathing cells induce less host astrocyte response and chondroitin sulphate proteoglycan expression than Schwann cells following transplantation into adult CNS white matter. Exp Neurol 184:237–246

Lakatos A, Smith PM, Barnett SC, Franklin RJ (2003b) Meningeal cells enhance limited CNS remyelination by transplanted olfactory ensheathing cells. Brain 126:598–609

Langford LA, Owens GC (1990) Resolution of the pathway taken by implanted Schwann cells to a spinal cord lesion by prior infection with a retrovirus encoding beta-galactosidase. Acta Neuropathol 80:514–520

Lankford KL, Imaizumi T, Honmou O, Kocsis JD (2002) A quantitative morphometric analysis of rat spinal cord remyelination following transplantation of allogenic Schwann cells. J Comp Neurol 443:259–274

Lankford KL, Sasaki M, Radtke C, Kocsis JD (2008) Olfactory ensheathing cells exhibit unique migratory, phagocytic, and myelinating properties in the X-irradiated spinal cord not shared by Schwann cells. Glia 56:1664–1678

Lavdas AA, Franceschini I, Dubois-Dalcq M, Matsas R (2006) Schwann cells genetically engineered to express PSA show enhanced migratory potential without impairment of their myelinating ability in vitro. Glia 53:868–878

Lavdas AA, Papastefanaki F, Thomaidou D, Matsas R (2008) Schwann cell transplantation for CNS repair. Curr Med Chem 15:151–160

Lee G, Kim H, Elkabetz Y, Al Shamy G, Panagiotakos G, Barberi T, Tabar V, Studer L (2007) Isolation and directed differentiation of neural crest stem cells derived from human embryonic stem cells. Nat Biotechnol 25:1468–1475

Lee G, Chambers SM, Tomishima MJ, Studer L (2010) Derivation of neural crest cells from human pluripotent stem cells. Nat Protoc 5:688–701

Levi AD (1996) Characterization of the technique involved in isolating Schwann cells from adult human peripheral nerve. J Neurosci Methods 68:21–26

Levi AD, Bunge RP (1994) Studies of myelin formation after transplantation of human Schwann cells into the severe combined immunodeficient mouse. Exp Neurol 130:41–52

Li HY, Say EH, Zhou XF (2007) Isolation and characterization of neural crest progenitors from adult dorsal root ganglia. Stem Cells 25:2053–2065

Lima C, Pratas-Vital J, Escada P, Hasse-Ferreira A, Capucho C, Peduzzi JD (2006) Olfactory mucosa autografts in human spinal cord injury: a pilot clinical study. J Spinal Cord Med 29:191–203, discussion 196–204

Lu J, Ashwell K. Olfactory ensheathing cells: their potential use for repairing the injured spinal cord. Spine (Phila Pa 1976). 2002 Apr 15;27(8):887–92

Lu P, Yang H, Culbertson M, Graham L, Roskams AJ, Tuszynski MH (2006) Olfactory ensheathing cells do not exhibit unique migratory or axonal growth-promoting properties after spinal cord injury. J Neurosci 26:11120–11130

Markakis EA, Sasaki M, Lankford KL, Kocsis JD (2009) Convergence of cells from the progenitor fraction of adult olfactory bulb tissue to remyelinating glia in demyelinating spinal cord lesions. PLoS One 4:e7260

Maro GS, Vermeren M, Voiculescu O, Melton L, Cohen J, Charnay P, Topilko P (2004) Neural crest boundary cap cells constitute a source of neuronal and glial cells of the PNS. Nat Neurosci 7:930–938

Marshall CT, Guo Z, Lu C, Klueber KM, Khalyfa A, Cooper NG, Roisen FJ (2005) Human adult olfactory neuroepithelial derived progenitors retain telomerase activity and lack apoptotic activity. Brain Res 1045:45–56

Marshall CT, Lu C, Winstead W, Zhang X, Xiao M, Harding G, Klueber KM, Roisen FJ (2006) The therapeutic potential of human olfactory-derived stem cells. Histol Histopathol 21:633–643

Martin D, Schoenen J, Delree P, Leprince P, Rogister B, Moonen G (1991) Grafts of syngenic cultured, adult dorsal root ganglion-derived Schwann cells to the injured spinal cord of adult rats: preliminary morphological studies. Neurosci Lett 124:44–48

Martino G, Franklin RJ, Van Evercooren AB, Kerr DA (2010) Stem cell transplantation in multiple sclerosis: current status and future prospects. Nat Rev Neurol 6:247–255

McKenzie IA, Biernaskie J, Toma JG, Midha R, Miller FD (2006) Skin-derived precursors generate myelinating Schwann cells for the injured and dysmyelinated nervous system. J Neurosci 26:6651–6660

McTigue DM, Horner PJ, Stokes BT, Gage FH (1998) Neurotrophin-3 and brain-derived neurotrophic factor induce oligodendrocyte proliferation and myelination of regenerating axons in the contused adult rat spinal cord. J Neurosci 18:5354–5365

Montero-Menei CN, Pouplard-Barthelaix A, Gumpel M, Baron-Van Evercooren A (1992) Pure Schwann cell suspension grafts promote regeneration of the lesioned septo-hippocampal cholinergic pathway. Brain Res 570:198–208

Mujtaba T, Mayer-Proschel M, Rao MS (1998) A common neural progenitor for the CNS and PNS. Dev Biol 200:1–15

Nagoshi N, Shibata S, Kubota Y, Nakamura M, Nagai Y, Satoh E, Morikawa S, Okada Y, Mabuchi Y, Katoh H, Okada S, Fukuda K, Suda T, Matsuzaki Y, Toyama Y, Okano H (2008) Ontogeny and multipotency of neural crest-derived stem cells in mouse bone marrow, dorsal root ganglia, and whisker pad. Cell Stem Cell 2:392–403

Ogier C, Bernard A, Chollet AM, LE Diquardher T, Hanessian S, Charton G, Khrestchatisky M, Rivera S (2006) Matrix metalloproteinase-2 (MMP-2) regulates astrocyte motility in connection with the actin cytoskeleton and integrins. Glia 54:272–284

Oudega M, Xu XM (2006) Schwann cell transplantation for repair of the adult spinal cord. J Neurotrauma 23:453–467

Papastefanaki F, Chen J, Lavdas AA, Thomaidou D, Schachner M, Matsas R (2007) Grafts of Schwann cells engineered to express PSA-NCAM promote functional recovery after spinal cord injury. Brain 130:2159–2174

Pearse DD, Sanchez AR, Pereira FC, Andrade CM, Puzis R, Pressman Y, Golden K, Kitay BM, Blits B, Wood PM, Bunge MB (2007) Transplantation of Schwann cells and/or olfactory ensheathing glia into the contused spinal cord: survival, migration, axon association, and functional recovery. Glia 55:976–1000

Plant GW, Christensen CL, Oudega M, Bunge MB (2003) Delayed transplantation of olfactory ensheathing glia promotes sparing/regeneration of supraspinal axons in the contused adult rat spinal cord. J Neurotrauma 20:1–16

Pomp O, Brokhman I, Ben-Dor I, Reubinoff B, Goldstein RS (2005) Generation of peripheral sensory and sympathetic neurons and neural crest cells from human embryonic stem cells. Stem Cells 23:923–930

Pomp O, Brokhman I, Ziegler L, Almog M, Korngreen A, Tavian M, Goldstein RS (2008) PA6-induced human embryonic stem cell-derived neurospheres: a new source of human peripheral sensory neurons and neural crest cells. Brain Res 1230:50–60

Radtke C, Aizer AA, Agulian SK, Lankford KL, Vogt PM, Kocsis JD (2009) Transplantation of olfactory ensheathing cells enhances peripheral nerve regeneration after microsurgical nerve repair. Brain Res 1254:10–17

Rutishauser U (2008) Polysialic acid in the plasticity of the developing and adult vertebrate nervous system. Nat Rev Neurosci 9:26–35

Rutkowski JL, Kirk CJ, Lerner MA, Tennekoon GI (1995) Purification and expansion of human Schwann cells in vitro. Nat Med 1:80–83

Saberi H, Moshayedi P, Aghayan HR, Arjmand B, Hosseini SK, Emami-Razavi SH, Rahimi-Movaghar V, Raza M, Firouzi M (2008) Treatment of chronic thoracic spinal cord injury patients with autologous Schwann cell transplantation: an interim report on safety considerations and possible outcomes. Neurosci Lett 443:46–50

Sailer MH, Hazel TG, Panchision DM, Hoeppner DJ, Schwab ME, McKay RD (2005) BMP2 and FGF2 cooperate to induce neural-crest-like fates from fetal and adult CNS stem cells. J Cell Sci 118:5849–5860

Santos-Silva A, Fairless R, Frame MC, Montague P, Smith GM, Toft A, Riddell JS, Barnett SC (2007) FGF/heparin differentially regulates Schwann cell and olfactory ensheathing cell interactions with astrocytes: a role in astrocytosis. J Neurosci 27:7154–7167

Sasaki M, Honmou O, Akiyama Y, Uede T, Hashi K, Kocsis JD (2001) Transplantation of an acutely isolated bone marrow fraction repairs demyelinated adult rat spinal cord axons. Glia 35:26–34

Sasaki M, Lankford KL, Zemedkun M, Kocsis JD (2004) Identified olfactory ensheathing cells transplanted into the transected dorsal funiculus bridge the lesion and form myelin. J Neurosci 24:8485–8493

Sasaki M, Lankford KL, Radtke C, Honmou O, Kocsis JD (2011) Remyelination after olfactory ensheathing cell transplantation into diverse demyelinating environments. Exp Neurol 229:88–98

Shirakabe K, Wakatsuki S, Kurisaki T, Fujisawa-Sehara A (2001) Roles of Meltrin beta/ADAM19 in the processing of neuregulin. J Biol Chem 276:9352–9358

Sieber-Blum M, Grim M, Hu YF, Szeder V (2004) Pluripotent neural crest stem cells in the adult hair follicle. Dev Dyn 231:258–269

Smith PM, Sim FJ, Barnett SC, Franklin RJ (2001) SCIP/Oct-6, Krox-20, and desert hedgehog mRNA expression during CNS remyelination by transplanted olfactory ensheathing cells. Glia 36:342–353

Szabo E, Rampalli S, Risueno RM, Schnerch A, Mitchell R, Fiebig-Comyn A, Levadoux-Martin M, Bhatia M (2010) Direct conversion of human fibroblasts to multilineage blood progenitors. Nature 468:521–526.

Takagi T, Ishii K, Shibata S, Yasuda A, Sato M, Nagoshi N, Saito H, Okano HJ, Toyama Y, Okano H, Nakamura M (2011) Schwann-spheres derived from injured peripheral nerves in adult mice – their in vitro characterization and therapeutic potential. PLoS One 6:e21497

Takahashi K, Yamanaka S (2006) Induction of pluripotent stem cells from mouse embryonic and adult fibroblast cultures by defined factors. Cell 126:663–676

Talbott JF, Cao Q, Enzmann GU, Benton RL, Achim V, Cheng XX, Mills MD, Rao MS, Whittemore SR (2006) Schwann cell-like differentiation by adult oligodendrocyte precursor cells following engraftment into the demyelinated spinal cord is BMP-dependent. Glia 54:147–159

Taveggia C, Zanazzi G, Petrylak A, Yano H, Rosenbluth J, Einheber S, Xu X, Esper RM, Loeb JA, Shrager P, Chao MV, Falls DL, Role L, Salzer JL (2005) Neuregulin-1 type III determines the ensheathment fate of axons. Neuron 47:681–694

Tetzlaff W, Okon EB, Karimi-Abdolrezaee S, Hill CE, Sparling JS, Plemel JR, Plunet WT, Tsai EC, Baptiste D, Smithson LJ, Kawaja MD, Fehlings MG, Kwon BK (2011) A systematic review of cellular transplantation therapies for spinal cord injury. J Neurotrauma 28:1611–1682

Toft A, Tomé M, Lindsay SL, Barnett SC, Riddell JS. Transplant-mediated repair properties of rat olfactory mucosal OM-I and OM-II sphere-forming cells. J Neurosci Res. 2012 Mar;90(3):619-31.

Toma JG, Akhavan M, Fernandes KJ, Barnabe-Heider F, Sadikot A, Kaplan DR, Miller FD (2001) Isolation of multipotent adult stem cells from the dermis of mammalian skin. Nat Cell Biol 3:778–784

Toma JG, McKenzie IA, Bagli D, Miller FD (2005) Isolation and characterization of multipotent skin-derived precursors from human skin. Stem Cells 23:727–737

Tomé M, Lindsay SL, Riddell JS, Barnett SC. Identification of nonepithelial multipotent cells in the embryonic olfactory mucosa. Stem Cells. 2009 Sep;27(9):2196–208.

Vierbuchen T, Ostermeier A, Pang ZP, Kokubu Y, Sudhof TC, Wernig M (2010) Direct conversion of fibroblasts to functional neurons by defined factors. Nature 463:1035–1041

Vincent AJ, West AK, Chuah MI (2005) Morphological and functional plasticity of olfactory ensheathing cells. J Neurocytol 34:65–80

Wang A, Tang Z, Park IH, Zhu Y, Patel S, Daley GQ, Li S (2011) Induced pluripotent stem cells for neural tissue engineering. Biomaterials 32:5023–5032

Weintraub H, Tapscott SJ, Davis RL, Thayer MJ, Adam MA, Lassar AB, Miller AD (1989) Activation of muscle-specific genes in pigment, nerve, fat, liver, and fibroblast cell lines by forced expression of MyoD. Proc Natl Acad Sci USA 86:5434–5438

Wewetzer K, Verdu E, Angelov DN, Navarro X (2002) Olfactory ensheathing glia and Schwann cells: two of a kind? Cell Tissue Res 309:337–345

Wewetzer K, Radtke C, Kocsis J, Baumgartner W (2011) Species-specific control of cellular proliferation and the impact of large animal models for the use of olfactory ensheathing cells and Schwann cells in spinal cord repair. Exp Neurol 229:80–87

Winstead W, Marshall CT, Lu CL, Klueber KM, Roisen FJ (2005) Endoscopic biopsy of human olfactory epithelium as a source of progenitor cells. Am J Rhinol 19:83–90

Wong CE, Paratore C, Dours-Zimmermann MT, Rochat A, Pietri T, Suter U, Zimmermann DR, Dufour S, Thiery JP, Meijer D, Beermann F, Barrandon Y, Sommer L (2006) Neural crest-derived cells with stem cell features can be traced back to multiple lineages in the adult skin. J Cell Biol 175:1005–1015

Wood PM (1976) Separation of functional Schwann cells and neurons from normal peripheral nerve tissue. Brain Res 115:361–375

Woodhoo A, Sommer L (2008) Development of the Schwann cell lineage: from the neural crest to the myelinated nerve. Glia 56:1481–1490

Woodhoo A, Sahni V, Gilson J, Setzu A, Franklin RJ, Blakemore WF, Mirsky R, Jessen KR (2007) Schwann cell precursors: a favourable cell for myelin repair in the Central Nervous System. Brain 130:2175–2185

Xiao M, Klueber KM, Lu C, Guo Z, Marshall CT, Wang H, Roisen FJ (2005) Human adult olfactory neural progenitors rescue axotomized rodent rubrospinal neurons and promote functional recovery. Exp Neurol 194:12–30

Xu Y, Liu Z, Liu L, Zhao C, Xiong F, Zhou C, Li Y, Shan Y, Peng F, Zhang C (2008) Neurospheres from rat adipose-derived stem cells could be induced into functional Schwann cell-like cells in vitro. BMC Neurosci 9:21

Yamauchi J, Miyamoto Y, Tanoue A, Shooter EM, Chan JR (2005) Ras activation of a Rac1 exchange factor, Tiam1, mediates neurotrophin-3-induced Schwann cell migration. Proc Natl Acad Sci USA 102:14889–14894

Zhou Y, Snead ML (2008) Derivation of cranial neural crest-like cells from human embryonic stem cells. Biochem Biophys Res Commun 376:542–547

Ziegler L, Grigoryan S, Yang IH, Thakor NV, Goldstein RS (2011) Efficient generation of schwann cells from human embryonic stem cell-derived neurospheres. Stem Cell Rev 7:394–403

Zujovic V, Bachelin C, Baron-Van Evercooren A (2007) Remyelination of the central nervous system: a valuable contribution from the periphery. Neuroscientist 13:383–391

Zujovic V, Thibaud J, Bachelin C, Vidal M, Coulpier F, Charnay P, Topilko P, Baron-Van Evercooren A (2010) Boundary cap cells are highly competitive for CNS remyelination: fast migration and efficient differentiation in PNS and CNS myelin-forming cells. Stem Cells 28:470–479

Zujovic V, Thibaud J, Bachelin C, Vidal M, Deboux C, Coulpier F, Stadler N, Charnay P, Topilko P, Baron-Van Evercooren A (2011) Boundary cap cells are peripheral nervous system stem cells that can be redirected into central nervous system lineages. Proc Natl Acad Sci USA 108:10714–10719

Chapter 7
Immune Modulation and Repair Following Neural Stem Cell Transplantation

Tamir Ben-Hur, Stefano Pluchino, and Gianvito Martino

7.1 Origin and Fate of Neural Stem/Precursor Cells

The extremely composite architecture of the central nervous system (CNS) can be viewed as the result of a number of subsequent cell divisions and precise cell-to-cell and cell-to-substrate interactions. This starts from a small number of undifferentiated intermediate progenitor cells in the developing neural tube, that assemble during the whole embryonic and one relatively short postnatal period (Noctor et al. 2007). This process, called neurogenesis, principally involves the highly heterogeneous population of neural stem and precursor cells (hereafter called NPCs) located in major CNS germinal regions. Within these CNS regions, NPCs undergo self-renewal and give

---

T. Ben-Hur (✉)
Department of Neurology, The Agnes Ginges Center for Human Neurogenetics,
Hadassah-Hebrew University Medical Center, PO Box 12000, Jerusalem 91120, Israel
e-mail: tamir@hadassah.org.il

S. Pluchino
CNS Repair Unit, Division of Neuroscience, San Raffaele Scientific Institute,
via Olgettina 58, 20132 Milan, Italy

Institute of Experimental Neurology, Division of Neuroscience,
San Raffaele Scientific Institute, via Olgettina 58, 20132 Milan, Italy

Department of Clinical Neurosciences, Cambridge Centre for Brain Repair
and Cambridge Stem Cell Initiative, E.D. Adrian Building, Forvie Site,
Robinson Way, Cambridge CB2 0PY, UK
e-mail: pluchino.stefano@hsr.it

G. Martino
Neuroimmunology Unit, Division of Neuroscience, San Raffaele Scientific
Institute, via Olgettina 58, 20132 Milan, Italy

Institute of Experimental Neurology, Division of Neuroscience, San Raffaele
Scientific Institute, via Olgettina 58, 20132 Milan, Italy
e-mail: martino.gianvito@hsr.it

rise to most neuronal and glial cell precursors that populate the developing CNS by a combination of centrifugal radial and tangential cell migration (Marin and Rubenstein 2003; Rakic 1990). At the end of embryonic life, upon CNS assembly and neuro/glial generation, the functional specificity of the CNS stem cell compartment is fixed by various molecular/cellular cues. This compartment is characterized by highly static properties that might both limit cell renewal and hamper brain repair following different types of tissue damages (reviewed in Bonfanti 2006).

In addition to widespread structural changes (e.g. modifications of synaptic contacts) reshaping neuronal circuits (Ito 2006), postnatal neuro- and glio-genesis persists in precise anatomical sites of the CNS, such as the subventricular zone (SVZ) of the lateral ventricles and the subgranular zone (SGZ) of the dentate gyrus (DG) of the hippocampus (Alvarez-Buylla and Garcia-Verdugo 2002; Bonfanti and Ponti 2008; Gage 2000). The SVZ of the adult mammalian brain more evidently retains embryonic features of primitive germinal layers, as it maintains direct contact with the ventricles and its neuronal precursors undergo long-distance migration to reach their final site of destination in the olfactory bulb (Lois and Alvarez-Buylla 1994). The rodent SVZ is made up of two main neural cell compartments: (a) newly generated, migrating neuroblasts, which form tangentially oriented chains towards the olfactory bulb (OB) (Doetsch and Alvarez-Buylla 1996; Lois et al. 1996) and (b) protoplasmic astrocytes organized to form longitudinally oriented *glial tubes* (Lois et al. 1996; Peretto et al. 1997). The neuroblasts, also referred to as type A cells (Lois et al. 1996), co-express β-tubulin, doublecortin, and poly-sialylated neural cell adhesion molecule (PSA-NCAM) and have an electron-dense cytoplasm (Menezes and Luskin 1994; Nacher et al. 2001). The astrocytes (type B cells) are glial fibrillary acidic protein (GFAP)$^+$, ramified cells with an electron lucent, watery cytoplasm. A subpopulation of type B cells has been indicated as the true neural stem cells, which would divide at a very slow rate in vivo (Doetsch et al. 1999). The third population, termed type C cells, with ultrastructural features intermediate between A and B cell types, consists of highly proliferative "transit amplifying" progenitor cells that bridge between slow proliferating stem cells and their progeny (Doetsch et al. 1997). The human SVZ possesses a highly complex organization with remarkable network of astrocytes and ependymal cells distributed within a total of four layers from the ventricular side throughout the brain parenchyma. This includes a monolayer of ependymal cells (Layer I), a hypocellular gap (Layer II), a ribbon of astrocytes (Layer III), and a final transitional zone (Layer IV) (Quinones-Hinojosa et al. 2006). Also, the vasculature has recently emerged as an important component of stem cell niches. Two recent seminal papers have in fact showed that the adult SVZ niche contains an extensive planar vascular plexus with highly specialized and unique properties. Within this context, dividing stem cells and their transit-amplifying progeny are tightly apposed to blood vessels both during homeostasis and regeneration. Bona fide stem cells are frequently in contact with the vasculature at sites that lack astrocyte endfeet and pericyte coverage (Mirzadeh et al. 2008; Tavazoie et al. 2008).

The DG of the hippocampus is a three-layered cortex made up of small neurons (*granules*), which form a 4–10 cell thick layer comprised between two fibre layers.

Hippocampal progenitor cells divide in the SGZ and generate a progeny, which gives rise to mature granule cells extending an axon within the mossy fibre pathway and reaching the Ammon's horn (Zhao et al. 2006). In the SGZ, both radial and horizontal astrocytes are present, the former being a special type of radial glia-like cell, similar to the type B cells of the SVZ. These glial cells can divide and give rise to the granule cell precursors (Seri et al. 2004). Nevertheless, transit amplifying cells—reminiscent of SVZ type C progenitors—have not been identified in the hippocampus, the intermediate-like cells (type D cells) being equivalent to SVZ type A cells. This partially addresses the apparent complexity in growing neurospheres from SGZ tissue, unless it is contaminated with portions of the dorsal–lateral wall of the lateral ventricle. Indeed, in comparison with the SVZ, the rate of adult hippocampal neurogenesis more evidently decreases with age (Kuhn et al. 1996), thus suggesting that the SGZ contains progenitors with limited self-renewal potential rather than true stem cells.

Other cell types of adult neurogenic areas, including the ependymal cells and endothelial cells both in the ventricular SVZ, are considered to be part of the adult stem cell niche of the forebrain and to originate from a modification of the primitive germinative layers (Shen et al. 2004). A portion of the caudal extension of the adult mouse SVZ no longer associated with an open ventricle but constricted between the hippocampus and the corpus callosum has been recently described as a distinct "subcallosal" zone (Seri et al. 2006). This region mainly produces oligodendrocyte precursors, thus confirming that adult neurogenic areas can generate different types of cells other than neurons.

Yet NPC features seem to be retained solely within persistent neurogenic sites, as astrocytes from other *non-germinal and non-neurogenic* CNS regions (e.g. cerebral cortex, cerebellum, spinal cord) can show multipotency in vitro only when isolated at early developmental stages, whereas the very same cells lose this property after the second postnatal week in mice (Laywell et al. 2000). These latter findings confirm that adult CNS regions other than prototypical germinal areas may possess certain intrinsic cellular/molecular factors—or peculiar anatomical components—capable of (*cell non-autonomous*) regulation of the maintenance of stem-like cells in specific (micro)environments in vivo. In this respect, the maintenance and differentiation of NPCs in CNS germinal niches is very likely relying also on their physical contact to the basal lamina (BL) that, acting as a scaffold, sequesters and/or modulates cytokines and growth factors derived from local cells (such as fibroblasts, macrophages, and pericytes) (Mercier et al. 2002). As such, type B cells in the SVZ are in close contact (interdigitated) with both the BL and the blood vessels, while in the SGZ, bursts of endothelial cell division are spatially and temporally related to clusters of neurogenesis (Palmer et al. 2000).

Current research reveals that different types of neurogenic events, in addition to those one(s) occurring more canonically within prototypical germinal regions, are being characterized. A deeper understanding of the (single vs. multiple) NPC origin and that of mechanisms allowing the retention of stem properties across pre- and postnatal development will be necessary for the design of new strategies for brain repair, aimed at modulating in situ the endogenous sources of neuronal and glial progenitor cells.

7.2 NPC Transplantation in Neurological Disorders

Recent advances in stem cell biology have raised great expectations that diseases of CNS may be ameliorated by cell transplantation. In particular, cell therapy has been studied for inducing efficient remyelination and/or tissue repair in both genetic myelin disorders and acquired inflammatory demyelinating disease models.

In genetically transmitted dysmyelinating diseases, hereditary defects lead to either a failure of myelination during development or to premature myelin breakdown. Here, large regions are demyelinated and depleted of competent glial and oligodendrocyte progenitor cells. Since the resident glial progenitor cell population is incapable of producing myelin in these conditions, then the transplantation of gene defect-free myelin-forming cells is the only possible strategy for achieving anatomic and functional myelin restoration (Duncan et al. 2011). To achieve this end, transplanted progenitors cells should be in sufficient numbers and competent for broad dispersal and extensive myelination. Importantly, for genetic dysmyelinating diseases, the issues at hand are focused mainly on the competence of transplanted cells, as they are expected to integrate into the highly permissive, normal developmental programme of the CNS. This leaves us essentially with choosing the optimal transplantable myelinating cell phenotype, of which there is no consensus yet. Different types of progenitor cells have been proposed and evaluated for this task, including oligodendroglial lineage cells (Duncan et al. 2011), neural stem cells, cells derived from embryonic stem cells, and non-CNS phenotypes, such as Schwann cells, olfactory ensheathing cells (reviewed in Ben-Hur and Goldman 2008), and boundary cap cells (Zujovic et al. 2010).

In acquired demyelinating diseases, the most common of which is multiple sclerosis (MS), the complex issues of cell therapy involve not only the optimal transplantable cell type but also the manipulation of the host CNS to allow the therapeutic actions of transplanted cells. The pathological hallmark of MS is the presence of highly heterogeneous, chronic inflammatory, demyelinating multifocal lesions within the CNS (Compston and Coles 2002; Dyment and Ebers 2002; Flugel et al. 2001; Lucchinetti et al. 2000; Noseworthy et al. 2000; Wingerchuk et al. 2001). The aetiology of MS is multifactorial and includes very complex interplay between environmental factors and susceptibility genes (Hemmer et al. 2002b; Lassmann et al. 2001; Lucchinetti et al. 1996; Martino and Hartung 1999). These factors trigger a cascade of events that engage the immune system, resulting in acute inflammatory injury of axons and glia, accompanied by frank demyelination (Akassoglou et al. 1998; Bjartmar et al. 1999; Kornek et al. 2000; Lassmann 2002; Trapp et al. 1998). The affected CNS demyelinated regions can undergo partial remyelination, leading to structural repair and recovery of function (Barkhof et al. 2003; Chang et al. 2002; Compston 1996; 1997; Prineas et al. 1993; Raine and Wu 1993). Attempts to regenerate myelin can be recognized pathologically in brains of MS patients by the existence of shadow plaques, which are partially remyelinated lesions (Lassmann et al. 1997; Prineas et al. 1993). However, remyelination in MS is typically incomplete and ultimately fails in the setting of recurrent episodes contributing to the progressive demyelination, gliosis, axonal damage, and neurodegeneration typically noted in MS (Blakemore et al. 2002; Franklin 2002; Franklin and Ffrench-Constant 2008). Several studies have indicated that axonal pathology is the best correlate of chronic neurological impairment in MS and its animal counterpart experimental autoimmune encephalomyelitis (EAE) (Bjartmar et al. 1999, 2000; De Stefano et al. 1998; Hemmer et al. 2002a; Steinman 2001; Wujek et al. 2002; Jackson et al. 2009; Papadopoulos et al. 2006). The sequential involvement of these processes underlies the clinical course, characterized by episodes of relapses, which after full remissions early in the course of disease eventually leave persistent deficits and finally deteriorate into a secondary chronic progressive phase. Thus, for MS cell therapies, the major expectation is that transplanted cells succeed where endogenous progenitors have eventually failed (Franklin 2002). Specifically, transplanted cells need to target the specific sites of disease, migrate and integrate in the host tissue, and survive in the CNS environment inflicted with inflammation. This adds crucial issues of timing, route of cell delivery, as well as long-term survival of grafted cells in the "*inhospitable*" adult CNS environment, which need to be critically considered when envisaging therapeutic cell transplants for MS.

Moreover, the choice of transplantable cells in MS depends on gaining insights as to the mechanisms by which they improve the clinical outcome of disease. Restorative neurotransplantation research for prototypical degenerative CNS disorders has focused until recently mainly on the potential of the neural graft to replace damaged or missing cell populations (Lindvall and Kokaia 2006; Lindvall et al. 2004).

Early work has demonstrated proof of principle about the remyelinating properties of various cell types, including oligodendrocyte progenitor cells (Blakemore and Crang 1988; Crang and Blakemore 1991; Crang et al. 1998), Schwann cells (Avellana-Adalid et al. 1998; Blakemore 1977; Kohama et al. 2001), olfactory nerve ensheathing cells (Franklin et al. 1996; Imaizumi et al. 1998, 2000), and multipotent NPCs (Ben-Hur et al. 2003b; Cummings et al. 2005; Eftekharpour et al. 2007; Keirstead et al. 1999; Pluchino et al. 2003). These properties have been mostly studied in focal demyelinated CNS lesions as well as in genetic CNS myelin disorders and, though to a lesser extent, in EAE (Ben-Hur and Goldman 2008; Martino and Pluchino 2006; Pluchino et al. 2004).

In parallel with the great expectations from the direct myelin repair potential of most the above summarized cell sources, recent studies have highlighted novel bystander neuroprotective and neurotrophic capabilities (alternative to direct cell replacement) by which transplanted NPCs may be remarkably beneficial to the host brain in experimental CNS disease models (reviewed in Martino and Pluchino 2006).

This chapter will specifically first focus on the basic properties of transplantable NPCs and then point at the various mechanisms regulating their therapeutic plasticity in experimental models of MS. Overall, we will discuss many of the practical considerations (and their biological basis) when designing a cell therapeutic approach in neurological diseases (Fig. 7.1).

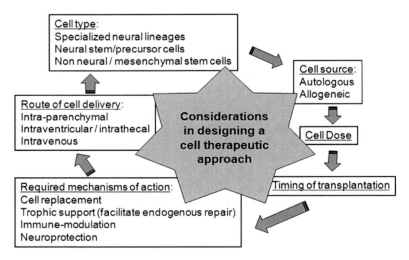

Fig. 7.1 There are multiple considerations when designing the clinical translation of cell therapy. These include choosing the relevant therapeutic target(s) for the candidate disease, choosing the proper cellular platform, timing of transplantation, and route of delivery. For example, congenital dysmyelinating diseases and age-related macular degeneration require specific cell types for cell-replacement therapy (by myelin-forming cells and retinal pigment epithelium cells) and their direct delivery into the target tissue. Spinal cord injury may enjoy from the trophic effects of transplanted cells to enhance endogenous repair systems, as well as from their immunomodulatory and direct regenerative properties. Parkinson's and Huntington's diseases are considered as candidates mainly for cell replacement, but early therapy with neural precursors may also have protective effects against disease progression. For multiple sclerosis, all four mechanisms of transplanted cell action may be beneficial and may be achieved to certain extents by both neural and non-neural cell platforms

7.2.1 The Route of Cell Delivery

The route of cell administration represents a major constraint for stem cell transplantation and appears to be very dependent on the site of the lesion in the CNS (focal vs. multifocal). MS is indeed a multifocal disease, and it would be then very unlikely to propose lesion-targeted injection of cells (Pluchino and Martino 2008). Moreover, it is often difficult to determine which of the multiple foci observed in the brain by non-invasive magnetic resonance imaging (MRI) may underscore clinical significance and whether they would eventually be amenable to effective remyelination upon cell therapy (Chen et al. 2007). Therefore, the ability of transplanted cells to migrate into inflamed brain areas, integrate, and differentiate is a crucial requisite to test, thus further highlighting the identification of the optimal time window and route of cell delivery as two key issues for cell transplantation in MS.

Transplanted multipotent NPCs have been shown to migrate and integrate in the developing rodent CNS and adopt cellular identity according to local and temporal cues (Brustle et al. 1995, 1998; Flax et al. 1998). Furthermore, the transplantation of both somatic and embryonic stem (ES) cell-derived oligodendrocyte progenitors has led to widespread myelination in the genetic dysmyelinating shiverer (*shi/shi*) mouse (Windrem et al. 2004, 2008; Yandava et al. 1999) and the myelin-deficient (*md*) rat models (Brustle et al. 1999; Hammang et al. 1997; Zhang et al. 1998, 1999). While earlier work suggests that the normal adult brain does not allow neither long-distance migration nor does it support the survival of focally transplanted neural cells (O'Leary and Blakemore 1997), more recent data has provided evidence that in some model systems, long-stage migration and survival of oligodendrocyte progenitors can occur (Buchet et al. 2011). Consequently, cell migration might be a major limiting factor for both endogenous remyelinating cells and transplanted lineage-committed myelin-forming cells in MS (Blakemore et al. 2000; Franklin and Blakemore 1997; Gensert and Goldman 1997). These and other observations have led to two major approaches for the delivery of therapeutic cells through biological fluids accessing multiple CNS areas in experimental MS models: the intracerebroventricular (i.c.) or intrathecal (i.t.) cell transplantation and the intravenous (i.v.) cell injection.

Following i.c. or i.t. transplantation of either neurospheres or single cell-dissociated NPCs, clinico-pathological amelioration of EAE in rodents has been achieved, with transplanted cells exhibiting targeted migration almost exclusively to the inflamed periventricular and perivascular white, but not grey, matter of the forebrain (Ben-Hur et al. 2003a, b, 2007; Bulte et al. 2003; Pluchino et al. 2003; Politi et al. 2007). Using this latter approach, the majority of transplanted NPCs have differentiated in vivo into glial cells (e.g. 30 % oligodendrocytes, 25 % astrocytes) (Ben-Hur et al. 2003b; Pluchino et al. 2003). Most white matter tracts that are involved in EAE and MS are in close proximity to the ventricular and spinal subarachnoid spaces (Brok et al. 2001; Bruck 2005). Therefore, the main advantage of these specific routes of cell delivery would be that transplanted NPCs disseminate throughout the ventricular and subarachnoid spaces, enabling their inflammation-driven migration to the white matter.

Several mediators of inflammation have been implicated in regulating NPC migration. The chemokine stromal-derived-factor (SDF)-1 α/CXCL12 and its exclusive receptor CXCR4 are important regulators of the migration of dentate granule cells (Bagri et al. 2002; Lu et al. 2002), sensory neurons (Belmadani et al. 2005), and cortical interneurons (Stumm et al. 2003) during development. In agreement with the knowledge that profound recapitulation of developmental processes takes place in the adult CNS during reparative regeneration (Martino 2004), the developmental chemokines SDF-1 α and monocyte chemoattractant protein (MCP)-1/CCL2, and their respective receptors CXCR4 and CCR2, have modulated the migration of SVZ-resident endogenous NPCs following experimental cerebral ischemia (Imitola et al. 2004; Yan et al. 2007). Also, tumour necrosis factor (TNF) α (Ben-Hur et al. 2003a), hepatocyte growth factor (HGF), transforming growth factor (TGF) β, and Rantes/CCL5 have also been implicated in NPC and oligodendrocyte progenitor cell self-renewal and migration, respectively, both in vivo and ex vivo after NPC grafting into hippocampal slice cultures (Belmadani et al. 2006; Lalive et al. 2005; Nicoleau et al. 2009; Pluchino et al. 2005). Recent transplant data has shown that transplanted oligodendrocyte progenitors migrate preferentially towards focal spinal cord lesions in which TGFβ expression is increased (Wang et al. 2011).

The i.c. and i.t. cell delivery routes may therefore allow therapeutic cells to accumulate as close as needed to the multiple damaged/inflamed areas in EAE/MS, without (or with very little) separating anatomical barriers. Moreover, recent studies suggest the transition of immune pathogenesis in MS from a peripheral initiated inflammatory injury during early phase of disease to a CNS compartmentalized immune and degenerative process during the chronic phase of disease that is localized mainly in subarachnoid space/pial surface (Howell et al. 2011). Thus, i.c./i.t. delivery may target the disease process directly.

An alternative route of cell delivery that has been recently suggested is the direct injection of NPCs into the blood stream (Pluchino et al. 2003, 2005; Pluchino et al. 2005). The exact molecular mechanism sustaining the significant homing capacity of i.v.-injected NPCs has been detailed in rodents with EAE as dependent on the expression of functional cell adhesion molecules (e.g. CD44), integrins (e.g. $\alpha 4$, $\beta 1$) (Campos et al. 2004, 2006; Leone et al. 2005; Pluchino et al. 2003, 2005), and chemokine receptors (e.g. CCR1, CCR2, CCR5, CXCR3, CXCR4) (Imitola et al. 2004; Ji et al. 2004; Pluchino et al. 2005) on NPC surface.

In addition to the ease of i.v. delivery, the main advantage of this latter protocol of NPC delivery is the mechanism of cell accumulation, initially described to target specifically the perivascular inflammatory foci in the CNS (Ben-Hur et al. 2003b; Pluchino et al. 2003, 2005). However, more recent evidence suggests that i.v.-injected NPCs may target and synergize with immune cells also outside the CNS, at the level of secondary lymphoid organs, such as the lymph nodes (Einstein et al. 2007) or the spleen (Lee et al. 2008), though this latter observation significantly supports the added therapeutic value transplantable NPCs capable of engaging multiple mechanisms of action within different CNS vs. non-CNS microenvironments in vivo (Einstein and Ben-Hur 2008; Martino and Pluchino 2006). It would also open new (still unsolved) questions about the overall safety and systemic effects of cell delivery routes allowing the dispersion of injected NPCs outside the CNS.

7.2.2 Tracking Transplanted Cells

To develop successful clinical cell-based therapies, it would be important to assess non-invasive monitoring of both cell fate and distribution. Nuclear medicine techniques, fluorescence imaging, and bioluminescence have been shown to provide valuable means for monitoring cell therapies in vivo. Positron emission tomography (PET) has been used in humans to evaluate viability and function of locally transplanted cells in the neurodegenerative disorders Parkinson's and Huntington's diseases (Brooks 2005; Mendez et al. 2005). Fluorescence imaging and bioluminescence have been successfully used to monitor focal stem cell transplantation in mice with cerebral infarcts (Kim et al. 2004). Unfortunately, all these techniques have shown poor spatial resolution, cannot be correlated to anatomical details, and have limited depth penetration (Jacobs and Cherry 2001). In contrast to these latter techniques, magnetic resonance imaging (MRI) is a more accurate means for cellular imaging (de Vries et al. 2005), allowing efficacious (cell) tracking over longer periods of time.

The recent development of super paramagnetic iron oxides (SPIOs) has further improved the sensitivity of cellular and molecular MRI (Modo et al. 2005). Magnetically labelled oligodendrocyte progenitors (Bulte et al. 1999; Franklin et al. 1999), neural progenitors (Ben-Hur et al. 2007; Bulte et al. 2001), and Schwann cells (Dunning et al. 2004) have been visualized with high magnetic field-strength, animal-dedicated MR scanners, when focally transplanted into preclinical rodent models of genetic, focal chemical, or inflammatory de/dysmyelination.

SPIO-labelled, bromodeoxyuridine (BrdU)-loaded NPCs have been transplanted i.c. into acute EAE in Lewis rats. Ex vivo MRI confirmed that while the transplant disseminated in the ventricular system of both naïve and EAE brains, widespread migration into white matter tracts occurred in EAE rats only. A good correlation was also found between the histological distribution of iron-labelled NPCs and the BrdU immunostaining, thus indicating that the magnetic label was retained by transplanted NPCs (and not transferred to neighbouring cells) in vivo (Bulte et al. 2003). Furthermore, proof of principle of the sensitive detection and quantification (e.g. down to 30 cells/region of interest) of i.v.-injected SPIO-labelled NPCs accumulating in the brain of mice with chronic EAE has also been provided by means of in vivo clinical-grade 3T MR imaging and ex vivo relaxometry (Politi et al. 2007). Also, serial MRI-based tracking of mouse adult and human ES cell-derived neural spheres showed specific (and early along the course of disease) migration of SPIO-labelled cells along white matter tracts after i.c. cell injection in chronic EAE mice (Ben-Hur et al. 2007), thus further suggesting that human ES-derived NPCs respond to inflammatory tissue signals similar to rodent cells, a prerequisite for their consideration as clinical vectors.

A recent report highlights that SPIO-labelled human adult mesenchymal stem cells (MSCs), implanted subcutaneously (s.c.) into nude mice, display preferential chondrogenic differentiation in vivo, as compared to implanted unlabeled MSCs (Farrell et al. 2008). On the other hand, SPIO labelling does not seem to change the migratory properties and lineage fate of NPCs both in vitro and in vivo (Cohen et al. 2010). Therefore, it would be necessary to evaluate whether contrast agent incorporation exerts any significant negative effects on the candidate cell population prior to use their in cell therapies.

More broadly though, these data suggest that the real-time MR monitoring of delivered cell therapeutics may become an important tool in evaluating the efficacy of transplant-based remyelination both experimentally, as well as in individual patients.

7.3 The Therapeutic Plasticity of NPCs

7.3.1 Cell Replacement

Although NPCs have been successfully transplanted in several preclinical models of neurological disorders, the remarkable functional recovery obtained upon NPC transplantation has generally showed little correlation with the absolute numbers of terminally differentiated neural cells originated by transplanted NPCs (see for review Martino and Pluchino 2006). As such, transplanted NPCs very scarcely differentiate into tyrosine hydroxylase (TH)-immunoreactive neurons, when transplanted in rodents with experimental Parkinson's disease (PD) (Ben-Hur et al. 2004; Ourednik et al. 2002; Rafuse et al. 2005; Richardson et al. 2005) or Huntington's disease (HD) (McBride et al. 2004; Ryu et al. 2004), despite significant behavioural improvement. Similarly, mice with spinal cord injury (SCI) (Fujiwara et al. 2004; Hofstetter et al. 2005), acute stroke (Chu et al. 2004), or intracerebral haemorrhage (Jeong et al. 2003; Lee et al. 2008) show clinical recovery regardless of transplanted NPCs acquiring preferential astroglial fate in vivo. Other evidence shows that NPCs injected into models of SCI do not differentiate at all into terminally differentiated neuronal cells (Cummings et al. 2005; Fujiwara et al. 2004; Lu et al. 2003; Teng et al. 2002). In EAE, very low differentiation of NPCs into myelin-forming oligodendrocytes is accompanied by neurophysiological evidence of axonal protection and remyelination (Ben-Hur et al. 2003b; Pluchino et al. 2003). In the very same context, much way over 20 % of transplanted cells reaching inflammatory demyelinated areas do not express differentiation markers (Pluchino et al. 2003). It is yet to be shown that transplanted cells can remyelinate efficiently in clinically relevant models of MS. This may depend both on grafting precursor cells that are already programmed to the oligodendroglial lineage (Ben-Hur et al. 1998; Hu et al. 2009) and the use of appropriate models of disease that allow remyelination to occur.

This scarce terminal differentiation and propensity to maintain an undifferentiated phenotype within the host tissue has suggested that transplanted NPCs might be therapeutic efficacious via a number of bystander mechanism(s) alternative to cell replacement.

7.3.2 (Tissue) Trophic Support

The poor repair capacity of the adult CNS results from a number of failing repair programmes, including the apparent inability of endogenous progenitors and stem cells to properly respond to disease states, to replace damaged cells and from lack of regenerative capacity of injured axons (Franklin 2002; Franklin and Ffrench-Constant 2008; Martino 2004; Pluchino et al. 2008).

NPC therapy is emerging as a mode of treatment that can enhance the host brain's ability to repair itself in both aspects (Einstein and Ben-Hur 2008; Pluchino and Martino 2008). Recent studies have focused also on several bystander tissue-protective and -trophic properties of transplanted NPCs. NPCs seeded on a synthetic biodegradable scaffold and grafted into the cord of hemi-sectioned rats induced significant clinical recovery while reducing the necrosis of the surrounding parenchyma and preventing extensive secondary cell loss, inflammation, and glial scar formation (Teng et al. 2002). Moreover, the NPC graft induced a permissive environment for axonal regeneration (Teng et al. 2002). Substantial endogenous reconstitution of the brain structural connectivity has been found following injection of NPCs in biodegradable scaffolds into regions of extensive brain degeneration caused by hypoxia (Park et al. 2002a, b) or following i.c. transplantation of NPCs after ischemia/reperfusion injury in mice (Capone et al. 2007). Transplanted NPCs have rescued endogenous dopaminergic neurons of the mesostriatal system in a PD model in rodents (Ourednik et al. 2002), while prevention of death of motor neurons has been observed when NPCs were transplanted in models of amyotrophic lateral sclerosis (ALS) (Ferrer-Alcon et al. 2007; Kerr et al. 2003; Suzuki et al. 2007). NPC-driven bystander tissue protection has generally led to a reduction of glial scar formation and increase of survival and/or functions of endogenous glial and neuronal progenitors surviving to the pathological insult.

NPC trophic effects relate in part to increased in vivo bioavailability of major neurotrophins (e.g. nerve growth factor [NGF], brain-derived neurotrophic factor [BDNF], ciliary neurotrophic factor [CNTF], glial-derived neurotrophic factor [GDNF]) (Chu et al. 2004; Einstein et al. 2006; Lu et al. 2003; Pluchino et al. 2003; Teng et al. 2002 and others). The multiple roles of neurotrophins as mediators in cell cycle regulation, cell survival, and differentiation during development and adulthood make them potential candidates for the regulation of endogenous NPC proliferation and differentiation following brain injury and for modulation of host environment into more permissive for regeneration. For example, neurotrophins secreted by transplanted NPCs help promote corticospinal axon growth after transplantation onto an organotypic co-culture system, containing dissected brain cortex and spinal cord (Kamei et al. 2007). Moreover, several neurotrophins that may be released by NPCs were shown to inhibit EAE. Insulin-like growth factor (IGF)-1 and glial growth factor (GGF)-2 are neurotrophic factors that promote survival and proliferation in the oligodendrocyte lineage (Barres et al. 1992; Canoll et al. 1996, 1999; Mason et al. 2000). Treatment with these factors was beneficial clinically and pathologically in animals with EAE (Akassoglou et al. 1998; Cannella et al. 1998; Yao et al. 1996). Interestingly, their effect was mediated not only by enhancing oligodendrocyte survival (Butzkueven et al. 2002; Linker et al. 2002) but also by decreasing neuroinflammation (Cannella et al. 1998; Flugel et al. 2001; Ruffini et al. 2001; Villoslada et al. 2000).

The broad (tissue) trophic support by transplanted stem cells has been studied extensively in models of spinal cord injury and attributed to multiple mechanisms. These include production of neurotrophic growth factors that enhance axonal regeneration (Lu et al. 2003); induction of matrix metalloproteinases that degrade the extracellular matrix and cell surface molecules that impede axonal regeneration, thus enabling axons to extend through the glial scar (Zhang et al. 2007); induction of angiogenesis in the lesioned tissue, which provides trophic support and enables tissue repair (Rauch et al. 2009); providing proper realignment and guidance to enable axonal regeneration along long fibre tracts (Pfeifer et al. 2004); and increasing remyelination in the lesion by both endogenous and graft-derived myelin-forming cells to enhance action potential conduction and limit secondary axonal degeneration (Keirstead et al. 2005). In addition, NPC transplantation has been shown to facilitate endogenous myelin repair following chronic cuprizone-induced demyelination, by induction of resident OPC proliferation and differentiation (Einstein et al. 2009).

Recent work has also indicated that transplanted stem cells, including NPCs, can enhance endogenous neurogenesis in certain physiological and pathological conditions (Hattiangady et al. 2007; Munoz et al. 2005). Mice exposed prenatally to opioids display impaired learning associated with reduced neurogenesis, and transplantation of NPCs improves learning functions, as well as host brain-derived neurogenesis in the DG of the hippocampus (Ben-Shaanan et al. 2008). A similar neurotrophic effect was also reported in physiological ageing. While neurogenesis in the DG declines severely by middle age, transplantation of NPCs stimulates the endogenous NPCs in the SGZ to produce new dentate granule cells (Hattiangady et al. 2007).

Thus, transplanted NPCs may enhance the adult CNS capacity to repair itself by restoring the ability of endogenous progenitors and stem cells to both respond properly to disease state and replace damaged CNS cells and the ability of severed axons to regenerate. Yet, a current and comprehensive illustration of whether these and other (tissue)trophic properties of NPCs are also relevant to EAE and MS is partially lacking.

7.3.3 Immune Modulation

Important progress towards the application of stem cell therapies in MS has been made recently with the observation that NPC transplantation attenuates the clinical course of acute (Einstein et al. 2003), chronic (Einstein et al. 2006; Pluchino et al. 2003) and relapsing (Pluchino et al. 2005; Pluchino et al. 2009) EAE in rodents and primates (Pluchino et al. 2009a). While dealing with the expected regenerative potential of stem cell transplantation as direct cell-replacement therapy, recent work has highlighted peculiar additional immune modulatory mechanisms by which transplanted NPCs exhibit therapeutic effects in preclinical models of MS. The first indication of novel (anti-inflammatory?) effect of NPCs was obtained when neurospheres were transplanted i.c. in acute spinal cord homogenate (SCH)-induced EAE Lewis rats (Einstein et al. 2003). These EAE rats show acute, reversible paralytic disease that is the result of disseminated CNS inflammation without demyelination or axonal injury. NPC transplantation in SCH-induced EAE Lewis rats attenuated the inflammatory brain process and clinical severity of disease (Einstein et al. 2003).

Subsequent studies examined the effect of NPC transplantation, upon either i.c. or i.v. cell injection, in the myelin oligodendrocyte glycoprotein (MOG)35-55-induced EAE in C57BL/6 mice. In this model, there is an acute paralytic disease due to a T cell-mediated autoimmune process that causes severe axonal injury and demyelination. Subsequently, the mice remain with fixed neurologic sequel, the severity of which is correlated with the extent of axonal loss (Wujek et al. 2002). NPC transplantation in MOG35-55-induced EAE mice attenuated the inflammatory process, rescued the endogenous pool of oligodendrocyte progenitor cells, reduced acute and chronic axonal injury and demyelination, and improved the overall clinical and neurophysiological performance of the mice (Einstein et al. 2006; Pluchino et al. 2003). Recent evidence has now suggested that this latter phenomenon may be dependent on the capacity of transplanted NPCs to engage multiple mechanisms of action within specific microenvironments in vivo (Martino and Pluchino 2006). Among a wide range of potential therapeutic actions, and in addition to the cell replacement and (tissue) trophic capacities (Pluchino et al. 2003), remarkable immune modulatory capacities are described for transplanted NPCs within specific CNS (Einstein et al. 2003, 2006; Pluchino et al. 2003, 2005; Swanborg 2001) and non-CNS areas (Einstein et al. 2007). As such, considerable proof that NPC-mediated bystander immune regulation may take place both in the CNS, at the level of the atypical perivascular niches (Pluchino et al. 2005), as well as in secondary lymphoid organs, such as the lymph nodes (Einstein et al. 2007) or the spleen (Lee et al. 2008), has recently been provided. However, the exact mechanisms by which transplanted NPCs attenuate CNS vs. peripheral inflammation are not yet clear.

One school of thought has first suggested an immune suppressive effect, by which NPCs induce apoptosis of Th1—but not Th2—cells selectively, via the inflammation-driven up-regulation of membrane expression of functional death receptor ligands (e.g. FasL, TRAIL, Apo3L) on NPCs (Pluchino et al. 2005). Alternatively, it has been suggested that NPCs inhibit T cell activation and proliferation by a non-specific, bystander immune suppressive action (Einstein et al. 2007). This notion emerged from co-culture experiments that showed a striking inhibition of the activation and proliferation of EAE-derived, as well as naive T cells by NPCs, following stimulation by various stimuli (Einstein et al. 2003, 2007). The suppressive effect of NPCs on T cells was accompanied by a significant suppression of pro-inflammatory cytokines, such as interleukin (IL)-2, TNFα, and IFNγ (Einstein et al. 2007). Moreover, NPCs inhibited multiple inflammatory signals, as exemplified by attenuation of T cell receptor-IL2- and IL6-mediated immune cell activation and/or proliferation (Fainstein et al. 2008). The relevance of such NPC/T cell interaction was first suggested when NPCs i.v. injected prior to EAE disease onset (e.g. at 8 days after the immunization) were transiently found in peripheral lymphoid organs, where they interacted with T cells to reduce their encephalitogenicity (Einstein et al. 2007). In this protocol i.v. NPC injection at an early time point, transplanted cells did not cross the blood–brain barrier, and their entire effect was mediated by peripheral immune suppression, resulting in reduced immune cell infiltration into the CNS and consequently milder CNS damage.

Recent studies have started addressing the role of individual molecular candidates in regulating this novel immune modulatory (or regulatory) capacity of transplanted NPCs in EAE. As such, when injected s.c. in a setup of passive cell vaccination in EAE mice, NPCs have shown remarkable capacity to target and synergize with immune cells in secondary lymphoid organs (but not in the CNS), where they stably (e.g. for more than 2 months after cell injection) modify the perivascular microenvironment. Within this context, surviving NPCs have hindered the activation of myeloid dendritic cells (DC) via a bone morphogenetic protein (BMP)-4-dependent mechanism, which was completely reverted by the BMP antagonist Noggin (Pluchino et al. 2009b; Pluchino et al. 2009).

Indeed, such a broad immune regulatory capacity has more recently been shown also for human somatic (Pluchino et al. 2009a) as well as human ES-cell-derived NPCs (Aharonowiz et al. 2008) after transplantation in non-human primates and mice with EAE, respectively. Again, the therapeutic effects of both these latter NPC sources was not related to graft or host-driven remyelination but was rather mediated by an immune regulatory mechanism that protected the CNS from immune-mediated injury.

Taken together, the results of experimental studies that we describe here have started to reveal the mechanisms by which neural stem cells act as immune regulators and in so doing create environments within the CNS that favour protection and repair. Such mechanistic insights are essential if the full therapeutic potential of CNS stem cells is to be achieved.

7.4 Conclusions

Since the first transplant of stem cells into the spinal cord of rodents in which an acute demyelinating lesion was induced (Liu et al. 2000), we have witnessed a spur of experimental cell-based transplantation approaches aimed at fostering biological and molecular mechanisms underlying CNS repair. Theories assuming that no (or very little) renewing potential is identified within the adult CNS have been contravened, new promising sources of myelinogenic cells for transplantation purposes (i.e. olfactory bulb ensheathing cells, adult and embryonic stem cells) have been characterized, and new cell-replacement strategies have been proposed and established. A better understanding of the dynamics of endogenous remyelination has been achieved, and insights concerning the process of remyelination driven by site-specific myelin-forming cell transplantation have been discovered. This has led to the first clinical trial—performed in MS patients—based on autologous Schwann cell transplantation into brain areas of autoimmune demyelination. This Phase I trial was carried out in three patients only, and biopsies of the site of transplantation apparently failed to identify transplanted Schwann. There were no repeats of adverse events associated with transplantation, yet the lack of publication of the results of the trial leaves unanswered questions. Other limitations of cell transplantation, which have not yet been overcome, include (a) the limited number of highly myelinating cells that can be grown in vitro and (b) the limited migratory capacity of myelinating cells once transplanted. Somatic stem cells might represent therefore an alternative and promising area of investigation with some potential in its essence.

As a matter of fact, new hopes have been raised by the encouraging preliminary results obtained by transplanting CNS-derived NPCs and bone marrow mesenchymal/stromal stem cells (BMSCs) in rodent models of MS (Aharonowiz et al. 2008; Ben-Hur et al. 2003b; Einstein et al. 2003, 2006, 2007; Gerdoni et al. 2007; Pluchino et al. 2003, 2005; Zappia et al. 2005). However, most of the results obtained so far using stem cells as therapeutic weapons for MS consistently challenge the sole and limited view that stem cells therapeutically work exclusively throughout cell

Table 7.1 Comparison between three candidate cell populations for transplantation in MS

Cell type	Bone marrow stromal cells	Adult neural stem cells	Embryonic stem cells
Cell source	Autologous	Allogeneic	Allogeneic[a]
Current availability for clinical use	Available immediately	Available banks, limited source	Under development
Safety	Safe[b]	Probably safe[c]	Under investigation[d]
Functional properties			
Integration in host CNS	No	Yes	Yes
Immunomodulation	Yes	Yes	Yes
Trophic effects	Yes	Yes	Yes
Remyelination	Probably no	Questionable	Yes

[a]Potentially autologous by developing induced pluripotent stem cells technology
[b]Possible risk of producing ectopic mesenchymal tissue
[c]Possible risk of neuroepithelial tumours
[d]Risk of teratomas

replacement. Indeed, the transplantation of the above different stem cell sources has promoted CNS repair via a number of previously unexpected bystander mechanisms, mainly exerted by undifferentiated stem cells releasing, at the different sites of inflammation and/or tissue damage, a milieu of (tissue) trophic and immune modulatory molecules whose release is temporally and spatially orchestrated by specific (micro)environmental cues (Table 7.1).

These molecules, acting in a paracrine/bystander fashion, are pleiotropic and redundant in nature and "constitutively" secreted by stem cells. Therefore, the (tissue) trophic and immune regulatory capacities we have been focussing within this chapter can easily be envisaged as highly peculiar stem cell signatures. These new stem cell properties may indeed help in explaining some of the data showing that different sources of somatic stem cells, other than CNS stem cells, even with low capabilities of neural (trans)differentiation efficiently promote CNS repair (reviewed in Martino and Pluchino 2006; Uccelli et al. 2008). In this view, the stem cell therapeutic plasticity can be viewed as the capacity of stem cells to adapt their fate and function(s) to specific environmental needs occurring as a result of different pathological conditions (Fig. 7.2). Initial clinical experience in a Phase I trial of autologous bone marrow-derived mesenchymal stem cell transplantation by both the intravenous and intrathecal routes is encouraging. First, the only adverse reaction observed was of transient meningeal irritation in some patients, and there were no long-term complications. Second, some of these chronic progressive patients that had failed all standard therapies showed clinical improvement following cell therapy. These results warrant further trials with power to detect therapeutic efficacy (Karussis et al. 2010).

While further studies are certainly required to assess the overall safety, efficacy, and in vivo therapeutic plasticity of NPCs, the great challenge for any future human application of NPC-based protocols in MS will be to develop more reliable and

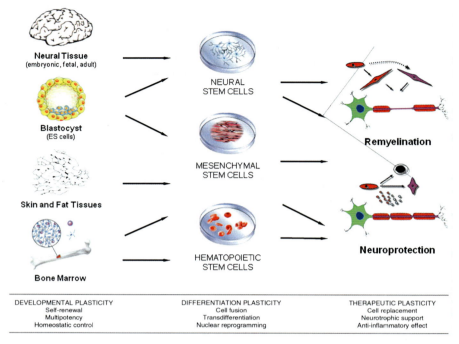

Fig. 7.2 Stem cells are plastic entities in their essence. Besides developmental and differentiation plasticity, the concept of "therapeutic plasticity" is now emerging. It can be viewed as the capacity of stem cells to adapt their fate and function(s) to specific environmental needs occurring as a result of different pathological conditions. Most of the results obtained so far using stem cells as therapeutic weapons for MS consistently challenge the sole and limited view that stem cells therapeutically work exclusively throughout cell replacement. Indeed, the transplantation of the above different stem cell sources has promoted CNS repair via a number of previously unexpected bystander mechanisms, mainly exerted by undifferentiated stem cells releasing, at the different sites of inflammation and/or tissue damage, a milieu of (tissue) trophic and immune modulatory molecules whose release is temporally and spatially orchestrated by specific (micro)environmental cues

reproducible approaches optimizing both (tissue) trophic and immune regulatory capacities of stem cells for functional and anatomical rescuing of myelin architecture in MS patients.

References

Aharonowiz M, Einstein O, Fainstein N, Lassmann H, Reubinoff B, Ben-Hur T (2008) Neuroprotective effect of transplanted human embryonic stem cell-derived neural precursors in an animal model of multiple sclerosis. PLoS One 3:e3145

Akassoglou K, Bauer J, Kassiotis G, Pasparakis M, Lassmann H, Kollias G, Probert L (1998) Oligodendrocyte apoptosis and primary demyelination induced by local TNF/p55TNF receptor signaling in the central nervous system of transgenic mice: models for multiple sclerosis with primary oligodendrogliopathy. Am J Pathol 153:801–813

Alvarez-Buylla A, Garcia-Verdugo JM (2002) Neurogenesis in adult subventricular zone. J Neurosci 22:629–634

Avellana-Adalid V, Bachelin C, Lachapelle F, Escriou C, Ratzkin B, Baron-Van Evercooren A (1998) In vitro and in vivo behaviour of NDF-expanded monkey Schwann cells. Eur J Neurosci 10:291–300

Bagri A, Gurney T, He X, Zou YR, Littman DR, Tessier-Lavigne M, Pleasure SJ (2002) The chemokine SDF1 regulates migration of dentate granule cells. Development 129:4249–4260

Barkhof F, Bruck W, De Groot CJ, Bergers E, Hulshof S, Geurts J, Polman CH, van der Valk P (2003) Remyelinated lesions in multiple sclerosis: magnetic resonance image appearance. Arch Neurol 60:1073–1081

Barres BA, Hart IK, Coles HS, Burne JF, Voyvodic JT, Richardson WD, Raff MC (1992) Cell death and control of cell survival in the oligodendrocyte lineage. Cell 70:31–46

Belmadani A, Tran PB, Ren D, Assimacopoulos S, Grove EA, Miller RJ (2005) The chemokine stromal cell-derived factor-1 regulates the migration of sensory neuron progenitors. J Neurosci 25:3995–4003

Belmadani A, Tran PB, Ren D, Miller RJ (2006) Chemokines regulate the migration of neural progenitors to sites of neuroinflammation. J Neurosci 26:3182–3191

Ben-Hur T, Goldman SA (2008) Prospects of cell therapy for disorders of myelin. Ann N Y Acad Sci 1142:218–249

Ben-Hur T, Rogister B, Murray K, Rougon G, Dubois-Dalcq M (1998) Growth and fate of PSA-NCAM+ precursors of the postnatal brain. J Neurosci 18:5777–5788

Ben-Hur T, Ben-Menachem O, Furer V, Einstein O, Mizrachi-Kol R, Grigoriadis N (2003a) Effects of proinflammatory cytokines on the growth, fate, and motility of multipotential neural precursor cells. Mol Cell Neurosci 24:623–631

Ben-Hur T, Einstein O, Mizrachi-Kol R, Ben-Menachem O, Reinhartz E, Karussis D, Abramsky O (2003b) Transplanted multipotential neural precursor cells migrate into the inflamed white matter in response to experimental autoimmune encephalomyelitis. Glia 41:73–80

Ben-Hur T, Idelson M, Khaner H, Pera M, Reinhartz E, Itzik A, Reubinoff BE (2004) Transplantation of human embryonic stem cell-derived neural progenitors improves behavioral deficit in Parkinsonian rats. Stem Cells 22

Ben-Hur T, van Heeswijk RB, Einstein O, Aharonowiz M, Xue R, Frost EE, Mori S, Reubinoff BE, Bulte JW (2007) Serial in vivo MR tracking of magnetically labeled neural spheres transplanted in chronic EAE mice. Magn Reson Med 57:164–171

Ben-Shaanan TL, Ben-Hur T, Yanai J (2008) Transplantation of neural progenitors enhances production of endogenous cells in the impaired brain. Mol Psychiatry 13:222–231

Bjartmar C, Yin X, Trapp BD (1999) Axonal pathology in myelin disorders. J Neurocytol 28:383–395

Bjartmar C, Kidd G, Mork S, Rudick R, Trapp BD (2000) Neurological disability correlates with spinal cord axonal loss and reduced N-acetyl aspartate in chronic multiple sclerosis patients. Ann Neurol 48:893–901

Blakemore WF (1977) Remyelination of CNS axons by Schwann cells transplanted from the sciatic nerve. Nature 266:68–69

Blakemore WF, Crang AJ (1988) Extensive oligodendrocyte remyelination following injection of cultured central nervous system cells into demyelinating lesions in adult central nervous system. Dev Neurosci 10:1–11

Blakemore WF, Gilson JM, Crang AJ (2000) Transplanted glial cells migrate over a greater distance and remyelinate demyelinated lesions more rapidly than endogenous remyelinating cells. J Neurosci Res 61:288–294

Blakemore WF, Chari DM, Gilson JM, Crang AJ (2002) Modelling large areas of demyelination in the rat reveals the potential and possible limitations of transplanted glial cells for remyelination in the CNS. Glia 38:155–168

Bonfanti L (2006) PSA-NCAM in mammalian structural plasticity and neurogenesis. Prog Neurobiol 80:129–164

Bonfanti L, Ponti G (2008) Adult mammalian neurogenesis and the New Zealand white rabbit. Vet J 175:310–331

Brok HP, Bauer J, Jonker M, Blezer E, Amor S, Bontrop RE, Laman JD, t Hart BA (2001) Non-human primate models of multiple sclerosis. Immunol Rev 183:173–185

Brooks DJ (2005) Positron emission tomography and single-photon emission computed tomography in central nervous system drug development. NeuroRx 2:226–236

Bruck W (2005) The pathology of multiple sclerosis is the result of focal inflammatory demyelination with axonal damage. J Neurol 252(Suppl 5):v3–v9

Brustle O, Maskos U, McKay RD (1995) Host-guided migration allows targeted introduction of neurons into the embryonic brain. Neuron 15:1275–1285

Brustle O, Choudhary K, Karram K, Huttner A, Murray K, Dubois-Dalcq M, McKay RD (1998) Chimeric brains generated by intraventricular transplantation of fetal human brain cells into embryonic rats. Nat Biotechnol 16:1040–1044

Brustle O, Jones KN, Learish RD, Karram K, Choudhary K, Wiestler OD, Duncan ID, McKay RD (1999) Embryonic stem cell-derived glial precursors: a source of myelinating transplants. Science 285:754–756

Buchet D, Garcia C, Deboux C, Nait-Oumesmar B, Baron-Van Evercooren A (2011) Human neural progenitors from different foetal forebrain regions remyelinate the adult mouse spinal cord. Brain 134:1168–1183

Bulte JW, Zhang S, van Gelderen P, Herynek V, Jordan EK, Duncan ID, Frank JA (1999) Neurotransplantation of magnetically labeled oligodendrocyte progenitors: magnetic resonance tracking of cell migration and myelination. Proc Natl Acad Sci USA 96(26): 15256–15261

Bulte JW, Douglas T, Witwer B, Zhang SC, Strable E, Lewis BK, Zywicke H, Miller B, van Gelderen P, Moskowitz BM, Duncan ID, Frank JA (2001) Magnetodendrimers allow endosomal magnetic labeling and in vivo tracking of stem cells. Nat Biotechnol 19:1141–1147

Bulte JW, Ben-Hur T, Miller BR, Mizrachi-Kol R, Einstein O, Reinhartz E, Zywicke HA, Douglas T, Frank JA (2003) MR microscopy of magnetically labeled neurospheres transplanted into the Lewis EAE rat brain. Magn Reson Med 50:201–205

Butzkueven H, Zhang JG, Soilu-Hanninen M, Hochrein H, Chionh F, Shipham KA, Emery B, Turnley AM, Petratos S, Ernst M, Bartlett PF, Kilpatrick TJ (2002) LIF receptor signaling limits immune-mediated demyelination by enhancing oligodendrocyte survival. Nat Med 8:613–619

Campos LS, Leone DP, Relvas JB, Brakebusch C, Fassler R, Suter U, Ffrench-Constant C (2004) Beta1 integrins activate a MAPK signalling pathway in neural stem cells that contributes to their maintenance. Development 131:3433–3444

Campos LS, Decker L, Taylor V, Skarnes W (2006) Notch, epidermal growth factor receptor, and beta1-integrin pathways are coordinated in neural stem cells. J Biol Chem 281:5300–5309

Cannella B, Hoban CJ, Gao YL, Garcia-Arenas R, Lawson D, Marchionni M, Gwynne D, Raine CS (1998) The neuregulin, glial growth factor 2, diminishes autoimmune demyelination and enhances remyelination in a chronic relapsing model for multiple sclerosis. Proc Natl Acad Sci USA 95:10100–10105

Canoll PD, Musacchio JM, Hardy R, Reynolds R, Marchionni MA, Salzer JL (1996) GGF/neuregulin is a neuronal signal that promotes the proliferation and survival and inhibits the differentiation of oligodendrocyte progenitors. Neuron 17:229–243

Canoll PD, Kraemer R, Teng KK, Marchionni MA, Salzer JL (1999) GGF/neuregulin induces a phenotypic reversion of oligodendrocytes. Mol Cell Neurosci 13:79–94

Capone C, Frigerio S, Fumagalli S, Gelati M, Principato MC, Storini C, Montinaro M, Kraftsik R, De Curtis M, Parati E, De Simoni MG (2007) Neurosphere-derived cells exert a neuroprotective action by changing the ischemic microenvironment. PLoS One 2:e373

Chang A, Tourtellotte WW, Rudick R, Trapp BD (2002) Premyelinating oligodendrocytes in chronic lesions of multiple sclerosis. N Engl J Med 346:165–173

Chen JT, Kuhlmann T, Jansen GH, Collins DL, Atkins HL, Freedman MS, O'Connor PW, Arnold DL (2007) Voxel-based analysis of the evolution of magnetization transfer ratio to quantify remyelination and demyelination with histopathological validation in a multiple sclerosis lesion. Neuroimage 36:1152–1158

Chu K, Kim M, Park KI, Jeong SW, Park HK, Jung KH, Lee ST, Kang L, Lee K, Park DK, Kim SU, Roh JK (2004) Human neural stem cells improve sensorimotor deficits in the adult rat brain with experimental focal ischemia. Brain Res 1016:145–153

Cohen ME, Muja N, Fainstein N, Bulte JW, Ben-Hur T (2010) Conserved fate and function of ferumoxides-labeled neural precursor cells in vitro and in vivo. J Neurosci Res 88:936–944

Compston A (1996) Remyelination of the central nervous system. Mult Scler 1:388–392

Compston A (1997) Remyelination in multiple sclerosis: a challenge for therapy. The 1996 European Charcot Foundation Lecture. Mult Scler 3:51–70

Compston A, Coles A (2002) Multiple sclerosis. Lancet 359:1221–1231

Crang AJ, Blakemore WF (1991) Remyelination of demyelinated rat axons by transplanted mouse oligodendrocytes. Glia 4:305–313

Crang AJ, Gilson J, Blakemore WF (1998) The demonstration by transplantation of the very restricted remyelinating potential of post-mitotic oligodendrocytes. J Neurocytol 27:541–553

Cummings BJ, Uchida N, Tamaki SJ, Salazar DL, Hooshmand M, Summers R, Gage FH, Anderson AJ (2005) Human neural stem cells differentiate and promote locomotor recovery in spinal cord-injured mice. Proc Natl Acad Sci USA 102:14069–14074

De Stefano N, Matthews PM, Fu L, Narayanan S, Stanley J, Francis GS, Antel JP, Arnold DL (1998) Axonal damage correlates with disability in patients with relapsing-remitting multiple sclerosis. Results of a longitudinal magnetic resonance spectroscopy study. Brain 121(Pt 8):1469–1477

de Vries IJ, Lesterhuis WJ, Barentsz JO, Verdijk P, van Krieken JH, Boerman OC, Oyen WJ, Bonenkamp JJ, Boezeman JB, Adema GJ, Bulte JW, Scheenen TW, Punt CJ, Heerschap A, Figdor CG (2005) Magnetic resonance tracking of dendritic cells in melanoma patients for monitoring of cellular therapy. Nat Biotechnol 23:1407–1413

Doetsch F, Alvarez-Buylla A (1996) Network of tangential pathways for neuronal migration in adult mammalian brain. Proc Natl Acad Sci USA 93:14895–14900

Doetsch F, Garcia-Verdugo JM, Alvarez-Buylla A (1997) Cellular composition and three-dimensional organization of the subventricular germinal zone in the adult mammalian brain. J Neurosci 17:5046–5061

Doetsch F, Caille I, Lim DA, Garcia-Verdugo JM, Alvarez-Buylla A (1999) Subventricular zone astrocytes are neural stem cells in the adult mammalian brain. Cell 97:703–716

Duncan ID, Kondo Y, Zhang SC (2011) The myelin mutants as models to study myelin repair in the leukodystrophies. Neurotherapeutics 8:607–624

Dunning MD, Lakatos A, Loizou L, Kettunen M, Ffrench-Constant C, Brindle KM, Franklin RJ (2004) Superparamagnetic iron oxide-labeled Schwann cells and olfactory ensheathing cells can be traced in vivo by magnetic resonance imaging and retain functional properties after transplantation into the CNS. J Neurosci 24:9799–9810

Dyment DA, Ebers GC (2002) An array of sunshine in multiple sclerosis. N Engl J Med 347:1445–1447

Eftekharpour E, Karimi-Abdolrezaee S, Wang J, El Beheiry H, Morshead C, Fehlings MG (2007) Myelination of congenitally dysmyelinated spinal cord axons by adult neural precursor cells results in formation of nodes of Ranvier and improved axonal conduction. J Neurosci 27:3416–3428

Einstein O, Ben-Hur T (2008) The changing face of neural stem cell therapy in neurologic diseases. Arch Neurol 65:452–456

Einstein O, Karussis D, Grigoriadis N, Mizrachi-Kol R, Reinhartz E, Abramsky O, Ben-Hur T (2003) Intraventricular transplantation of neural precursor cell spheres attenuates acute experimental allergic encephalomyelitis. Mol Cell Neurosci 24:1074–1082

Einstein O, Grigoriadis N, Mizrachi-Kol R, Reinhartz E, Polyzoidou E, Lavon I, Milonas I, Karussis D, Abramsky O, Ben-Hur T (2006) Transplanted neural precursor cells reduce brain inflammation to attenuate chronic experimental autoimmune encephalomyelitis. Exp Neurol 198:275–284

Einstein O, Fainstein N, Vaknin I, Mizrachi-Kol R, Reihartz E, Grigoriadis N, Lavon I, Baniyash M, Lassmann H, Ben-Hur T (2007) Neural precursors attenuate autoimmune encephalomyelitis by peripheral immunosuppression. Ann Neurol 61:209–218

Einstein O, Friedman-Levi Y, Grigoriadis N, Ben-Hur T (2009) Transplanted neural precursors enhance host brain-derived myelin regeneration. J Neurosci 29:15694–15702

Fainstein N, Vaknin I, Einstein O, Zisman P, Ben Sasson SZ, Baniyash M, Ben-Hur T (2008) Neural precursor cells inhibit multiple inflammatory signals. Mol Cell Neurosci 39:335–341

Farrell E, Wielopolski P, Pavljasevic P, van Tiel S, Jahr H, Verhaar J, Weinans H, Krestin G, O'Brien FJ, van Osch G, Bernsen M (2008) Effects of iron oxide incorporation for long term cell tracking on MSC differentiation in vitro and in vivo. Biochem Biophys Res Commun 369(4):1076–81

Ferrer-Alcon M, Winkler-Hirt C, Perrin FE, Kato AC (2007) Grafted neural stem cells increase the life span and protect motoneurons in pmn mice. Neuroreport 18:1463–1468

Flax JD, Aurora S, Yang C, Simonin C, Wills AM, Billinghurst LL, Jendoubi M, Sidman RL, Wolfe JH, Kim SU, Snyder EY (1998) Engraftable human neural stem cells respond to developmental cues, replace neurons, and express foreign genes. Nat Biotechnol 16:1033–1039

Flugel A, Berkowicz T, Ritter T, Labeur M, Jenne DE, Li Z, Ellwart JW, Willem M, Lassmann H, Wekerle H (2001) Migratory activity and functional changes of green fluorescent effector cells before and during experimental autoimmune encephalomyelitis. Immunity 14:547–560

Franklin RJ (2002) Why does remyelination fail in multiple sclerosis? Nat Rev 3:705–714

Franklin RJ, Blakemore WF (1997) To what extent is oligodendrocyte progenitor migration a limiting factor in the remyelination of multiple sclerosis lesions? Mult Scler 3:84–87

Franklin RJ, Ffrench-Constant C (2008) Remyelination in the CNS: from biology to therapy. Nat Rev 9:839–855

Franklin RJ, Gilson JM, Franceschini IA, Barnett SC (1996) Schwann cell-like myelination following transplantation of an olfactory bulb-ensheathing cell line into areas of demyelination in the adult CNS. Glia 17:217–224

Franklin RJ, Blaschuk KL, Bearchell MC, Prestoz LL, Setzu A, Brindle KM, ffrench-Constant C (1999) Magnetic resonance imaging of transplanted oligodendrocyte precursors in the rat brain. Neuroreport 10(18):3961–3965

Fujiwara Y, Tanaka N, Ishida O, Fujimoto Y, Murakami T, Kajihara H, Yasunaga Y, Ochi M (2004) Intravenously injected neural progenitor cells of transgenic rats can migrate to the injured spinal cord and differentiate into neurons, astrocytes and oligodendrocytes. Neurosci Lett 366:287–291

Gage FH (2000) Mammalian neural stem cells. Science 287:1433–1438

Gensert JM, Goldman JE (1997) Endogenous progenitors remyelinate demyelinated axons in the adult CNS. Neuron 19:197–203

Gerdoni E, Gallo B, Casazza S, Musio S, Bonanni I, Pedemonte E, Mantegazza R, Frassoni F, Mancardi G, Pedotti R, Uccelli A (2007) Mesenchymal stem cells effectively modulate pathogenic immune response in experimental autoimmune encephalomyelitis. Ann Neurol 61:219–227

Hammang JP, Archer DR, Duncan ID (1997) Myelination following transplantation of EGF-responsive neural stem cells into a myelin-deficient environment. Exp Neurol 147:84–95

Hattiangady B, Shuai B, Cai J, Coksaygan T, Rao MS, Shetty AK (2007) Increased dentate neurogenesis after grafting of glial restricted progenitors or neural stem cells in the aging hippocampus. Stem Cells 25:2104–2117

Hemmer B, Archelos JJ, Hartung HP (2002a) New concepts in the immunopathogenesis of multiple sclerosis. Nat Rev 3:291–301

Hemmer B, Cepok S, Nessler S, Sommer N (2002b) Pathogenesis of multiple sclerosis: an update on immunology. Curr Opin Neurol 15:227–231

Hofstetter CP, Holmstrom NA, Lilja JA, Schweinhardt P, Hao J, Spenger C, Wiesenfeld-Hallin Z, Kurpad SN, Frisen J, Olson L (2005) Allodynia limits the usefulness of intraspinal neural stem cell grafts; directed differentiation improves outcome. Nat Neurosci 8:346–353

Howell OW, Reeves CA, Nicholas R, Carassiti D, Radotra B, Gentleman SM, Serafini B, Aloisi F, Roncaroli F, Magliozzi R, Reynolds R (2011) Meningeal inflammation is widespread and linked to cortical pathology in multiple sclerosis. Brain 134:2755–2771

Hu BY, Du ZW, Li XJ, Ayala M, Zhang SC (2009) Human oligodendrocytes from embryonic stem cells: conserved SHH signaling networks and divergent FGF effects. Development 136:1443–1452

Imaizumi T, Lankford KL, Waxman SG, Greer CA, Kocsis JD (1998) Transplanted olfactory ensheathing cells remyelinate and enhance axonal conduction in the demyelinated dorsal columns of the rat spinal cord. J Neurosci 18:6176–6185

Imaizumi T, Lankford KL, Kocsis JD (2000) Transplantation of olfactory ensheathing cells or Schwann cells restores rapid and secure conduction across the transected spinal cord. Brain Res 854:70–78

Imitola J, Raddassi K, Park KI, Mueller FJ, Nieto M, Teng YD, Frenkel D, Li J, Sidman RL, Walsh CA, Snyder EY, Khoury SJ (2004) Directed migration of neural stem cells to sites of CNS injury by the stromal cell-derived factor 1alpha/CXC chemokine receptor 4 pathway. Proc Natl Acad Sci USA 101:18117–18122

Ito M (2006) Cerebellar circuitry as a neuronal machine. Prog Neurobiol 78:272–303

Jackson SJ, Lee J, Nikodemova M, Fabry Z, Duncan ID (2009) Quantification of myelin and axon pathology during relapsing progressive experimental autoimmune encephalomyelitis in the Biozzi ABH mouse. J Neuropathol Exp Neurol 68:616–625

Jacobs RE, Cherry SR (2001) Complementary emerging techniques: high-resolution PET and MRI. Curr Opin Neurobiol 11:621–629

Jeong SW, Chu K, Jung KH, Kim SU, Kim M, Roh JK (2003) Human neural stem cell transplantation promotes functional recovery in rats with experimental intracerebral hemorrhage. Stroke 34:2258–2263

Ji JF, He BP, Dheen ST, Tay SS (2004) Expression of chemokine receptors CXCR4, CCR2, CCR5 and CX3CR1 in neural progenitor cells isolated from the subventricular zone of the adult rat brain. Neurosci Lett 355:236–240

Kamei N, Tanaka N, Oishi Y, Hamasaki T, Nakanishi K, Sakai N, Ochi M (2007) BDNF, NT-3, and NGF released from transplanted neural progenitor cells promote corticospinal axon growth in organotypic cocultures. Spine 32:1272–1278

Karussis D, Karageorgiou C, Vaknin-Dembinsky A, Gowda-Kurkalli B, Gomori JM, Kassis I, Bulte JW, Petrou P, Ben-Hur T, Abramsky O, Slavin S (2010) Safety and immunological effects of mesenchymal stem cell transplantation in patients with multiple sclerosis and amyotrophic lateral sclerosis. Arch Neurol 67:1187–1194

Keirstead HS, Ben-Hur T, Rogister B, O'Leary MT, Dubois-Dalcq M, Blakemore WF (1999) Polysialylated neural cell adhesion molecule-positive CNS precursors generate both oligodendrocytes and Schwann cells to remyelinate the CNS after transplantation. J Neurosci 19:7529–7536

Keirstead HS, Nistor G, Bernal G, Totoiu M, Cloutier F, Sharp K, Steward O (2005) Human embryonic stem cell-derived oligodendrocyte progenitor cell transplants remyelinate and restore locomotion after spinal cord injury. J Neurosci 25:4694–4705

Kerr DA, Llado J, Shamblott MJ, Maragakis NJ, Irani DN, Crawford TO, Krishnan C, Dike S, Gearhart JD, Rothstein JD (2003) Human embryonic germ cell derivatives facilitate motor recovery of rats with diffuse motor neuron injury. J Neurosci 23:5131–5140

Kim DE, Schellingerhout D, Ishii K, Shah K, Weissleder R (2004) Imaging of stem cell recruitment to ischemic infarcts in a murine model. Stroke 35:952–957

Kohama I, Lankford KL, Preiningerova J, White FA, Vollmer TL, Kocsis JD (2001) Transplantation of cryopreserved adult human Schwann cells enhances axonal conduction in demyelinated spinal cord. J Neurosci 21:944–950

Kornek B, Storch MK, Weissert R, Wallstroem E, Stefferl A, Olsson T, Linington C, Schmidbauer M, Lassmann H (2000) Multiple sclerosis and chronic autoimmune encephalomyelitis: a comparative quantitative study of axonal injury in active, inactive, and remyelinated lesions. Am J Pathol 157:267–276

Kuhn HG, Dickinson-Anson H, Gage FH (1996) Neurogenesis in the dentate gyrus of the adult rat: age-related decrease of neuronal progenitor proliferation. J Neurosci 16:2027–2033

Lalive PH, Paglinawan R, Biollaz G, Kappos EA, Leone DP, Malipiero U, Relvas JB, Moransard M, Suter T, Fontana A (2005) TGF-beta-treated microglia induce oligodendrocyte precursor cell chemotaxis through the HGF-c-Met pathway. Eur J Immunol 35:727–737

Lassmann H (2002) Mechanisms of demyelination and tissue destruction in multiple sclerosis. Clin Neurol Neurosurg 104:168–171

Lassmann H, Bruck W, Lucchinetti C, Rodriguez M (1997) Remyelination in multiple sclerosis. Mult Scler 3:133–136

Lassmann H, Bruck W, Lucchinetti C (2001) Heterogeneity of multiple sclerosis pathogenesis: implications for diagnosis and therapy. Trends Mol Med 7:115–121

Laywell ED, Rakic P, Kukekov VG, Holland EC, Steindler DA (2000) Identification of a multipotent astrocytic stem cell in the immature and adult mouse brain. Proc Natl Acad Sci USA 97:13883–13888

Lee ST, Chu K, Jung KH, Kim SJ, Kim DH, Kang KM, Hong NH, Kim JH, Ban JJ, Park HK, Kim SU, Park CG, Lee SK, Kim M, Roh JK (2008) Anti-inflammatory mechanism of intravascular neural stem cell transplantation in haemorrhagic stroke. Brain 131:616–629

Leone DP, Relvas JB, Campos LS, Hemmi S, Brakebusch C, Fassler R, Ffrench-Constant C, Suter U (2005) Regulation of neural progenitor proliferation and survival by beta1 integrins. J Cell Sci 118:2589–2599

Lindvall O, Kokaia Z (2006) Stem cells for the treatment of neurological disorders. Nature 441:1094–1096

Lindvall O, Kokaia Z, Martinez-Serrano A (2004) Stem cell therapy for human neurodegenerative disorders-how to make it work. Nat Med 10(Suppl):S42–S50

Linker RA, Maurer M, Gaupp S, Martini R, Holtmann B, Giess R, Rieckmann P, Lassmann H, Toyka KV, Sendtner M, Gold R (2002) CNTF is a major protective factor in demyelinating CNS disease: a neurotrophic cytokine as modulator in neuroinflammation. Nat Med 8:620–624

Liu S, Qu Y, Stewart TJ, Howard MJ, Chakrabortty S, Holekamp TF, McDonald JW (2000) Embryonic stem cells differentiate into oligodendrocytes and myelinate in culture and after spinal cord transplantation. Proc Natl Acad Sci USA 97:6126–6131

Lois C, Alvarez-Buylla A (1994) Long-distance neuronal migration in the adult mammalian brain. Science 264:1145–1148

Lois C, Garcia-Verdugo JM, Alvarez-Buylla A (1996) Chain migration of neuronal precursors. Science 271:978–981

Lu M, Grove EA, Miller RJ (2002) Abnormal development of the hippocampal dentate gyrus in mice lacking the CXCR4 chemokine receptor. Proc Natl Acad Sci USA 99:7090–7095

Lu P, Jones LL, Snyder EY, Tuszynski MH (2003) Neural stem cells constitutively secrete neurotrophic factors and promote extensive host axonal growth after spinal cord injury. Exp Neurol 181:115–129

Lucchinetti CF, Bruck W, Rodriguez M, Lassmann H (1996) Distinct patterns of multiple sclerosis pathology indicates heterogeneity on pathogenesis. Brain Pathol 6:259–274

Lucchinetti C, Bruck W, Parisi J, Scheithauer B, Rodriguez M, Lassmann H (2000) Heterogeneity of multiple sclerosis lesions: implications for the pathogenesis of demyelination. Ann Neurol 47:707–717

Marin O, Rubenstein JL (2003) Cell migration in the forebrain. Annu Rev Neurosci 26:441–483

Martino G (2004) How the brain repairs itself: new therapeutic strategies in inflammatory and degenerative CNS disorders. Lancet Neurol 3:372–378

Martino G, Hartung HP (1999) Immunopathogenesis of multiple sclerosis: the role of T cells. Curr Opin Neurol 12:309–321

Martino G, Pluchino S (2006) The therapeutic potential of neural stem cells. Nat Rev 7:395–406

Mason JL, Ye P, Suzuki K, D'Ercole AJ, Matsushima GK (2000) Insulin-like growth factor-1 inhibits mature oligodendrocyte apoptosis during primary demyelination. J Neurosci 20:5703–5708

McBride JL, Behrstock SP, Chen EY, Jakel RJ, Siegel I, Svendsen CN, Kordower JH (2004) Human neural stem cell transplants improve motor function in a rat model of Huntington's disease. J Comp Neurol 475:211–219

Mendez I, Sanchez-Pernaute R, Cooper O, Vinuela A, Ferrari D, Bjorklund L, Dagher A, Isacson O (2005) Cell type analysis of functional fetal dopamine cell suspension transplants in the striatum and substantia nigra of patients with Parkinson's disease. Brain 128:1498–1510

Menezes JR, Luskin MB (1994) Expression of neuron-specific tubulin defines a novel population in the proliferative layers of the developing telencephalon. J Neurosci 14:5399–5416

Mercier F, Kitasako JT, Hatton GI (2002) Anatomy of the brain neurogenic zones revisited: fractones and the fibroblast/macrophage network. J Comp Neurol 451:170–188

Mirzadeh Z, Merkle FT, Soriano-Navarro M, Garcia-Verdugo JM, Alvarez-Buylla A (2008) Neural stem cells confer unique pinwheel architecture to the ventricular surface in neurogenic regions of the adult brain. Cell Stem Cell 3:265–278

Modo M, Hoehn M, Bulte JW (2005) Cellular MR imaging. Mol Imaging 4:143–164

Munoz JR, Stoutenger BR, Robinson AP, Spees JL, Prockop DJ (2005) Human stem/progenitor cells from bone marrow promote neurogenesis of endogenous neural stem cells in the hippocampus of mice. Proc Natl Acad Sci USA 102:18171–18176

Nacher J, Crespo C, McEwen BS (2001) Doublecortin expression in the adult rat telencephalon. Eur J Neurosci 14:629–644

Nicoleau C, Benzakour O, Agasse F, Thiriet N, Petit J, Prestoz L, Roger M, Jaber M, Coronas V (2009) Endogenous hepatocyte growth factor is a niche signal for subventricular zone neural stem cell amplification and self-renewal. Stem Cells 27:408–419

Noctor SC, Martinez-Cerdeno V, Kriegstein AR (2007) Neural stem and progenitor cells in cortical development. Novartis Found Symp 288:59–73, discussion 73–78, 96–98

Noseworthy JH, Lucchinetti C, Rodriguez M, Weinshenker BG (2000) Multiple sclerosis. N Engl J Med 343:938–952

O'Leary MT, Blakemore WF (1997) Oligodendrocyte precursors survive poorly and do not migrate following transplantation into the normal adult central nervous system. J Neurosci Res 48:159–167

Ourednik J, Ourednik V, Lynch WP, Schachner M, Snyder EY (2002) Neural stem cells display an inherent mechanism for rescuing dysfunctional neurons. Nat Biotechnol 20:1103–1110

Palmer TD, Willhoite AR, Gage FH (2000) Vascular niche for adult hippocampal neurogenesis. J Comp Neurol 425:479–494

Papadopoulos D, Pham-Dinh D, Reynolds R (2006) Axon loss is responsible for chronic neurological deficit following inflammatory demyelination in the rat. Exp Neurol 197:373–385

Park KI, Ourednik J, Ourednik V, Taylor RM, Aboody KS, Auguste KI, Lachyankar MB, Redmond DE, Snyder EY (2002a) Global gene and cell replacement strategies via stem cells. Gene Ther 9:613–624

Park KI, Teng YD, Snyder EY (2002b) The injured brain interacts reciprocally with neural stem cells supported by scaffolds to reconstitute lost tissue. Nat Biotechnol 20:1111–1117

Peretto P, Merighi A, Fasolo A, Bonfanti L (1997) Glial tubes in the rostral migratory stream of the adult rat. Brain Res Bull 42:9–21

Pfeifer K, Vroemen M, Blesch A, Weidner N (2004) Adult neural progenitor cells provide a permissive guiding substrate for corticospinal axon growth following spinal cord injury. Eur J Neurosci 20:1695–1704

Pluchino S, Martino G (2008) The therapeutic plasticity of neural stem/precursor cells in multiple sclerosis. J Neurol Sci 265:105–110

Pluchino S, Quattrini A, Brambilla E, Gritti A, Salani G, Dina G, Galli R, Del Carro U, Amadio S, Bergami A, Furlan R, Comi G, Vescovi AL, Martino G (2003) Injection of adult neurospheres induces recovery in a chronic model of multiple sclerosis. Nature 422:688–694

Pluchino S, Furlan R, Martino G (2004) Cell-based remyelinating therapies in multiple sclerosis: evidence from experimental studies. Curr Opin Neurol 17:247–255

Pluchino S, Zanotti L, Rossi B, Brambilla E, Ottoboni L, Salani G, Martinello M, Cattalini A, Bergami A, Furlan R, Comi G, Constantin G, Martino G (2005) Neurosphere-derived multipotent precursors promote neuroprotection by an immunomodulatory mechanism. Nature 436:266–271

Pluchino S, Muzio L, Imitola J, Deleidi M, Alfaro-Cervello C, Salani G, Porcheri C, Brambilla E, Cavasinni F, Bergamaschi A, Garcia-Verdugo JM, Comi G, Khoury SJ, Martino G (2008) Persistent inflammation alters the function of the endogenous brain stem cell compartment. Brain 131:2564–2578

Pluchino S, Gritti A, Blezer E, Amadio S, Brambilla E, Borsellino G, Cossetti C, Del Carro U, Comi G, t Hart B, Vescovi A, Martino G (2009a) Human neural stem cells ameliorate autoimmune encephalomyelitis in non-human primates. Ann Neurol 66:343–354

Pluchino S, Zanotti L, Brambilla E, Rovere-Querini P, Capobianco A, Alfaro-Cervello C, Salani G, Cossetti C, Borsellino G, Battistini L, Ponzoni M, Doglioni C, Garcia-Verdugo JM, Comi G, Manfredi AA, Martino G (2009b) Immune regulatory neural stem/precursor cells protect from central nervous system autoimmunity by restraining dendritic cell function. PLoS One 4:e5959

Politi LS, Bacigaluppi M, Brambilla E, Cadioli M, Falini A, Comi G, Scotti G, Martino G, Pluchino S (2007) Magnetic-resonance-based tracking and quantification of intravenously injected neural stem cell accumulation in the brains of mice with experimental multiple sclerosis. Stem Cells 25:2583–2592

Prineas JW, Barnard RO, Kwon EE, Sharer LR, Cho ES (1993) Multiple sclerosis: remyelination of nascent lesions. Ann Neurol 33:137–151

Quinones-Hinojosa A, Sanai N, Soriano-Navarro M, Gonzalez-Perez O, Mirzadeh Z, Gil-Perotin S, Romero-Rodriguez R, Berger MS, Garcia-Verdugo JM, Alvarez-Buylla A (2006) Cellular composition and cytoarchitecture of the adult human subventricular zone: a niche of neural stem cells. J Comp Neurol 494:415–434

Rafuse VF, Soundararajan P, Leopold C, Robertson HA (2005) Neuroprotective properties of cultured neural progenitor cells are associated with the production of sonic hedgehog. Neuroscience 131:899–916

Raine CS, Wu E (1993) Multiple sclerosis: remyelination in acute lesions. J Neuropathol Exp Neurol 52:199–204

Rakic P (1990) Principles of neural cell migration. Experientia 46:882–891

Rauch MF, Hynes SR, Bertram J, Redmond A, Robinson R, Williams C, Xu H, Madri JA, Lavik EB (2009) Engineering angiogenesis following spinal cord injury: a coculture of neural progenitor and endothelial cells in a degradable polymer implant leads to an increase in vessel density and formation of the blood-spinal cord barrier. Eur J Neurosci 29:132–145

Richardson RM, Broaddus WC, Holloway KL, Fillmore HL (2005) Grafts of adult subependymal zone neuronal progenitor cells rescue hemiparkinsonian behavioral decline. Brain Res 1032:11–22

Ruffini F, Furlan R, Poliani PL, Brambilla E, Marconi PC, Bergami A, Desina G, Glorioso JC, Comi G, Martino G (2001) Fibroblast growth factor-II gene therapy reverts the clinical course and the pathological signs of chronic experimental autoimmune encephalomyelitis in C57BL/6 mice. Gene Ther 8:1207–1213

Ryu JK, Kim J, Cho SJ, Hatori K, Nagai A, Choi HB, Lee MC, McLarnon JG, Kim SU (2004) Proactive transplantation of human neural stem cells prevents degeneration of striatal neurons in a rat model of Huntington disease. Neurobiol Dis 16:68–77

Seri B, Garcia-Verdugo JM, Collado-Morente L, McEwen BS, Alvarez-Buylla A (2004) Cell types, lineage, and architecture of the germinal zone in the adult dentate gyrus. J Comp Neurol 478:359–378

Seri B, Herrera DG, Gritti A, Ferron S, Collado L, Vescovi A, Garcia-Verdugo JM, Alvarez-Buylla A (2006) Composition and organization of the SCZ: a large germinal layer containing neural stem cells in the adult mammalian brain. Cereb Cortex 16(Suppl 1):i103–i111

Shen Q, Goderie S, Jin L, Karanth N, Sun Y, Abramova N, Vincent P, Pumiglia K, Temple S (2004) Endothelial cells stimulate self-renewal and expand neurogenesis of neural stem cells. Science 304(5675):1338–1340

Steinman L (2001) Myelin-specific CD8 T cells in the pathogenesis of experimental allergic encephalitis and multiple sclerosis. J Exp Med 194:F27–F30

Stumm RK, Zhou C, Ara T, Lazarini F, Dubois-Dalcq M, Nagasawa T, Hollt V, Schulz S (2003) CXCR4 regulates interneuron migration in the developing neocortex. J Neurosci 23: 5123–5130

Suzuki M, McHugh J, Tork C, Shelley B, Klein SM, Aebischer P, Svendsen CN (2007) GDNF secreting human neural progenitor cells protect dying motor neurons, but not their projection to muscle, in a rat model of familial ALS. PLoS One 2:e689

Swanborg RH (2001) Experimental autoimmune encephalomyelitis in the rat: lessons in T-cell immunology and autoreactivity. Immunol Rev 184:129–135

Tavazoie M, Van der Veken L, Silva-Vargas V, Louissaint M, Colonna L, Zaidi B, Garcia-Verdugo JM, Doetsch F (2008) A specialized vascular niche for adult neural stem cells. Cell Stem Cell 3:279–288

Teng YD, Lavik EB, Qu X, Park KI, Ourednik J, Zurakowski D, Langer R, Snyder EY (2002) Functional recovery following traumatic spinal cord injury mediated by a unique polymer scaffold seeded with neural stem cells. Proc Natl Acad Sci USA 99:3024–3029

Trapp BD, Peterson J, Ransohoff RM, Rudick R, Mork S, Bo L (1998) Axonal transection in the lesions of multiple sclerosis. N Engl J Med 338:278–285

Uccelli A, Moretta L, Pistoia V (2008) Mesenchymal stem cells in health and disease. Nat Rev Immunol 8:726–736

Villoslada P, Hauser SL, Bartke I, Unger J, Heald N, Rosenberg D, Cheung SW, Mobley WC, Fisher S, Genain CP (2000) Human nerve growth factor protects common marmosets against autoimmune encephalomyelitis by switching the balance of T helper cell type 1 and 2 cytokines within the central nervous system. J Exp Med 191:1799–1806

Wang Y, Piao JH, Larsen EC, Kondo Y, Duncan ID (2011) Migration and remyelination by oligodendrocyte progenitor cells transplanted adjacent to focal areas of spinal cord inflammation. J Neurosci Res 89(11):1737–1746

Windrem MS, Nunes MC, Rashbaum WK, Schwartz TH, Goodman RA, McKhann G 2nd, Roy NS, Goldman SA (2004) Fetal and adult human oligodendrocyte progenitor cell isolates myelinate the congenitally dysmyelinated brain. Nat Med 10:93–97

Windrem MS, Schanz SJ, Guo M, Tian GF, Washco V, Stanwood N, Rasband M, Roy NS, Nedergaard M, Havton LA, Wang S, Goldman SA (2008) Neonatal chimerization with human glial progenitor cells can both remyelinate and rescue the otherwise lethally hypomyelinated shiverer mouse. Cell Stem Cell 2:553–565

Wingerchuk DM, Lucchinetti CF, Noseworthy JH (2001) Multiple sclerosis: current pathophysiological concepts. Lab Invest 81:263–281

Wujek JR, Bjartmar C, Richer E, Ransohoff RM, Yu M, Tuohy VK, Trapp BD (2002) Axon loss in the spinal cord determines permanent neurological disability in an animal model of multiple sclerosis. J Neuropathol Exp Neurol 61:23–32

Yan YP, Sailor KA, Lang BT, Park SW, Vemuganti R, Dempsey RJ (2007) Monocyte chemoattractant protein-1 plays a critical role in neuroblast migration after focal cerebral ischemia. J Cereb Blood Flow Metab 27:1213–1224

Yandava BD, Billinghurst LL, Snyder EY (1999) "Global" cell replacement is feasible via neural stem cell transplantation: evidence from the dysmyelinated shiverer mouse brain. Proc Natl Acad Sci USA 96:7029–7034

Yao DL, Liu X, Hudson LD, Webster HD (1996) Insulin-like growth factor-I given subcutaneously reduces clinical deficits, decreases lesion severity and upregulates synthesis of myelin proteins in experimental autoimmune encephalomyelitis. Life Sci 58:1301–1306

Zappia E, Casazza S, Pedemonte E, Benvenuto F, Bonanni I, Gerdoni E, Giunti D, Ceravolo A, Cazzanti F, Frassoni F, Mancardi G, Uccelli A (2005) Mesenchymal stem cells ameliorate experimental autoimmune encephalomyelitis inducing T-cell anergy. Blood 106:1755–1761

Zhang SC, Lundberg C, Lipsitz D, O'Connor LT, Duncan ID (1998) Generation of oligodendroglial progenitors from neural stem cells. J Neurocytol 27:475–489

Zhang SC, Ge B, Duncan ID (1999) Adult brain retains the potential to generate oligodendroglial progenitors with extensive myelination capacity. Proc Natl Acad Sci USA 96:4089–4094

Zhang Y, Klassen HJ, Tucker BA, Perez MT, Young MJ (2007) CNS progenitor cells promote a permissive environment for neurite outgrowth via a matrix metalloproteinase-2-dependent mechanism. J Neurosci 27:4499–4506

Zhao C, Teng EM, Summers RG Jr, Ming GL, Gage FH (2006) Distinct morphological stages of dentate granule neuron maturation in the adult mouse hippocampus. J Neurosci 26:3–11

Zujovic V, Thibaud J, Bachelin C, Vidal M, Coulpier F, Charnay P, Topilko P, Baron-Van Evercooren A (2010) Boundary cap cells are highly competitive for CNS remyelination: fast migration and efficient differentiation in PNS and CNS myelin-forming cells. Stem Cells 28:470–479

Chapter 8
Axonal Protection with Sodium Channel Blocking Agents in Models of Multiple Sclerosis

Joel A. Black, Kenneth J. Smith, and Stephen G. Waxman

8.1 Introduction

Axonal pathology in multiple sclerosis (MS) has been recognized for nearly 150 years (Charcot 1868), but it is within the last decade that the loss of axons has come to be increasingly appreciated as a major contributor to nonremitting disability in MS (DeStefano et al. 1998; Bjartmar et al. 2000; Dutta and Trapp 2007). Early studies demonstrated significant axonal damage and loss within acute MS plaques (Ferguson et al. 1997; Trapp et al. 1998; Bitsch et al. 2000; Bjartmar et al. 2000; Lovas et al. 2000; DeStefano et al. 2001; DeLuca et al. 2004), but the loss continues to occur, albeit at an attenuated rate, in chronic inactive plaques (Kornek et al. 2000; Kuhlmann et al. 2002). Correspondingly, a positive correlation between the prevalence of damaged axons and inflammatory activity has been observed in some MS lesions (Ferguson et al. 1997; Trapp et al. 1998; Kornek et al. 2000; Bitsch et al. 2001), although axonal loss within normal-appearing white matter (NAWM), which

J.A. Black • S.G. Waxman (✉)
Department of Neurology, Yale University School of Medicine,
PO Box 20818, New Haven, CT 06520, USA

Paralyzed Veterans of America/United Spinal Association,
Center for Neuroscience and Regeneration Research,
Yale University School of Medicine,
West Haven, CT, USA

Rehabilitation Research Center, VA Connecticut Healthcare System,
West Haven, CT 06516, USA
e-mail: stephen.waxman@yale.edu

K.J. Smith
Department of Neuroinflammation, The Institute of Neurology (Queen Square),
University College London, 1 Wakefield Street, London, WC1N 1PJ, UK

has limited immune cell activity, has also been reported (Ganter et al. 1999; Evangelou et al. 2000; Bjartmar et al. 2001; Narayanan et al. 2006). Delineation of the mechanisms responsible for axonal degeneration in MS is a daunting challenge. Yet, especially in light of the key role that axonal loss plays in the progression of disability, there has been considerable interest over the last several years in identifying neuroprotective therapies that can ameliorate axonal injury and degeneration in neuroinflammatory disorders.

A series of in vitro studies in the early 1990s provided evidence that voltage-gated sodium channels can participate in Ca^{2+}-mediated damage of central white matter axons following anoxic injury (Stys et al. 1991, 1992a; Waxman et al. 1994). These studies demonstrated that, as a result of anoxia-triggered impairment of axonal Na^+/K^+ activity, persistent sodium current flowing through tetrodotoxin (TTX)-sensitive sodium channels can drive reverse operation of the Na^+/Ca^{2+} exchanger (NCX), leading to importation of high levels of Ca^{2+} that can progress to irreversible axonal injury. Consistent with a role for sodium channels in the pathogenesis of axonal injury, sodium channel blocking agents, TTX, phenytoin, carbamazepine, and quaternary anesthetics, were shown to prevent the development of irreversible dysfunction of axons in white matter in vitro following anoxia (Stys et al. 1992b; Fern et al. 1993). Subsequently, sodium channel blockers were also shown to exert protective effects on axons exposed to trauma and ischemic injuries (Agrawal and Fehlings 1996; Imaizumi et al. 1997; Teng and Wrathall 1997; Rosenberg et al. 1999; Hewitt et al. 2001; Schwartz and Fehlings 2001; Hains et al. 2004; Hains and Waxman 2005; Kaptanoglu et al. 2005).

The findings in anoxic injury were extended and applied to neuroinflammatory demyelinating disease in experiments that examined the effects of nitric oxide on axons. It was reasoned that because nitric oxide was present in raised concentrations in MS lesions (Bo et al. 1994; Brosnan et al. 1994; Smith and Lassmann 2002) and because nitric oxide is a potent inhibitor of mitochondrial metabolism (reviewed in Brown 2007; Soane et al. 2007), MS lesions might experience a similar energy deficit as that resulting from anoxia (Kapoor et al. 2000, 2003). If so, axons at a site of inflammation might be expected to accumulate dangerous levels of sodium ions. Indeed, it seemed that this problem might especially affect demyelinated axons because it had been found that such axons were rendered inherently vulnerable to sodium accumulation by their acquisition of a persistent inward sodium current at the site of demyelination (Kapoor et al. 1997). Nitric oxide had been demonstrated to be a potent inhibitor of axonal conduction, particularly in demyelinated axons (Redford et al. 1997), and it could cause axonal degeneration at higher concentrations, or, especially, if the energy requirements of the axons was increased by sustained electrical activity at physiological frequencies (Kapoor et al. 1999; Smith et al. 2001). If neuroinflammatory lesions resembled anoxic lesions, it seemed possible that sodium channel blocking agents might again be beneficial, acting via the same mechanism involving sodium accumulation and reverse operation of the NCX (Kapoor et al. 2000, 2003). This possibility was tested in experiments that found that sodium channel blocking agents were potent in protecting axons from degeneration mediated by nitric oxide: flecainide and lidocaine/lignocaine were employed as blocking agents by Kapoor et al. (2000, 2003), and tetrodotoxin was employed by Garthwaite et al. (2002). Flecainide was especially valuable in the protection of axons from degeneration induced by the combination of NO and impulse activity (Kapoor et al. 2000, 2003). The line of reasoning regarding the involvement of sodium channels in the cascade leading to axonal degeneration was then extended to the study of animal models more closely allied to MS, which is the focus of this chapter.

The idea that neuroinflammatory lesions in MS may suffer from a lack of energy has accumulated additional support in recent years from observations on MS tissue itself. Notably, some MS lesions (Pattern III) have a hypoxia-like morphological appearance (Aboul-Enein and Lassmann 2005), and there is mounting evidence that there can be a substantive mitochondrial dysfunction within MS CNS tissue (Dutta et al. 2006; Mahad et al. 2008). If so, it seems increasingly likely that axons may be damaged in MS by sodium accumulation and consequent reverse operation of the NCX, in which case a therapeutic approach employing partial sodium channel blockade is especially indicated.

This chapter presents work from our laboratories that have examined the effects of sodium channel blocking agents on disease progression in rodent models of neuroinflammatory lesions, including experimental autoimmune encephalomyelitis (EAE), a disease that is widely utilized to model aspects of MS. The sodium channel blocking agents utilized in our studies—phenytoin, carbamazepine, flecainide, and lamotrigine—are well-characterized in the clinical setting, with flecainide commonly used to treat cardiac arrhythmias (Wit and Rosen 1983) and phenytoin, carbamazepine and lamotrigine employed as anticonvulsants and/or in pain management (Rogowski and Porter 1990; Steiner et al. 1994; Pöllmann and Feneberg 2008). The results summarized here demonstrate that these sodium channel blocking agents provide robust protection of spinal cord axons, preserve action potential conduction, significantly diminish immune cell infiltration, and attenuate neurological deficits in EAE. Results from these studies provided a rationale for planning and implementing clinical studies utilizing sodium channel blocking agents in patients with MS, and several clinical trials examining the efficacy of sodium channel blockade in ameliorating clinical disability in MS are currently ongoing.

8.2 Sodium Channel Blockade Improves EAE

8.2.1 Clinical Status

EAE induced in rodent species has been used extensively to model different aspects of MS (Steinman and Zamvil 2006). Depending on the species/strain and immunization protocol, differing clinical courses of MS patients (i.e., progressive, relapsing–remitting) are, at least partially, mimicked in these EAE models (Brown and McFarlin 1981; Zamvil et al. 1985; Gold et al. 2006). In the studies of Lo et al. (2003) and Black et al. (2007), C57/Bl6 mice injected with 33-55 peptide of rat

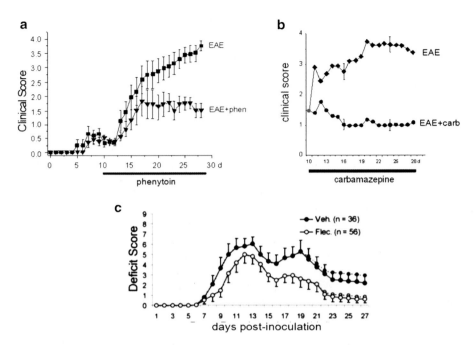

Fig. 8.1 Sodium channel blocking agents improve neurological status of rodents with EAE. (**a, b**) Clinical scores (mean ± SEM) are shown for untreated C57/Bl6 mice with progressive EAE (*inverted triangles* in **a** and *filled diamonds* in **b**); both phenytoin (*filled boxes* in **a**) and carbamazepine (*filled circles* in **b**) treatment of mice with progressive EAE significantly improved the clinical course of the disease. Oral administration of phenytoin or carbamazepine was started on day 10 following MOG inoculation, as indicated by the *horizontal bar*, and continued until the end of the study at day 28. (**c**) The progression of chronic-relapsing EAE in DA rats followed a relapsing disease course (*filled circles*). Flecainide treatment commencing at 7 days post spinal cord homogenate inoculation reduced the severity of the disease course (*filled circles*) (modified and reprinted with permission from Lo et al. 2003, Bechtold et al. 2004, Black et al. 2007)

myelin oligodendrocyte glycoprotein (MOG) developed a disease onset that was evident 7–10 days postinjection and a progression of disability that plateaued around 19–22 days postinjection (Fig. 8.1a). In contrast, Bechtold (2004) and Bechtold et al. (2002, 2004, 2006) studied dark agouti (DA) rats inoculated with syngeneic spinal cord homogenate that exhibited a chronic-relapsing form of EAE (Fig. 8.1c), in which neurological symptoms became evident approximately 1 week following injection, with relapses occurring over the next 2 weeks.

In both progressive (C57/Bl6 mice) and chronic-relapsing (DA rats) forms of rodent EAE, administration of sodium channel blocking agents significantly improved the clinical course of the disease (Lo et al. 2003; Black et al. 2006, 2007; Bechtold et al. 2002, 2004, 2006). Treatment of C57/Bl6 mice with phenytoin or carbamazepine commencing on day 10 following MOG injection (using an oral treatment that resulted in phenytoin and carbamazepine serum levels within the human therapeutic range) significantly improved the clinical course in the treated mice compared with untreated mice, as assessed from day 13 until the end of the study at day 28 (Fig. 8.1a, b; Lo et al. 2003; Black et al. 2007). Likewise, flecainide administered to DA rats, with EAE by twice daily injections beginning either 7 days following (Fig. 8.1c) or 3 days prior to inoculation, significantly improved the mean peak deficit and terminal deficit scores compared with vehicle-treated rats (Bechtold et al. 2004). Lamotrigine treatment in DA rats provided more limited clinical improvement in the acute course of EAE compared with untreated rats but significantly improved the terminal deficit in the lamotrigine-treated rats in comparison with vehicle-treated EAE rats (Bechtold et al. 2006).

These studies demonstrated the efficacy of sodium channel blockade as an intervention that improves the neurological status in EAE for the duration of drug administration (up to 28 days) following induction of EAE. Subsequently, it was shown that phenytoin administration significantly improved the clinical course of EAE in C57/Bl6 mice when administered for up to 180 days, which was the endpoint of this study (Black et al. 2006). Taken together, these observations clearly established sodium channel blocking agents as providing significant improvement in the neurological status of rodent EAE and suggested their use as therapeutic agents in neuroinflammatory disorders.

8.2.2 Axonal Protection

Early studies in EAE demonstrated a loss of axons (Raine and Cross 1989; White et al. 1992; Kornek et al. 2000; Onuki et al. 2001; Wujek et al. 2002), which was correlated with the accumulation of nonremitting disability (Wujek et al. 2002). To determine whether the improved neurological status observed in EAE with sodium channel blocking agents was accompanied by protection of axons from degeneration, we assessed the magnitude of axonal degeneration in spinal cords of treated *versus* untreated mice and rats with EAE. Our results with the two distinct EAE models, and with three different sodium channel blocking agents, demonstrated consistent protection of axons in these animals. In C57/Bl6 mice, untreated EAE was accompanied by significant loss of axons within the dorsal corticospinal tract (CST, a descending tract) and the dorsal funiculus (DF, ascending tracts) (Fig. 8.2a; Black et al. 2007), with an approximately 60% loss of axons in the CST and a 40% loss in the DF compared to control mice, as assessed at 28–30 days post-MOG inoculation (Fig. 8.2a; Lo et al. 2003). Interestingly, in a separate study, axonal loss in the CST and DF at 180 days post-MOG inoculation were similar to those at 28–30 days post-MOG injection (Black et al. 2007), indicating that axonal degeneration in this murine EAE model occurs primarily during the initial phase of the disease. In contrast to untreated mice with EAE, treatment with phenytoin at clinically relevant levels significantly attenuated the loss of axons in mice with EAE at 28–30 days post-MOG inoculation, such that there was only a 28% loss of axons in the corticospinal tract and a 17% loss in the dorsal funiculus compared

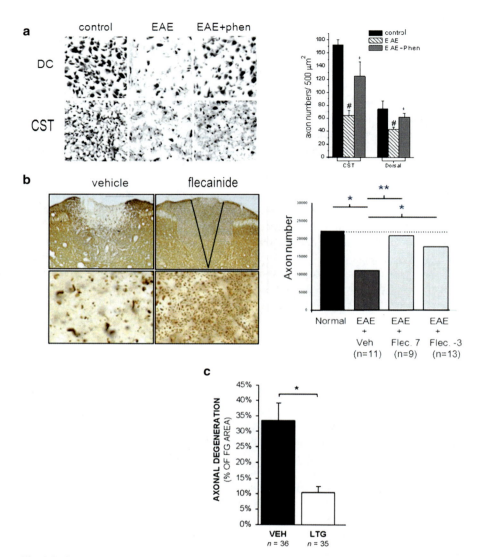

Fig. 8.2 Sodium channel blocking agents reduce the loss of spinal cord axons in EAE. (**a**) There was a substantial loss of axons in the dorsal funiculus (DF) and dorsal corticospinal tract (CST) in untreated EAE compared to control mice. Phenytoin-treated mice with EAE (EAE+phen) exhibited substantial increases in axons in the DF and CST compared to untreated EAE mice. Quantification of CST and DC axons (axon numbers/500 µm^2) from control, untreated EAE, and phenytoin-treated EAE mice demonstrates a significant ($p<0.05$) reduction of axon counts in the CST and DC in untreated EAE (*hash symbols*) compared to untreated controls. Phenytoin treatment of EAE (*asterisks*) results in a significant ($p<0.05$) increase in axons within the CST and DC compared to untreated EAE. (**b**) Sections of dorsal columns immunolabeled with neurofilament showed substantial loss of axons in vehicle-treated DA rats with EAE. In comparison, flecainide administration commencing at day 7 postinoculation greatly reduced the loss of axons. Quantification of the number of axon within a selected portion of the dorsal columns

with control values (Fig. 8.2a). Significantly, the axonal protection afforded by phenytoin treatment persisted during longer treatment periods, for at least 180 days (Black et al. 2007).

Similar protective effects of sodium channel blockade were seen in DA rats with chronic-relapsing EAE treated with flecainide or lamotrigine. DA rats inoculated with spinal cord homogenate exhibited prominent axonal loss and ongoing axonal degeneration, particularly within the medial region of the dorsal columns (fasciculus gracilis) (Fig. 8.2b, c; Bechtold et al. 2004), that was closely associated with the peak deficit scores exhibited by these rats. DA rats with EAE that had peak scores ≥8 (on 15-point scale) displayed an approximately 40% loss of axons relative to control rats. In contrast, both flecainide (Bechtold et al. 2002, 2004) and lamotrigine (Bechtold et al. 2006) treatments provided significant protection of spinal cord axons when administered to DA rats with EAE (Fig. 8.2b, c). In comparison with the number of axons in normal rats, flecainide treatment commencing either 3 days prior to or 7 days following homogenate injection provided near complete axonal protection, with axonal counts remaining at $98\% \pm 20\%$ and $83\% \pm 14\%$ of control values, respectively (Bechtold et al. 2004). Similar results were obtained with lamotrigine treatment of DA rats with EAE, such that, when assaying degeneration as a percentage of total area measured, lamotrigine-treated rats exhibited axonal degeneration in only 9% of the area, while untreated rats displayed degeneration in 31% of the area (Bechtold et al. 2006). The circulating concentration of lamotrigine (mean 11.5 µg/ml) in EAE was within the clinical therapeutic range (1–15 µg/ml) for the treatment of epilepsy. Flecainide therapy was also found to be effective in axonal protection in rats with a related disease, namely, experimental autoimmune neuritis, a model of Guillain–Barré syndrome (Bechtold et al. 2005).

8.2.3 Functional Protection of Axons

To assess whether the preservation of axons observed with morphological methods in EAE with sodium channel blocking therapy was accompanied by preservation of function in the axons, we performed recordings of the compound action potential (CAP) from the dorsal surface of spinal cords of untreated and sodium channel blocker-treated mice and rats with EAE. Our results in both rodent species with EAE demonstrated that treatment with sodium channel blocking agents provides

Fig. 8.2 (continued) (shown by the "V" in the upper right illustration) of normal, vehicle-treated EAE, and EAE rats treated with flecainide commencing at day 7 or -3 demonstrates a significant loss of axons in untreated EAE dorsal columns, with significant protection of axons mediated by flecainide therapy. $*p < 0.05$, $**p < 0.01$. (c) Histogram showing the magnitude of axonal degeneration in severely diseased DA rats with chronic-relapsing EAE. Lamotrigine treatment significantly reduced axonal degeneration in DA rats with CR-EAE compared with vehicle-treated rats with EAE ($*p < 0.050$) (modified and reprinted with permission from Lo et al. 2003, Bechtold et al. 2004, 2006, Black et al. 2007)

increased numbers of functional axons (greater CAPs) compared with untreated EAE, consistent with the attenuation of axonal loss and improvement of clinical status in these animals.

C57/Bl6 mice with EAE displayed clear improvements in CAP areas when administered phenytoin compared with untreated mice with EAE. The maximal CAP area in untreated mice with EAE was significantly attenuated compared with control mice (0.038 mV ms in EAE vs. 0.16 mV ms in controls; Fig. 8.3a; Lo et al. 2003). In contrast, phenytoin-treated mice with EAE exhibited maximal CAP areas (0.13 mV ms) that were significantly greater than those in untreated mice with EAE (0.038 mV ms) and that approached control values (0.16 mV ms). In addition, assessment of mean conduction velocities demonstrated significant improvement in conduction in phenytoin-treated versus untreated mice with EAE (control: 11.6 ± 2.1 m/s; EAE: 2.3 ± 1.7 m/s; EAE+phenytoin: 13.2 ± 1.1 m/s; Lo et al. 2003).

A similar protective effect on axonal conduction was seen in DA rats with chronic-relapsing EAE. Normal DA rats exhibited supramaximal-stimulated monophasic CAP areas of around 50–55 µV ms (Fig. 8.3b; Bechtold et al. 2004, 2006). The CAP areas of DA rats with EAE were significantly reduced by 40–60% (20–30 µV ms). Consistent with the observed preservation of axons achieved with treatment with sodium channel blocking agents, both flecainide (Bechtold et al. 2002, 2004) and lamotrigine (Bechtold et al. 2006) administration to DA rats with EAE significantly increased the CAP areas compared with vehicle-treated DA rats with EAE (Fig. 8.3b). Flecainide treatment commencing either -3 days prior to or 7 days following spinal cord homogenate inoculation significantly improved the CAP areas (-3-day flecainide: 28 µV ms vs. untreated 21 µV ms; 7-day flecainide: 37 µV ms vs. untreated 20 µV ms; Bechtold et al. 2004). Lamotrigine also preserved axonal conduction in DA rats with EAE, increasing the CAP area to 43 µV ms compared with 31 µV ms for untreated rats (Bechtold et al. 2006).

8.2.4 Inflammatory Infiltrates

The onset and progression of EAE is associated with infiltration of immune cells into the CNS (Zamvil and Steinman 1990; Gold et al. 2006). Interestingly, sodium channels are expressed by immune cells, including T lymphocytes (Gallin 1991; Lai et al. 2000; Fraser et al. 2004), microglia (Korotzer and Cotman 1992; Norenberg et al. 1994; Eder 1998; Craner et al. 2005; Black et al. 2009) and macrophages (Carrithers et al. 2007). Sodium currents are suggested to contribute to T lymphocyte activation and co-stimulation (Khan and Poisson 1999; Lai et al. 2000), while phagocytosis by microglia and macrophages and migration by microglia in vitro are attenuated by sodium channel blockade (Craner et al. 2005; Carrithers et al. 2007; Black et al. 2009). Figure 8.4 shows a significant reduction in phagocytosis by LPS-activated microglia in vitro when treated with either TTX or phenytoin (Black et al. 2009). Carrithers et al. (2007, 2009) showed that Nav1.5 sodium channels in

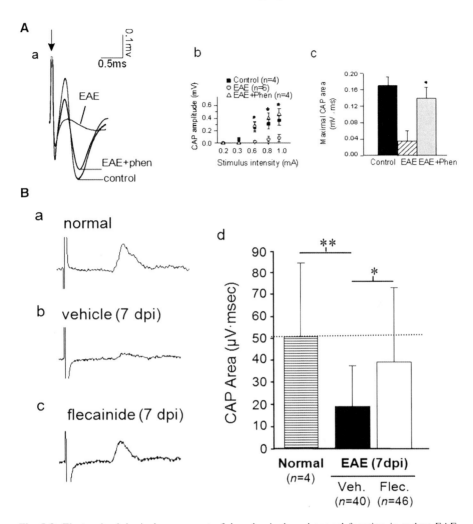

Fig. 8.3 Electrophysiological assessment of dorsal spinal cord axonal function in rodent EAE. (**A**) (**a**) Superimposed CAP traces from representative control, untreated EAE, and phenytoin-treated C57/Bl6 mice with EAE. The typical biphasic wave seen in controls is highly attenuated in untreated EAE but is restored in phenytoin-treated EAE. (**b**) Mean CAP amplitudes (±SEM, mV) in control, untreated EAE, and phenytoin-treated EAE at different stimulating current intensities ($*p<0.01$, phenytoin-treated EAE compared to untreated EAE). (**c**) Mean supramaximal CAP area (±SEM, mV×ms) in control, untreated EAE, and phenytoin-treated EAE ($*p<0.05$ phenytoin-treated EAE compared to untreated EAE). (**B**) Representative records from normal (**a**) and CR-EAE DA rats treated with vehicle (**b**) or flecainide (**c**) from 7 days postinoculation. (**d**) The area of the CAP was significantly reduced from normal, but this reduction was largely prevented by flecainide treatment. This protection was statistically significant with therapy from 7 dpi ($*p<0.05$, $**p<0.01$) (modified and reprinted with permission from Lo et al. 2003, Bechtold et al. 2004)

Fig. 8.4 Sodium channel blocking agents inhibit microglial phagocytosis. (**a**) In astrocyte-conditioned medium (ACM), few latex beads (*green*) are phagocytosed by CD11b+ microglia (*red*); in contrast, LPS-activated microglia exhibit increased numbers of phagocytosed latex beads (colocalized latex beads and microglia are *yellow*). Both phenytoin and TTX attenuate latex bead phagocytosis by LPS-activated microglia. (**b**) Quantification of phagocytic activity demonstrates a nearly threefold increase in latex bead phagocytosis with LPS activation compared to microglia maintained in ACM only. Phagocytic activity is significantly reduced with phenytoin and TTX treatment in LPS-activated microglia (modified and reprinted with permission from Black et al. 2009)

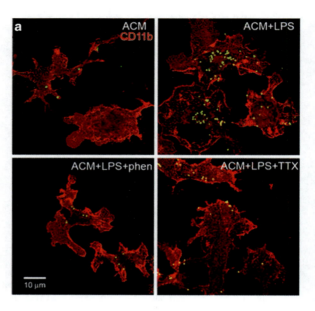

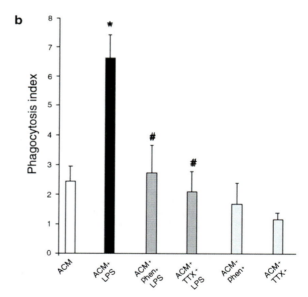

endosomes provide a route for Na+ ion efflux that offsets proton influx during endosomal acidification, a late stage of phagocytosis in macrophages, and demonstrated a role of Nav1.6 sodium channels in controlling macrophage motility via regulation of podosome formation. Microglial migration in vitro is also significantly attenuated by exposure to TTX or phenytoin (Fig. 8.5; Black et al. 2009). Similarly, there is evidence for a role of Nav1.6 in microglial migration (Black et al. 2009).

Fig. 8.5 Sodium channel blockade attenuates microglial migration. (**a**) Few microglia migrate through the trans-well-membrane pores in ACM only; in contrast, ATP activates a large number of microglia to migrate. Both phenytoin and TTX reduce the number of migrating ATP-activated microglia. (**b**) Quantification of microglial migration demonstrates a nearly fourfold increase in migrating cells following ATP activation compared to nonactivated microglia. Both phenytoin and TTX (0.3 μM) decrease the number of ATP-activated migrating microglia by ~50% (modified and reprinted with permission from Black et al. 2009)

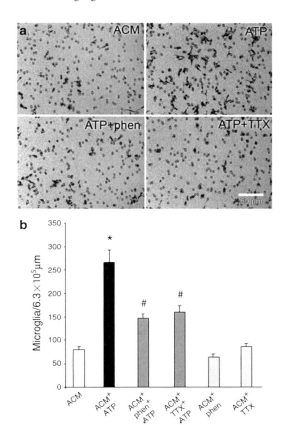

Craner et al. (2005) demonstrated that treatment with phenytoin reduces the inflammatory infiltrate with the CNS by 75%. Black et al. (2007) more recently carried out additional studies on the effect of treatment with sodium channel blocking agents on immune cell infiltration and/or functions in EAE. Spinal cord sections of control C57/Bl6 mice labeled for CD45 (leukocyte common antigen), which provides a marker for most immune cells (Sedgwick et al. 1991), exhibited relatively few CD45+ cells (Fig. 8.6). In contrast, untreated mice with EAE mice at 28 days post-MOG injection displayed substantial infiltration of CD45+ immune cells in dorsal, ventral, and lateral funiculi (Fig. 8.6). Mice with EAE that were administered phenytoin (Fig. 8.6) or carbamazepine, however, showed substantially reduced numbers of CD45+ cells in their spinal cords (Black et al. 2007).

Sodium channel blockade with flecainide also reduced spinal inflammation significantly in DA rats with EAE (Bechtold 2004). Quantification of ED-1 immunoreactivity (marking activated macrophages/microglia) in the spinal cords of the rats revealed that flecainide therapy effected a significant reduction in this measure of inflammation during the relapse stage of EAE (flecainide: 6.8% ± 3.0 of spinal cord area; vehicle: 11.5% ± 4.8; $p<0.05$). The expression of the inducible form of nitric

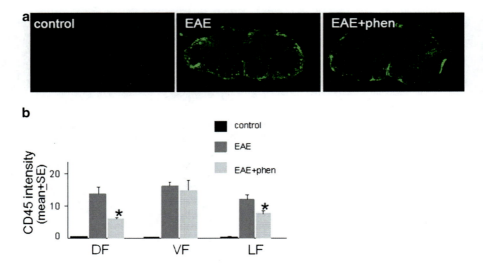

Fig. 8.6 Phenytoin attenuates immune cell infiltration in C57/Bl6 mice with EAE. (**a**) Low magnification images of cross-sections through lumbar spinal cords labeled for CD45 are shown for control, EAE 35 days postinjection (35 day EAE), and EAE+phenytoin 35 days post-injection (35 day EAE+phen) mice. There is substantial infiltration of CD45$^+$ immune cells in spinal cords of EAE; phenytoin treatment attenuates the infiltration of immune cells. (**b**) Quantification of CD45$^+$ intensity in dorsal (DF), ventral (VF) and lateral (LF) funiculi of control, EAE, and EAE+phen mice at 35 days post-injection. Phenytoin treatment significantly (*$p<0.05$) reduces CD45 intensity in DF and LF in EAE+phen mice compared to untreated EAE mice (modified and reprinted with permission from Black et al. 2007)

oxide synthase during the relapse period of EAE was also significantly reduced in the rats treated with flecainide (Fig 8.7; 0.4% ± 1.0), when compared with vehicle-treated animals (2.0% ± 2.2, $p<0.05$). In this context, it is interesting that lamotrigine also reduced lymph node cell proliferation in response to phytohemagglutinin (Makowska et al. 2004). Together with the results obtained in mice, these observations suggest that immunomodulation of immune cell functions by sodium channel blocking agents may contribute to the improvement of the clinical course of EAE by these agents.

8.3 Withdrawal of Sodium Channel Blockers Can Exacerbate EAE

8.3.1 Neurological Status

The studies described above demonstrated that continued administration of sodium channel blocking agents to mice and rats with EAE can significantly improve the course of the disease; however, whether these protective effects persist following withdrawal of treatment was not known. To examine this question, Black et al.

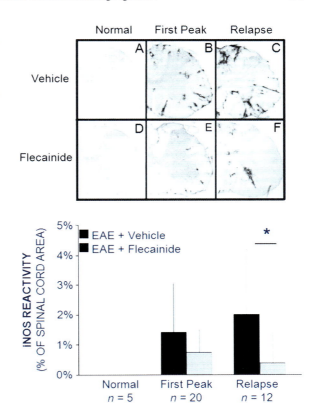

Fig. 8.7 Expression of iNOS in the spinal cords of rats with CR-EAE. iNOS⁺ cells were not observed in the spinal cords of normal DA rats (A) but were clearly visible during the first peak of CR-EAE (B). The number of iNOS⁺ cells in the spinal cord increased during the relapse phase of the disease (C). iNOS was not observed in normal rats treated with flecainide (D), but it was observed in the spinal cords of flecainide-treated CR-EAE rats during both the first peak (E) and the relapse (F). However, the magnitude of iNOS reactivity in the rats treated with flecainide was significantly reduced in comparison with rats treated with vehicle (DA Bechtold, P Hassoon and KJ Smith, unpublished observations—modified from Bechtold 2004)

(2007) treated mice with EAE with phenytoin or carbamazepine commencing on day 10 post-MOG inoculation and then withdrew treatment on day 28 (Black et al. 2007). As demonstrated in Fig. 8.8, withdrawal of phenytoin was accompanied by a rapid (within 2 days) exacerbation of the clinical status of mice with EAE; EAE mice with continuous phenytoin treatment did not exhibit exacerbation. The acute withdrawal of phenytoin was associated with a substantial death rate; 59% of the mice died within 11 days following the withdrawal, with most of the deaths occurring within 48 h of the cessation of treatment (Fig. 8.8a). Like withdrawal of phenytoin from mice with EAE, abrupt cessation of carbamazepine treatment was accompanied by acute worsening of the clinical symptoms of EAE, although the death rate following withdrawal was substantially less (Fig. 8.8b; 7.7%; Black et al. 2007). For EAE mice that survived the exacerbation following sodium channel blocker withdrawal, the clinical scores worsened substantially and rapidly approached those of untreated mice with EAE at similar time points. Notably, there were no ill effects observed following withdrawal of phenytoin or carbamazepine from wild-type (non-EAE) mice. Given the findings in EAE, it is interesting that patients with MS can be weaned from sodium channel blocking agents with no major (or perhaps any) detectable worsening of symptoms (see below).

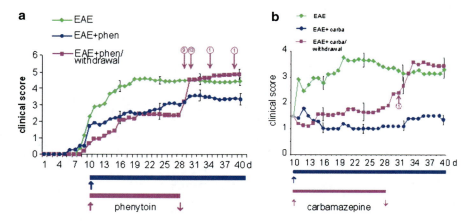

Fig. 8.8 Withdrawal of sodium channel blocking agents exacerbates EAE. (**a**) Withdrawal of phenytoin increases clinical dysfunction in mice with EAE. Mean clinical scores are shown for C57/Bl6 mice with untreated EAE (*green diamonds*), EAE+phenytoin (*blue circles*) and EAE+phenytoin/withdrawal (*magenta squares*). Phenytoin treatment is indicated by *blue* (continuous treatment) and *magenta* (withdrawal) *bars*. EAE+phenytoin mice exhibit significantly lower clinical scores than untreated mice on all days after day 12 (asterisks omitted for clarity). Withdrawal of phenytoin at day 28 results in rapid worsening of clinical scores. Numbers above magenta arrows indicate number of deaths at each post-withdrawal timepoint. No deaths occurred in EAE and EAE+phenytoin groups of mice from day 28 to 40. (**b**) Mean clinical scores are shown for C57/Bl6 mice with untreated EAE (*green diamonds*), EAE+carba (*blue circles*), and EAE+carbamazepine/withdrawal (*magenta squares*) commencing on day 10. Carbamazepine treatment is indicated by *blue* (continuous treatment) and *magenta* (withdrawal) *bars*. EAE+carbamazepine mice (*blue circles*) exhibit significantly lower clinical scores than untreated mice on all days after day 12. Withdrawal of carbamazepine at day 28 results in acute worsening of clinical scores. Number above *magenta arrow* indicates number of deaths at post-withdrawal timepoint. No deaths occurred in EAE and EAE+carbamazepine groups of mice from day 28 to 40 (modified and reprinted with permission from Black et al. 2007)

8.3.2 Inflammatory Rebound

What is/are the underlying mechanism(s) responsible for exacerbation of EAE following abrupt withdrawal of sodium channel blockers? Since loss of axons has been correlated with accumulation of nonremitting disability in rodent EAE (Wujek et al. 2002), it was conceivable that the observed worsening in EAE following phenytoin or carbamazepine withdrawal was due to massive injury to axons. Quantitative studies, however, demonstrated that there was not a significant difference in the number of axons within dorsal columns of mice with EAE maintained on phenytoin and those in which phenytoin was withdrawn 7–10 days prior (Black et al. 2007). While it is possible that axonal injury had occurred but had not progressed to a quantifiable loss of axons, because Wallerian degeneration of CNS axons can take weeks to become manifest (McDonald 1972), the histological results did not provide evidence for a massive loss of axons.

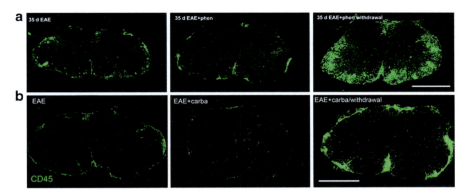

Fig. 8.9 Withdrawal of sodium channel blocking agents results in increased immune cell infiltrate. (**a**) Low magnification images of cross-sections through lumbar spinal cords labeled for CD45 are shown for EAE 35 days postinjection (35 day EAE), EAE + phenytoin 35 days postinjection (35 day EAE + phen), and 35 day EAE + phenytoin/withdrawal (35 day EAE + phen/withdrawal) mice. Withdrawal of phenytoin is accompanied by a substantial CD45$^+$ infiltrate into the spinal cord. *Scale bar*, 500 μm. (**b**) Low magnification images of cross-sections through lumbar spinal cords labeled for CD45 are shown for EAE 35 days postinjection (EAE), EAE + carbamazepine 35 days postinjection (EAE + carba), and 35 day EAE + carbamazepine/withdrawal (EAE + carba/withdrawal) mice. Withdrawal of carbamazepine is accompanied by a substantial CD45$^+$ cell infiltrate into the spinal cord; the immune cell infiltrate is substantially greater following carbamazepine withdrawal than in untreated EAE (modified and reprinted with permission from Black et al. 2007)

To determine whether inflammation could have contributed to worsening following sodium channel blocker withdrawal, Black et al. (2007) next directed attention to the extent of immune infiltrate into the spinal cords of mice with EAE following phenytoin and carbamazepine withdrawal (Fig. 8.9). As described above, CD45$^+$ cells, which were generally not observed in control mice, were clearly present in the dorsal, lateral, and ventral funiculi of the spinal cord in untreated mice with EAE at 35 days postinjection. When assessed during treatment with phenytoin (Fig. 8.9a) or carbamazepine (Fig. 8.9b) at day 28 postinoculation for EAE, the inflammatory infiltrate was suppressed. Surprisingly, EAE mice treated with phenytoin or carbamazepine from day 10 to day 28 followed by withdrawal at day 28 exhibited a prominent infiltration of inflammatory cells in all funiculi of the spinal cord (Fig. 8.9a, b; Black et al. 2007). Immunocytochemical analysis showed increased numbers of mature T lymphocytes labeled with anti-CD3, helper-inducer T lymphocytes labeled with anti-CD4, cytotoxic/suppressor T lymphocytes labeled with anti-CD8, and activated microglia/macrophages labeled with anti-CD11b and CD68 in the inflammatory infiltrate after phenytoin withdrawal (Black et al. 2007). Additional studies with flow cytometry demonstrated quantitative differences in the inflammatory infiltrate between untreated EAE mice at day 35 postinoculation and EAE mice in which phenytoin was withdrawn at day 28, prior to study at day 35 (Black et al. 2007). There was a 2.5-fold increase in the number of CD3$^+$ cells

(T lymphocytes) and a 3.5-fold increase in the number of CD11b$^+$ cells(microglia/macrophages) from the spinal cords and brains of EAE mice following phenytoin withdrawal as compared with untreated EAE mice (Black et al. 2007).

8.4 How Do Sodium Channel Blockers Provide Axonal Protection in EAE?

The results described above demonstrate that treatment with sodium channel blocking agents improves neurological status, reduces axonal loss, preserves functional axons, and attenuates inflammatory infiltrate in rodent EAE. Significantly, similar results were obtained with different EAE models (C57/Bl6 mice and DA rats) and with four different sodium channel blockers, phenytoin, carbamazepine, flecainide, and lamotrigine. While at present the underlying mechanism by which sodium channel blockade provides neuroprotection in EAE is not well understood, current evidence is suggestive of two independent converging pathways, one a direct action of sodium channel blockade on axons and one modulatory role on immune cell activity.

8.5 Sodium Channel Blockade Has a Direct Effect on Axons

The protective effects of sodium channel blockers on optic nerve axons in vitro following anoxic injury, with resultant energy depletion and impaired Na$^+$/K$^+$-ATPase activity, have been attributed to direct blockade of a persistent axonal sodium influx (Stys et al. 1993). Electron microprobe analysis has, in fact, demonstrated a continuous rise in intracellular [Na$^+$] paralleled by a rise in [Ca^{2+}] levels within anoxic myelinated axons (LoPachin and Stys 1995). Block of this sodium influx prevents reversal of the Na$^+$/Ca^{2+} exchanger (NCX) and the resultant import of damaging levels of Ca^{2+} and activation of degeneration pathways (Stys et al. 1992a, b; Waxman et al. 1992; Fern et al. 1993). Importantly, this protective action of sodium channel blockade is seen in an in vitro model of acute anoxia (Stys et al. 1991,1992a), where inflammation plays very little, if any, role, supporting the idea of a direct neuroprotective action of sodium channel blockers on axons.

Furthermore, earlier in this report we noted that mitochondria are likely to be inhibited in inflammatory MS lesions by nitric oxide, and that nitric oxide can cause axonal degeneration, especially if the metabolic demands of the axons are increased by impulse activity (Smith et al. 2001). We also described that axons could be protected from the degeneration by therapy with the sodium channel blocking agents, flecainide, lignocaine, and tetrodotoxin (Kapoor et al. 2000, 2003; Garthwaite et al. 2002). In these experiments with nitric oxide, there was no known involvement of inflammatory cells, and so again the beneficial effects of the sodium channel blocking agents was presumably mediated by direct effects on the nerve fibers themselves. Taken together, the evidence suggests that sodium channel blocking agents contribute *directly* to the preservation of axons in EAE by limiting sodium influx in axons, thereby preventing reverse operation of NCX and the initiation of a Ca^{2+}-mediated injury cascade in axons. Sodium channel blocking agents have also been shown to protect axons in vivo and in vitro in models of traumatic white matter injury (Agrawal and Fehlings 1996; Imaizumi et al. 1997; Teng and Wrathall 1997; Rosenberg et al. 1999; Hains et al. 2004), consistent with this role of sodium channels in the axonal injury cascade.

8.6 Sodium Channel Blockade Has a Direct Effect on Immune Cells

It is also possible that sodium channel blockers might *indirectly* protect axons by their actions on immune cells. Sodium channels have been shown to be expressed by macrophages and microglia (Norenberg et al. 1994; Schmidtmayer et al. 1994; Korotzer and Cotman 1992; Eder 1998; Craner et al. 2005; Black et al. 2007; Carrithers et al. 2007) and some T lymphocytes (DeCoursey et al. 1985; Gallin 1991; Roselli et al. 2006; Black et al. 2007), and blockade of sodium channels in immune cells has functional consequences (see Roselli et al. 2006). For instance, in vitro application of the sodium channel blockers TTX and phenytoin attenuates the phagocytotic activity of macrophages and microglia (Craner et al. 2005; Carrithers et al. 2007; Black et al. 2009), the ATP-stimulated migration of microglia (Black et al. 2009), and podosome formation, which is important for motility in macrophages (Carrithers et al. 2009). In addition, in vitro exposure of natural killer cells to clinical levels of phenytoin significantly depresses α-interferon-augmented cytotoxicity (Margaretten et al. 1987). Phenytoin also has an inhibitory effect on the release of IL-1α, IL-1β, and TNFα from cultured microglia (Black et al. 2009). Phenytoin administered in vivo has been reported to depress humoral and cellular immune responses, the activity of killer cells, and IFN-γ production by $CD4^+$ and $CD8^+$ lymphocytes (Andrade-Mena et al. 1994; Okamoto et al. 1988; Yamada et al. 2000). As discussed above, sodium channel blocking agents ameliorate immune cell infiltration within spinal cords in C57/Bl6 mice and DA rats with EAE (Bechtold et al. 2004; Craner et al. 2005; Black et al. 2007). These observations are consistent with the findings of Okada et al. (2001) in which mice administered phenytoin and challenged with keyhole limpet hemocyanin exhibited a Th2 immune response, in contrast to the Th1 response of untreated mice. These data strongly support the conclusion that in vivo administration of sodium channel blocking agents to mice and rats with EAE has an anti-inflammatory action, which could result in reduced axonal loss. Thus, sodium channel blockers may protect axons in EAE via two independent, converging mechanisms, one acting directly on axons and limiting sodium influx and consequent reverse (Ca^{2+} importing) Na^+/Ca^{2+} exchange and the second acting on immune cells to ameliorate the inflammatory response, thereby limiting the immune-mediated attack on axons.

Irrespective of the exact mechanisms by which sodium channel blocking agents exert their actions, it is clear that continuous treatment with these agents enhances clinical status and provides axonal preservation in EAE. Abrupt withdrawal of phenytoin and carbamazepine, however, results in acute exacerbation of EAE in C57/Bl6 mice, with substantial mortality. It is unclear, at this time, whether similar exacerbation would occur with additional sodium channel blocking agents or in different EAE models. Since accumulating evidence indicates that sodium channel blocking agents have functional effects on immune cell activity as well as axons, questions about the long-term effects of sodium channel blockade with agents such as phenytoin, carbamazepine and lamotrigine and about the effects of withdrawal of these agents, in neuroinflammatory disorders should be addressed. It can already be reported, however, that although an exacerbation is observed upon abruptly withdrawing sodium channel blocking agents from animals with EAE, no major clinical rebound has been observed on tapered withdrawal of the trial medication (lamotrigine or placebo) in patients with multiple sclerosis who have been treated in a current clinical trial for axonal protection (R Kapoor, D Miller, R Hughes, J Furby, T Hayton, and K Smith, unpublished observations). Whether a small effect occurs upon withdrawal will require knowledge to be obtained only upon the eventual unblinding of the clinical and imaging outcomes. The question of whether sodium channel blocking agents will provide long-term benefit in human neuroinflammatory disease clearly merits careful study (see Waxman 2008; Kapoor 2008), and indeed clinical trials to explore this possibility are underway in London (lamotrigine in secondary progressive MS), Philadelphia (topiramate in addition to Avonex in relapsing–remitting MS), and San Francisco (riluzole in addition to Avonex in patients with clinically isolated syndromes).

8.7 Conclusions

In this chapter, we have reviewed studies from our laboratories that demonstrated neuroprotective actions of sodium channel blocking agents, including flecainide, phenytoin, carbamazepine and lamotrigine, in multiple rodent models of neuroinflammation/demyelination, which are widely utilized to mimic aspects of multiple sclerosis (MS). Sodium channel blockade provides robust protection of optic nerve, and spinal cord axons, preserves action potential conduction, significantly diminishes immune cell infiltration and attenuates neurological deficits. However, abrupt withdrawal of sodium channel blockers can result in acute exacerbation of EAE, with increased mortality and immune cell infiltration, although no adverse effects were observed upon tapered withdrawal in clinical trial in MS. Blockade of sodium channels may provide neuroprotection by two converging pathways: (1) by limiting reverse operation of the axonal sodium–calcium exchanger (NCX), preventing import of damaging levels of Ca^{2+} and (2) by attenuating multiple effector functions of activated microglia. Our studies reviewed here provide a rationale for cautious planning and implementation of clinical studies utilizing sodium channel blocking agents in patients with MS.

Acknowledgments Research in the authors' laboratories is supported by funds from the National Multiple Sclerosis Society (RG 1912), the Medical Research Service and Rehabilitation Research Service, Department of Veterans Affairs, the Brain Research Trust, the European Union (NeuroproMiSe), the Medical Research Council (UK), and the Multiple Sclerosis Society of G.B. and N.I.

References

Aboul-Enein F, Lassmann H (2005) Mitochondrial damage and histotoxic hypoxia: a pathway of tissue injury in inflammatory brain disease? Acta Neuropathol 109:49–55

Agrawal SK, Fehlings MG (1996) Mechanisms of secondary injury to spinal cord axons in vitro: role of Na^+, Na^+-K^+-ATPase, the Na^+-H^+ exchanger, and the Na^+-Ca^{2+} exchanger. J Neurosci 16:545–552

Andrade-Mena CE, Sardo-Olmedo JAJ, Ramirez-Lizardo EJ (1994) Effects of phenytoin administration on murine immune function. J Neuroimmunol 50:3–7

Bechtold DA (2004) Axonal protection in experimental models of inflammatory demyelinating disease. PhD Thesis, University of London

Bechtold DA, Kapoor R, Smith KJ (2002) Axonal protection mediated by flecainide therapy in experimental inflammatory demyelinating disease. J Neurol 249(suppl 1):204

Bechtold DA, Kapoor R, Smith KJ (2004) Axonal protection using flecainide in experimental autoimmune encephalomyelitis. Ann Neurol 55:607–616

Bechtold DA, Yue X, Evans RM, Davies M, Gregson NA, Smith KJ (2005) Axonal protection in experimental autoimmune neuritis by the sodium channel blocking agent flecainide. Brain 128: 18–28

Bechtold DA, Miller SJ, Dawson AC, Sun Y, Kapoor R, Berry D, Smith KJ (2006) Axonal protection achieved in a model of multiple sclerosis using lamotrigine. J Neurol 253:1542–1551

Bitsch A, Schuchardt J, Bunkowski S, Kuhlmann T, Brück W (2000) Acute axonal injury in multiple sclerosis. Correlation with demyelination and inflammation. Brain 123:1174–1183

Bitsch A, Kuhlmann T, Stadelmann C, Lassmann H, Lucchinetti C, Brück W (2001) A longitudinal MRI study of histopathologically defined hypointense multiple sclerosis lesions. Ann Neurol 49(6):793–796

Bjartmar C, Kidd G, Mork S, Rudick R, Trapp BD (2000) Neurological disability correlates with spinal cord axonal loss and reduced N-acetyl-aspartate in chronic multiple sclerosis patients. Ann Neurol 48:893–901

Bjartmar C, Kinkel RP, Kidd G, Rudick RA, Trapp BD (2001) Axonal loss in normal-appearing white matter in a patient with acute MS. Neurology 57(7):1248–1252

Black JA, Liu S, Hains BC, Saab CY, Waxman SG (2006) Longterm protection of central axons with phenytoin in monophasic and chronic-relapsing EAE. Brain 129:3196–3208

Black JA, Liu S, Carrithers M, Carrithers LM, Waxman SG (2007) Exacerbation of experimental autoimmune encephalomyelitis after withdrawal of phenytoin and carbamazepine. Ann Neurol 67:21–33

Black JA, Liu S, Waxman SG (2009) Sodium channel activity modulates multiple functions in microglia. Glia 57(10):1072–1081

Bo L, Dawson TM, Wesselingh S, Mork S, Choi S, Kong PA, Hanley D, Trapp BD (1994) Induction of nitric oxide synthetase in demyelinating regions of multiple sclerosis. Ann Neurol 36: 78–786

Brosnan CF, Battistini L, Raine C, Dickson DW, Casadevall A (1994) Reactive nitrogen intermediates in human neuropathology: an overview. Dev Neurosci 16:152–161

Brown GC (2007) Nitric oxide and mitochondria. Front Biosci 12:1024–1033

Brown AM, McFarlin DE (1981) Relapsing experimental allergic encephalomyelitis in the SJL/J mouse. Lab Invest 45(3):278–284

Carrithers MD, Dib-Hajj S, Carrithers LM, Tokmoulina G, Pypaert M, Jonas EA, Waxman SG (2007) Expression of the voltage-gated sodium channel Nav1.5 in the macrophage late endosome regulates endosomal acidification. J Immunol 178:7822–7832

Carrithers MD, Chatterjee G, Carrithers LM, Offoha R, Iheagwara U, Rahner R, Graham M, Waxman SG (2009) Regulation of podosome formation in macrophages by a splice variant of the sodium channel SCN8A. J Biol Chem 284(12):8114–8126

Charcot M (1868) Histologie de la sclerose en plaques. Gaz Hosp 141(554–5):557–558

Craner MC, Damarjian TG, Liu S, Hains BC, Lo AC, Black JA, Newcombe J, Cuzner ML, Waxman SG (2005) Sodium channels contribute to microglia/macrophage activation and function in EAE and MS. Glia 49:220–229

DeCoursey TE, Chandy KG, Gupta S et al (1985) Voltage-dependent ion channels in T-lymphocytes. J Neuroimmunol 10:71–95

DeLuca GC, Ebers GC, Esiri MM (2004) The extent of axonal loss in the long tracts in hereditary spastic paraplegia. Neuropathol Appl Neurobiol 30(6):576–584

DeStefano N, Matthews PM, Fu L, Narayanan S, Stanley J, Francis GS, Antel JP, Arnold DL (1998) Axonal damage correlates with disability in patients with relapsing-remitting multiple sclerosis. Results of a longitudinal magnetic resonance spectroscopy study. Brain 121:1469–1477

DeStefano N, Narayanan S, Francis GS, Arnaoutelis R, Tartaglia MC, Antel JP, Matthews PM, Arnold DL (2001) Evidence of axonal damage in the early stages of multiple sclerosis and its relevance to disability. Arch Neurol 58(1):65–70

Dutta R, Trapp BD (2007) Pathogenesis of axonal and neuronal damage in multiple sclerosis. Neurology 68(22 Suppl 3):S22–S31

Dutta R, McDonough J, Yin X, Peterson J, Chang A, Torres T, Gudz T, Macklin WB, Lewis DA, Fox RJ, Rudick R, Mirnics K, Trapp BD (2006) Mitochondrial dysfunction as a cause of axonal degeneration in multiple sclerosis patients. Ann Neurol 59:478–489

Eder C (1998) Ion channels in microglia (brain macrophages). Am J Physiol 275:C327–C342

Evangelou N, Esiri MM, Smith S, Palace J, Matthews PM (2000) Quantitative pathological evidence for axonal loss in normal appearing white matter in multiple sclerosis. Ann Neurol 47(3): 391–395

Ferguson B, Matyszak MK, Esiri MM, Perry VH (1997) Axonal damage in acute multiple sclerosis lesions. Brain 120:393–399

Fern R, Ransom BR, Stys PK, Waxman SG (1993) Pharmacological protection of CNS white matter during anoxia: actions of phenytoin, carbamazepine, and diazepam. J Pharmacol Exp Ther 266:1549–1555

Fraser SP, Diss JKJ, Lloyd LJ et al (2004) T-lymphocyte invasiveness: control by voltage-gated Na$^+$ channel activity. FEBS Lett 569:191–194

Gallin EK (1991) Ion channels in leukocytes. Physiol Rev 71:775–811

Ganter P, Prince C, Esiri MM (1999) Spinal cord axonal loss in multiple sclerosis: a post-mortem study. Neuropathol Appl Neurobiol 25(6):459–467

Garthwaite G, Goodin DA, Batchelor AM, Leeming K, Garthwaite J (2002) Nitric oxide toxicity in CNS white matter: an in vitro study using rat optic nerve. Neuroscience 109:145–155

Gold R, Linington C, Lassmann H (2006) Understanding pathogenesis and therapy of multiple sclerosis via animal models: 70 years of merits and culprits in experimental autoimmune encephalomyelitis research. Brain 129:1953–1971

Hains BC, Waxman SG (2005) Neuroprotection by sodium channel blockade with phenytoin in an experimental model of glaucoma. Invest Ophthalmol Vis Sci 46(11):4164–4169

Hains BC, Saab CY, Lo AC, Waxman SG (2004) Sodium channel blockade with phenytoin protects spinal cord axons, enhances axonal conduction, and improves functional motor recovery after contusion SCI. Exp Neurol 188:365–377

Hewitt KE, Stys PK, Lesiuk HJ (2001) The use-dependent sodium channel blocker mexiletine is neuroprotective against global ischemic injury. Brain Res 898(2):281–287

Imaizumi T, Kocsis JD, Waxman SG (1997) Anoxic injury in the rat spinal cord: pharmacological evidence for multiple steps in Ca^{2+}-dependent injury of the dorsal columns. J Neurotrauma 14: 299–311

Kapoor R (2008) Sodium channel blockers and neuroprotection in multiple sclerosis using lamotrigine. J Neurol Sci 274(1–2):54–56

Kapoor R, Li R-G, Smith KJ (1997) Slow sodium-dependent potential oscillations contribute to ectopic firing in mammalian demyelinated axons. Brain 120:647–652

Kapoor R, Davies M, Smith KJ (1999) Temporary axonal conduction block and axonal loss in inflammatory neurological disease: a potential role for nitric oxide? Oxidative/Energy Metabolism in Neurodegenerative Disorders. Ann NY Acad Sci 893:304–308

Kapoor R, Blaker PA, Hall SM, Davies M, Smith KJ (2000) Protection of axons from degeneration resulting from exposure to nitric oxide. Rev Neurol (Paris) 156:3S67

Kapoor R, Davies M, Blaker PA, Hall SM, Smith KJ (2003) Blockers of sodium and calcium entry protect axons from nitric-oxide mediated degeneration. Ann Neurol 53:174–180

Kaptanoglu E, Solaroglu I, Surucu HS, Akbiyik F, Beskonakli E (2005) Blockade of sodium channels by phenytoin protects ultrastructure and attenuates lipid peroxidation in experimental spinal cord injury. Acta Neurochi (Wien) 147:405–412

Khan N-A, Poisson JP (1999) 5-HT_3 receptor-channels coupled with Na^+ influx in human T-cells: role in T cell activation. J Neuroimmunol 99:53–60

Kornek B, Storch MK, Weissert R, Wallstroem E, Stefferl A, Olsson T, Linington C, Schmidbauer M, Lassmann H (2000) Multiple sclerosis and chronic autoimmune encephalomyelitis: a comparative quantitative study of axonal injury in active, inactive, and remyelinated lesions. Am J Pathol 157:267–276

Korotzer AR, Cotman CW (1992) Voltage-gated currents expressed by rat microglia in culture. Glia 6:81–88

Kuhlmann T, Lingfeld G, Bitsch A, Schuchardt J, Bruck W (2002) Acute axonal damage in multiple sclerosis is most extensive in early disease stages and decreases over time. Brain 125:2202–2212

Lai Z-F, Chen Y-Z, Nishimura Y, Nishi K (2000) An amiloride-sensitive and voltage-dependent Na^+ channel in an HLA-DR-restricted human T cell clone. J Immunol 165:83–90

Lo AC, Saab CY, Black JA, Waxman SG (2003) Phenytoin protects spinal cord axons and preserves axonal conduction and neurological function in a model of neuroinflammation in vivo. J Neurophysiol 90:3566–3571

LoPachin RM Jr, Stys PK (1995) Elemental composition and water content of rat optic nerve myelinated axons and glial cells: effects of in vitro anoxia and reoxygenation. J Neurosci 15(10):6735–6746

Lovas G, Szilagyi N, Majtenyi K, Palkovits M, Komoly S (2000) Axonal changes in chronic demyelinated cervical spinal cord plaques. Brain 123:308–317

Mahad D, Ziabreva I, Lassmann H, Turnbull D (2008) Mitochondrial defects in acute multiple sclerosis lesions. Brain 34:577–589

Makowska A, Bechtold DA, Sajic M, Gregson NA, Hughes RA, Smith KJ (2004) Sodium channel blocking agents affect T-cell function. J Neuroimmunol 154:88 (abstract)

Margaretten NC, Hincks JR, Warren RP, Coulombe RA Jr (1987) Effects of phenytoin and carbamazepine on human natural killer cell activity and genotoxicity *in vitro*. Toxicol Appl Pharmacol 87:10–17

McDonald WI (1972) The time course of conduction failure during degeneration of a central tract. Exp Brain Res 14(5):550–556

Narayanan S, Francis SJ, Sled JG, Santos AC, Antel S, Levesque I, Brass S, Lapierre Y, Sappey-Marinier D, Pike GB, Arnold DL (2006) Axonal injury in the cerebral normal-appearing white matter of patients with multiple sclerosis is related to concurrent demyelination in lesions but not to concurrent demyelination in normal-appearing white matter. Neuroimage 29:637–642

Norenberg W, Illes P, Gebicke-Haerter PJ (1994) Sodium channels in isolated human brain macrophages (microglia). Glia 10:165–172

Okada K, Sugiura T, Kuroda E, Tsuji S, Yamashita U (2001) Phenytoin promotes Th2 type immune response in mice. Clin Exp Immunol 124:406–413

Okamoto Y, Shimizu K, Tamura K, Miyao Y, Yamada M, Tsuda N, Matsui Y, Mogami H (1988) Effects of phenytoin on cell-mediated immunity. Cancer Immunol Immunother 26:176–179

Onuki M, Ayers MM, Bernard CC, Orian JM (2001) Axonal degeneration is an early pathological feature in autoimmune-mediated demyelination in mice. Microsc Res Tech 52(6):731–739

Pöllmann W, Feneberg W (2008) Current management of pain associated with multiple sclerosis. CNS Drugs 22(4):291–324

Raine CS, Cross AH (1989) Axonal dystrophy as a consequence of long-term demyelination. Lab Invest 60(5):714–725

Redford EJ, Kapoor R, Smith KJ (1997) Nitric oxide donors reversibly block axonal conduction: demyelinated axons are especially susceptible. Brain 120:2149–2157

Rogowski MA, Porter RJ (1990) Antiepileptic drugs: pharmacological mechanisms and clinical efficacy with consideration of promising developmental stage compounds. Pharmacol Rev 42: 223–286

Roselli F, Livrea P, Jirillo E (2006) Voltage-gated sodium channel blockers as immunomodulators. Rec Pat CNS Drug Discov 1:83–91

Rosenberg LJ, Teng YD, Wrathall JR (1999) Effects of the sodium channel blocker tetrodotoxin on acute white matter pathology after experimental contusive spinal cord injury. J Neurosci 19:6122–6133

Schmidtmayer J, Jacobsen C, Miksch G, Sievers J (1994) Blood monocytes and spleen macrophages differentiate into microglia-like cells on monolayers of astrocytes: membrane currents. Glia 12:259–267

Schwartz G, Fehlings MG (2001) Evaluation of the neuroprotective effects of sodium channel blockers after spinal cord injury: improved behavioral and neuroanatomical recovery with riluzole. J Neurosurg 94(2 Suppl):245–256

Sedgwick JD, Schwender S, Imrich H (1991) Isolation and direct characterization of resident microglial cells from the normal and inflamed central nervous system. Proc Natl Acad Sci USA 88:7438–7442

Smith KJ, Lassmann H (2002) The role of nitric oxide in multiple sclerosis. Lancet Neurol 1: 232–241

Smith KJ, Kapoor R, Hall SM, Davies M (2001) Electrically active axons degenerate when exposed to nitric oxide. Ann Neurol 49:470–476

Soane L, Kahraman S, Kristian T, Fiskum G (2007) Mechanisms of impaired mitochondrial energy metabolism in acute and chronic neurodegenerative disorders. J Neurosci Res 85:3407–3415

Steiner TJ, Silveira C, Yuan AWC, North Thames Lamictal Study Group (1994) Comparison of lamotrigine (Lamictal) and phenytoin in newly diagnosed epilepsy. Epilepsia 35:61

Steinman L, Zamvil SS (2006) How to successfully apply animal studies in experimental allergic encephalomyelitis to research on multiple sclerosis. Ann Neurol 60(1):12–21

Stys PK, Waxman SG, Ransom BR (1991) Na^+-Ca^{2+} exchanger mediates Ca2+ influx during anoxia in mammalian central nervous system white matter. Ann Neurol 30(3):375–380

Stys PK, Ransom BR, Waxman SG (1992a) Tertiary and quaternary local anesthetics protect CNS white matter from anoxic injury at concentrations that do not block excitability. J Neurophysiol 67(1):236–240

Stys PK, Waxman SG, Ransom BR (1992b) Ionic mechanisms of anoxic injury in mammalian CNS white matter: role of Na^+ channels and Na^+-Ca^{2+} exchanger. J Neurosci 12:430–439

Stys PK, Sontheimer H, Ransom BR, Waxman SG (1993) Non-inactivating TTX-sensitive Na^+ conductance in rat optic nerves. Proc Natl Acad Sci USA 90:6976–6980

Teng YD, Wrathall JR (1997) Local blockade of sodium channels by tetrodotoxin ameliorates tissue loss and long-term functional deficits resulting from experimental spinal cord injury. J Neurosci 17:4359–4366

Trapp BD, Peterson J, Ransohoff RM, Rudick R, Mörk S, Bö L (1998) Axonal transection in the lesions of multiple sclerosis. N Engl J Med 338(5):278–285

Waxman SG (2008) Sodium channels and neuroprotection in multiple sclerosis: current status. Nat Clin Pract Neurol 4(3):159–169

Waxman SG, Black JA, Stys PK, Ransom BR (1992) Ultrastructural concomitants of anoxic injury and early post-anoxic recovery in rat optic nerve. Brain Res 574(1–2):105–119

Waxman SG, Black JA, Ransom BR, Stys PK (1994) Anoxic injury of rat optic nerve: ultrastructural evidence for coupling between Na+ influx and Ca(2+)-mediated injury in myelinated CNS axons. Brain Res 644(2):197–204

White SR, Black PC, Samathanam GK, Paros KC (1992) Prazosin suppresses development of axonal damage in rats inoculated for experimental allergic encephalomyelitis. J Neuroimmunol 39:211–218

Wit AL, Rosen MR (1983) Pathophysiologic mechanisms of cardiac arrhythmias. Am Heart J 106(4 Pt 2):798–811

Wujek JR, Bjartmar C, Richer E, Ransohoff RM, Yu M, Tuohy VK, Trapp BD (2002) Axon loss in the spinal cord determines permanent neurological disability in an animal model of multiple sclerosis. J Neuropathol Exp Neurol 61:23–32

Yamada M, Ohkawa M, Tamura K et al (2000) Anticonvulsant-induced suppression of IFN-γ production by lymphocytes obtained from cervical lymph nodes in glioma-bearing mice. J Neurooncol 47:125–132

Zamvil SS, Steinman L (1990) The T lymphocyte in experimental allergic encephalomyelitis. Ann Rev Immunol 8:579–621

Zamvil SS, Nelson PA, Mitchell DJ, Knobler RL, Fritz RB, Steinman L (1985) Encephalitogenic T cell clones specific for myelin basic protein. An unusual bias in antigen recognition. J Exp Med 162(6):2107–2124

Chapter 9
Effects of Current Medical Therapies on Reparative and Neuroprotective Functions in Multiple Sclerosis

Jack P. Antel and Veronique E. Miron

9.1 Introduction

9.1.1 Pathology and Repair in Multiple Sclerosis Lesions

Multiple sclerosis (MS) is most frequently clinically characterized by an initial relapsing–remitting phase that evolves into a secondary progressive course; these phases can also be ongoing concurrently. Each of these clinical disease aspects reflects distinctive pathologic substrates, each of which may require its own therapeutic strategies. The pathologic correlate of clinical relapses is the development of new demyelinating lesions within the CNS that feature an inflammatory infiltrate comprised of cells of the adaptive (lymphocytes) and innate (myeloid cells) systems that have transgressed the blood–brain barrier (BBB) and/or the subarachnoid space/brain barrier (Kivisakk et al. 2003). Axonal transections are a further feature of the acute lesions (Trapp et al. 1998). Magnetic resonance imaging (MRI) studies indicate that multiple such lesions can occur without apparent clinical symptoms (Engell 1989). Recovery from relapses reflects the combined effects of multiple factors including resolution of inflammation, axonal adaptation to demyelination, remyelination, and cerebral reorganization. The pathologic correlates of the later progressive disease phase are even more complex involving changes both within initial

---

J.P. Antel (✉)
Neuroimmunology Unit, Montreal Neurological Institute, McGill University,
Montreal, QC, Canada H3A 2B4
e-mail: jack.antel@mcgill.ca

V.E. Miron
Scottish Centre for Regenerative Medicine, The University of Edinburgh,
5 Little France Drive, Edinburgh EH16 4UU, UK

Table 9.1 Basis for effects of current medical therapies on neuroprotection and repair in multiple sclerosis

Indirect effects	Active (consequences of molecular changes in immune cells exposed to systemic therapy)	Molecules expressed by modulated immune cells could mediate injury or promote protection/repair of OLs/neurons
		Effects of modulated immune cells on CNS microenvironment (astrocytes and microglia) with consequences for OLs/neurons
	Passive (deletion of immune cells or blocking CNS access)	Loss of molecular mechanisms involved in clearance of tissue injury (e.g., macrophages)
		Loss of physiological signals involved in repair and protection (neurotrophins)
Direct effects		Direct interactions with OLs/neurons
		Impact on CNS microenvironment (microglia and astrocytes) with consequences for OLs/neurons

lesions and in normal appearing white matter (NAWM) (Fu et al. 1998). Intralesional changes include further loss of oligodendrocytes (OLs) and axons, enhanced gliosis, and apparent failure of remyelination. Such lesions are dominated by innate immune cells (microglia/macrophages) rather than lymphocytes (Revesz et al. 1994). Observed changes in NAWM include continued decrease in axonal density and activated glial cells (microglia, astrocytes). There is also an apparent increase of lesions within the gray matter over time characterized by loss of neurons and demyelination without marked inflammation (Kidd et al. 1999). Ongoing remyelination is more apparent in recent rather than more chronic lesions and may be more prevalent in gray matter than in white matter lesions (Prineas and Connell 1979; Raine and Wu 1993; Stadelmann and Bruck 2008). Disease progression is thus likely to involve variable contributions by ongoing tissue injury, degeneration of previously injured OLs and neurons, and failure to initiate or sustain repair mechanisms (Table 9.1).

9.1.2 Assessment of Effects of Multiple Sclerosis-Directed Immunomodulatory Therapies on Neuroprotection and Repair

Given the complex pathologies of MS, the effects of current and emerging therapies for MS on reparative and neuroprotective functions are likely to reflect their impact on the injury and recovery aspects of the disease. The impact on these aspects need not be congruent as these two phases of the disease may be differentially affected by a particular drug. The emphasis of this chapter will be on agents that are primarily designed to act on the immune aspects of the disease process, as only these are currently approved or in advanced clinical trials. Although considerable data exist regarding the potential benefits of therapies that would be considered to impact directly on neuroprotection and/or repair in animal models of MS, particularly experimental autoimmune encephalomyelitis (EAE), their translation into the clinical MS field has not yet occurred. Examples of these include trophic factors and hormones such as prolactin, estrogens, and progesterone. One notes that at least some of these "neurobiological" directed therapies such as the neurohormones are already known to have effects on the immune system that may or may not be of benefit for MS (Luger et al. 1996).

The modern therapeutic era of multiple sclerosis began in the early 1990s with the completion of double-blind clinical trials using interferon β (IFNβ) and glatiramer acetate (GA) (Bornstein et al. 1987; Jacobs et al. 1995). These showed partial but sufficiently significant reductions in frequency and severity of disease relapses which resulted in regulatory approval of these agents. The clinical effects of these agents were associated with even more significant decreases in the number of new inflammatory lesions in the CNS, as measured by magnetic resonance imaging (MRI) techniques. MRI studies with GA further documented that the extent of initial injury, as judged by the irreversibility of initial lesions (formation of persistent "black holes"), was reduced by the therapy. Magnetic resonance (MR) spectroscopy studies indicate that IFNβ therapy can result in an increase in N-acetyl aspartate (NAA) signal, a measure of axonal integrity (Narayanan et al. 2001).

Natural history studies from the pretherapeutic era indicated that the frequency of relapses early in the disease course was linked with subsequent development of disability. The relatively short duration (2 years) of the blinded phase of the pivotal clinical trials makes it difficult to conclude whether there was an effect on disability that would reflect mechanisms other than reduced initial tissue injury, such as enhanced repair and delayed neural degeneration. The major clinical and imaging challenge of how to evaluate the effects of therapies on each of these disease aspects (injury, repair, and neuroprotection) is discussed in a separate chapter. An ongoing challenge in both MS and the animal model EAE is to define the extent to which capacity for repair is linked to sites and extent of the initial injury. There is strong documentation that later disease progression can occur in MS in the absence of measurable continued clinical relapses or MRI-defined new lesion formation (Phadke and Best 1983; Gronseth and Ashman 2000; Mews et al. 1998). Long-term follow-up studies, although limited by methodologic problems, suggest that early and prolonged use of the currently approved agents IFNβ and GA in patients with the relapsing form of MS reduces long-term disease progression (Kappos et al. 2006a, b; Miller et al. 2008). However, clinical trials with these immunomodulatory agents in established progressive phases of MS, either primary or secondary progressive, did not lead to their approval for these phases of disease. This is likely consequent to the marked decrease in inflammation within lesions in these phases compared to earlier stages.

9.1.3 Indirect and Direct Effects of Immunomodulatory Therapies on Neural Cells

Interferon beta (IFNβ) and glatiramer acetate (GA) are currently applied therapies which are systemically administered yet not recognized to cross the BBB, indicating that their primary effects are on the systemic immune system and/or on the BBB. Since neither agent completely abrogates continued immune cell trafficking across the BBB, they may still impact on neuroprotection and repair processes within the CNS by modulating the properties of systemic immune cells that may access the brain parenchyma. Such modulation would involve induction of secreted or cell surface molecules that could promote or impede processes of CNS tissue injury and repair. We will refer to these as *active indirect* mechanisms of action on neuroprotective or repair mechanisms (Table 9.1); specific mechanisms that apply to each of the agents are discussed later in this chapter. Active indirect mechanisms would also apply to evolving immune-directed therapeutic strategies aimed at selectively modulating the properties of disease-relevant immune components, particularly proinflammatory autoreactive T cells. Examples of the "immune deviation" approach include immunizing with native or altered myelin components or their encoding DNAs (Bar-Or et al. 2007; Bielekova et al. 2000) (Table 9.2).

The ongoing development of therapeutic agents in MS has continued to focus on agents that act on the systemic immune system or its capacity to access the CNS. A large family of such agents is comprised of monoclonal antibodies (mabs). Natalizumab, which recognizes the adhesion molecule very late antigen 4 (VLA-4) (also known as $\alpha_4\beta_1$-integrin), markedly reduces relapse frequency and MRI activity (Miller et al. 2003).VLA-4 expression on the surface of endothelial cells is required for binding of activated T lymphocytes to the BBB and subsequent transmigration into the CNS. The blockade of this interaction by natalizumab prevents T cell infiltration into the brain parenchyma of treated MS patients and is thought to underlie its anti-inflammatory properties (Keszthelyi et al. 1996; Yednock et al. 1992). Despite its effects on reducing new lesion formation, clinical trials attempting to show that the therapy-enhanced rate of recovery from relapses were inconclusive (O'Connor et al. 2004). Additional VLA-4 antagonists/inhibitors, such as Firategrast and TV-1102, are being evaluated as potential therapies for MS (Hohlfeld et al. 2011). More recently evaluated monoclonal antibody therapies include daclizumab which blocks the interleukin (IL)-2 receptor (Rose et al. 2007), rituximab and ofatumumab which recognize cluster designation (CD) 20 on B cell lineage cells (Hauser et al. 2008; Buttmann 2010), and alemtuzumab that recognizes CD52 expressed by most lymphoid cells (Coles et al. 2008). Other agents include an oral therapy cladribine, the purine analog which disrupts DNA synthesis and repair in lymphocytes (Giovannoni et al. 2010), and the immunomodulators teriflunomide and ibudilast which inhibit proliferation and activation of lymphocytes (Barkhof et al. 2010; Claussen and Korn 2012). The high degree of efficacy of some of these agents in reducing immune activity or eliminating immune cell trafficking into the CNS raises the issue of whether there could be interruption of physiological immune

Table 9.2 Indirect and direct effects of clinically applied therapies on neuroprotection and repair

	Indirect effects		Direct effects	
	Active	Passive	OLs/neurons	CNS microenvironment
IFNβ	– Immune deviation (altered cytokine profile) – Induction of death receptors on immune cells (fas, TRAIL) – Induction of neurotrophic and growth factors in immune astrocytic, and endothelial cells	– Reduced immune cell trafficking to CNS – Removal of endogenously produced IFNβ by IFNβ-directed antibodies		
GA	– Immune deviation (Th2 bias) that alter microglia antigen presentation – Induction of neurotrophic factors – Increased "protective immunity"			
Antibodies-immune cell and adhesion molecule reactive	– Immune deviation (i.e., reduction in immune cell trafficking across BBB)	– Reduction of immune mechanisms involved in tissue remodeling and regeneration	– Germline antibodies promote remyelination (i.e., IgM)	
Chemotherapeutic agents		– Reduction of immune mechanisms involved in tissue remodeling and regeneration	– Direct cytotoxicity of OLs/neurons	
Myelin vaccines	– Immune deviation			

(continued)

Table 9.2 (continued)

	Indirect effects		Direct effects	
	Active	Passive	OLs/neurons	CNS microenvironment
Minocycline		– Impairment of microglia/macrophage function		– Inhibition of proinflammatory mediator production by microglia and astrocytes – Inhibition of astrocyte-mediated glutamate sequestration
Statins	– Immune deviation	– Reduced immune cell trafficking to CNS	– Modulation of cytoskeletal and survival properties of progenitors, mature OLs, and neurons – Experimental data indicate effects are dose- and treatment duration-dependent	
FTY 720 (fingolimod)		– Reduction of entry of naïve and central memory T cells and B cells into CNS	– Modulation of cytoskeletal and survival properties of progenitors and mature OLs – Experimental data indicate effects are dose- and treatment duration-dependent	
IFNγ		– Dose-dependent ER stress response confers protection – Dose-dependent enhancement of progenitor differentiation and remyelination		

function within the CNS. We refer to these as *passive indirect* mechanisms. The balance between potential benefit and toxicity of these agents will determine whether they receive regulatory approval and their use in clinical practice (Tables 9.1, 9.2).

The physiological effects of the immune system are usually considered in context of controlling infection or tumor formation; in this chapter, we focus on the role of the immune constituents in neuroprotection/repair. Inflammation has been shown to be important for the activation of progenitor cells in repair in animal models of demyelination with regard to inducing an activated progenitor phenotype, clearance of inhibitory myelin debris by reactive macrophages, and secretion of molecules that can influence repair (Chari et al. 2006; Glezer et al. 2006; Kotter et al. 2006). Activated immune cells and glia, which can be found in MS lesions, can produce chemokines, cytokines, and growth factors which can promote oligodendrocyte progenitor migration, proliferation, and differentiation and induce remyelination (Arnett et al. 2001; Vela et al. 2002; Maysami et al. 2006; Zhang et al. 2006). Schwartz and colleagues have shown that administration of autoreactive T cells enhances recovery of injured CNS tissue, a process referred to as "protective autoimmunity" (Barouch and Schwartz 2002). MBP-reactive T cells have the potential to be neuroprotective via secretion of BDNF upon activation (Kerschensteiner et al. 1999). In addition, the proinflammatory cytokines IFNγ and tumor necrosis factor alpha (TNFα) are shown to have neuroprotection/repair-related properties in experimental models in which these molecules and/or their receptors have been genetically deleted (Arnett et al. 2001; Rodriguez et al. 2003). Administration of IFNγ in vitro or in vivo in the EAE and cuprizone demyelination models activates an adaptive endoplasmic reticulum (ER) stress response program termed the "integrated stress response" (ISR) that protects mature and oligodendrocyte lineage progenitors against subsequent damage (Lin et al. 2008). However, in small clinical trials, systemic administration of IFNγ exacerbated disease in MS patients, as did anti-TNF antibodies and soluble TNF receptor molecules (Panitch et al. 1987; van Oosten et al. 1996; Gomez-Gallego et al. 2008). Together, these findings suggest that one needs to consider the consequences of therapeutic interventions on the dual functions of immune constituents as promoters/effectors of tissue injury in MS as well as potential contributors to tissue protection/repair (Table 9.2).

The question of the extent of access of systemically administered monoclonal antibodies to the CNS remains open, especially in cases of MS with disrupted BBB. Thus unresolved is the effect that the monoclonal antibody-based therapeutic agents mentioned above may have on adaptive (T cell, B cell) or innate immune activity ongoing within the CNS compartment. The receptors targeted by these monoclonal antibodies are not expressed by primary neural cells. Although immunoglobulin access to the CNS is usually considered as being relatively limited, "therapeutic vaccination" with spinal cord homogenates in experimental CNS closed injury models results in high levels of antimyelin antibodies in the CNS. Experience with CNS amyloid-directed therapies using active immunization regimens and systemic administration of antibody indicate that sufficient antibody can access the CNS to lead to removal of significant amounts of the protein. Sufficient amounts of systemically administered naturally occurring IgM antibodies are able to access the CNS of experimental animals so as to promote remyelination (Bieber et al. 2002).

There are now immune-directed therapies entering clinical use or under investigation in MS that can access the CNS, and thus, one needs to consider their direct effects on neural cells in addition to effects on immune cells within the CNS. Chemotherapeutic agents (cyclophosphamide, mitoxantrone) that have broad cytotoxic effects have been, and continue to be, used for MS patients with worsening disease judged to be unresponsive to the currently approved immunomodulatory agents mentioned previously (Table 9.2). High-dose chemotherapy with agents that can access the CNS is shown by in vivo imaging to be associated with enhanced loss of brain tissue (Freedman 2007). There are newly emerging BBB-permeable therapies that were propelled into MS clinical trials due to observations of their anti-inflammatory properties in experimental systems such as EAE, without extensive data on direct CNS effects. The sphingosine-1-phosphate receptor agonists, including the recently approved FTY720 and the more selective agonist BAF312, presents an example of CNS-accessible agents that act via specific receptors that are broadly expressed within the immune system and the CNS. Lipophilic statins have now been evaluated in clinical trials in relapsing–remitting forms of MS and as a means to prevent development of clinically definite MS in patients with clinical isolated syndromes (CIS) (Vollmer et al. 2004; Menge et al. 2005) (Table 9.2). Some newly emerging oral therapies, including fumarate acid esters (i.e., BG00012), lamotrigine, and laquinimod, have been described as having neuroprotective properties and have been assessed clinically in small-scale exploratory trials. Many of these agents that can access the CNS have primarily been used for non-MS clinical applications, yet their established safety profiles relate to target organs other than the CNS.

9.2 Therapeutic Agents with Indirect Effects on Neural Cells (Fig. 9.1)

9.2.1 Interferon Beta

Interferon beta (IFNβ) (also known as Betaseron, Avonex, and Rebif) is a cytokine that is produced by components of the systemic immune system and also within the CNS by astrocytes (Javed and Reder 2006; Boutros et al. 1997). The finding that IFNβ is most clinically effective in the early, inflammation-driven stages of MS suggests primary effects via immunomodulation. IFNβ induces changes in expression of several hundred genes in immune cells that are relevant for cell activation, migration, proliferation, and effector functions (Javed and Reder 2006). As regards the latter, ligands that are members of the tumor necrosis factor (TNF) superfamily, namely, Fas and TNF-related apoptosis induced ligand (TRAIL), are upregulated by IFNβ (Greil et al. 2003). Both can induce caspase-dependent signaling in target cells that leads to programmed cell death. Fas- and TRAIL-deficient animals develop autoim-

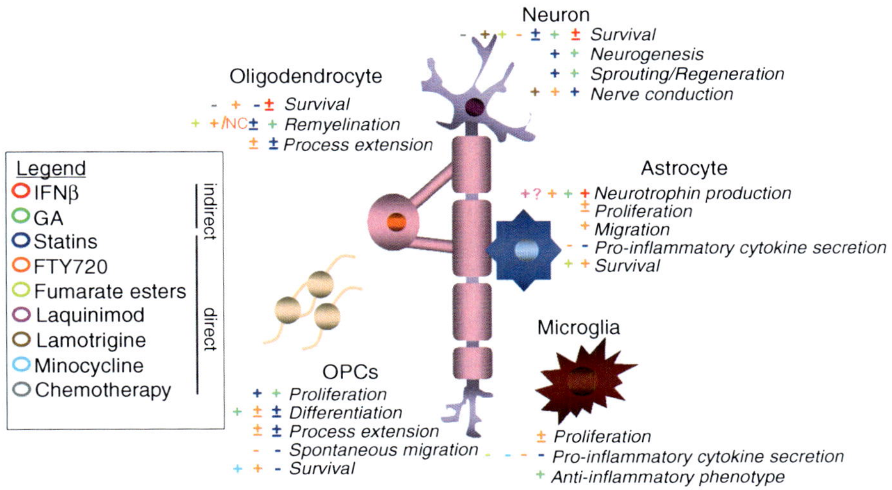

Fig. 9.1 Summary of the effects of MS-directed therapies on neural cell properties important for repair. Listed are properties or responses of neural cells including oligodendrocytes, oligodendrocyte progenitor cells (OPCs), neurons, astrocytes, and microglia. Included are effects on neural cells that may be indirect [i.e., from BBB-impermeable interferon (IFN)-β (*red*) and glatiramer acetate (GA; *green*)] or direct [i.e., from BBB-permeable statins (*blue*), FTY720 (*orange*), fumarate esters (*yellow*), laquinimod (*pink*), lamotrigine (*brown*), minocycline (*turquoise*), chemotherapy (*gray*)]. A promoting/beneficial effect is represented by *plus symbol*, a negative/detrimental effect is represented by *minus symbol*, a situation where both positive and negative effects have been observed is represented by *plus or minus symbol*, no effect/no change is represented by (NC), and where an effect is speculated is represented by *question mark*

mune disorders. The extent of Fas and TRAIL upregulation in lymphocytes has been correlated with a favorable disease course in MS (Wandinger et al. 2003). However, the same TRAIL-dependent mechanisms are also shown to underlie lymphocyte-mediated injury of neural cells in slice culture systems (Aktas et al. 2007). IFNβ would have the potential to worsen neural injury by upregulating the expression of TRAIL and Fas ligands on lymphocytes while the receptor counterparts are concurrently upregulated on oligodendroglia under proinflammatory conditions, as observed following exposure to IFNγ in vitro and in MS lesions in vivo (Wosik et al. 2003). Furthermore, upregulation of p53 in human adult CNS-derived OLs in vitro results in increased Fas and TRAIL receptor expression and enhanced susceptibility to ligand-mediated toxicity (Wosik et al. 2003). Increased levels of p53 expression might be expected to occur in response to an array of insults and has been observed in OLs in MS lesions that featured prominent OL loss (Wosik et al. 2003).

Whether IFNβ therapy could enhance repair by sparing neural cells from immune-mediated damage is unclear based on recent animal studies. IFNβ administration to EAE-afflicted animals can dampen or exacerbate clinical symptoms when disease is induced by Th1 versus Th17 lymphocytes, respectively (Axtell et al. 2010). Coadministration with estrogen receptor ligand β decreases EAE severity while preserving axonal densities in the spinal cord (Du et al. 2010), and IFNβ enhances regeneration of axons in a sciatic nerve crush model (Zanon et al. 2010). IFNβ can also induce neurotrophic factors in a number of cell types. Peripheral blood mononuclear cells (MNCs) derived from MS patients are reported to produce neurotrophic factors during IFNβ therapy (Caggiula et al. 2006; Yoshimura et al. 2010). Exposing human brain endothelial cells (HBECs) to MNCs pretreated in vitro with IFNβ, or to MNCs derived from IFNβ-treated patients, induces nerve growth factor (NGF) production by these cells (Biernacki et al. 2005). It has not been determined whether this represents the pro or mature form of the growth factor; the former may have deleterious effects for neural cells (Hempstead and Salzer 2002). IFNβ itself did not induce NGF production by HBECs (Biernacki et al. 2005). The induction of NGF production by HBECs was only observed when MNCs were derived from MS patients at the early onset phase of the disease and correlated inversely with EDSS and MRI lesion burden (Biernacki et al. 2005), suggesting a possible protective rather than injury-mediating function.

In vitro studies have demonstrated that IFNβ can directly influence neural cell properties. NGF is induced in astrocytes following exposure to IFNβ (Boutros et al. 1997). In addition, IFNβ decreases adult mouse neural progenitor apoptosis under death-inducing conditions in vitro (Hirsch et al. 2009) and decreases neurosphere proliferation (Lum et al. 2009). However the systemically administered recombinant cytokine is unlikely to reach high enough levels in the CNS to stimulate this process. IFNβ is also endogenously produced by astrocytes; its activity can be inhibited in vitro by anti-IFNβ antibodies that develop in some IFNβ-treated patients (Shapiro et al. 2006). The titers of the antibodies in serum are sufficiently high in some IFNβ-treated patients such that some access to the CNS would be predicted based on an expected 1:300 ratio between levels of serum and CSF antibodies. Corticosteroid administration at the onset of IFNβ therapy decreases the risk of development of these neutralizing antibodies (Zarkou et al. 2010). Together, this literature suggests that IFNβ therapy has the potential to indirectly influence neural cell properties by modulating peripheral immune components directly or by inducing anti-IFNβ antibodies which interfere with the regulatory function of endogenous IFNβ produced within the CNS. This can potentially have either advantageous or deleterious consequences for neural function or repair.

9.2.2 Glatiramer Acetate

Glatiramer acetate (GA, also known as copaxone or copolymer-1) is a synthetic random polymer of four amino acids (glutamine, lysine, alanine, tyrosine) that are found in the same molar ratio as in MBP. The protective/reparative effects of GA have been linked with the demonstrated capacity of this agent to induce a shift in polarization of GA-reactive T cells from a Th1 to a Th2 phenotype in the periphery which is maintained in the CNS (Aharoni et al. 2000). Th2 polarized lymphocytes can alter the antigen presentation properties of microglia/macrophages, resulting in the subsequent polarization of naïve T cells into a Th2 rather than Th1 phenotype (Kim et al. 2004). CNS injuries have been suggested to demonstrate a shift toward a Th2 cytokine profile. This has been hypothesized to be a necessary component of repair in the CNS subsequent to injury by promoting neural regeneration and preventing autoimmune processes (Hendrix and Nitsch 2007), supporting the concept of "protective immunity" (Kipnis and Schwartz 2002). Th2-polarized lymphocytes can migrate across an endothelial barrier in vitro more effectively than Th1-biased lymphocytes (Biernacki et al. 2001) and may therefore be expected to more readily impact neural cell properties. To be established is whether "immune deviated" myelin antigen-reactive lymphocytes induced by antigen-specific therapies show similar protective/repair properties as GA-reactive T cells.

GA administration to MS patients results in fewer "black holes" on MRI scans, which can be interpreted as less axonal damage or enhanced repair (Bitsch et al. 2001). Systemic administration of GA has been shown to reduce the extent of neuronal injury in a number of experimental models including CNS trauma, neurodegenerative, and autoimmune diseases (Kipnis et al. 2000; Kipnis and Schwartz 2002; Schori et al. 2001; Aharoni et al. 2003; Benner et al. 2004). For instance, GA enhanced OPC numbers and promoted remyelination in a rodent spinal cord demyelination model (Skihar et al. 2009) and prevented demyelination and neuronal loss, stimulated neurogenesis and remyelination, and encouraged axonal sprouting/regeneration in relapse-remitting and chronic EAE (Aharoni et al. 2005, 2008, 2011). The protective/repair effects of GA-reactive T cells within the CNS, whether completely related to their Th2 properties or not, could be mediated directly on the target cell or indirectly via modulation of disease-relevant properties of glial cells. Both in vitro and in vivo studies have demonstrated that GA induces production of neurotrophins such as BDNF, NT3, NT-4 (Aharoni et al. 2003; Ziemssen et al. 2002; Skihar et al. 2009), growth factors such as IGF-1 (Skihar et al. 2009), and cytokines such as TGFβ (Aharoni et al. 2003, 2005). These can be predicted to protect neural cells against insults (Riley et al. 2004), rescue degenerating neurons (Riley et al. 2004), and affect regeneration of mature OLs and their progenitors (Althaus 2004). Administration of GA to EAE-afflicted animals at various clinical stages resulted in a sustained augmentation of levels of neurotrophins in infiltrating lymphocytes, neurons, astrocytes, and newly migrated neural progenitors within the lesion resulting in attenuation of axonal damage (Aharoni et al. 2003). GA also increases growth and neurotrophic factor production in remyelinating lesions of the rodent spinal cord, and by GA-reactive lymphocytes in vitro, resulting in enhanced repair and OPC specification/differentiation, respectively (Aharoni et al. 2008; Skihar et al. 2009; Ziemssen et al. 2002; Zhang et al. 2010).

Other glial cell functions that could be influenced by transmigrated immune cells in GA-treated patients would include the capacity to produce or remove potential injury mediators (such as glutamate and free radicals), production of trophic molecules, and expression of signaling molecules that modulate migration, differentiation, and myelin production by progenitor cells. Some have suggested that GA-carrying dendritic cells have the potential to cross the BBB (Liu et al. 2007), raising the potential of thus far uninvestigated direct effects of GA on neural cells.

9.3 Therapeutic Agents with Direct Effects on Neural Cells

9.3.1 Chemotherapeutic Agents

A number of cytotoxic agents have been used typically for patients who fail conventional therapies. Examples of such therapies include cyclophosphamide, mitoxantrone, and cladribine. Subcutaneous administration of cladribine (2-chlorodeoxyadenosine) is associated with decreased levels of proinflammatory cytokines and chemokines in the CSF of treated MS patients in remission, possibly indicating a potential effect on CNS cells in addition to peripheral suppression (Bartosik-Psujek et al. 2004). A strategy being adopted from other autoimmune disorders is to administer such agents as initial induction therapies to be followed by immunomodulatory therapy. Such therapies can induce initial loss of brain volume that cannot be attributed to eliminating inflammation. This was observed in the Canadian immunoablative therapy protocol, in which patients underwent serial MRI studies before and after the immunoablation phase that included administration of CNS-accessible agents (busulfan), followed by infusion of autologous bone marrow-derived stem cells (Freedman 2007).

9.3.2 Minocycline

Minocycline is a semisynthetic tetracycline that has been in clinical use for non-neurologic indications (e.g., acne) that readily crosses the BBB. The agent has been shown to alleviate CNS pathology in a number of neurodegenerative and inflammatory animal models including stroke, spinal cord injury, Parkinson's disease, and multiple sclerosis (Arvin et al. 2002; Brundula et al. 2002; Popovic et al. 2002; Wu et al. 2002; Lee et al. 2003; Fox et al. 2005; Nikodemova et al. 2007, 2010). Mechanisms held to account for this include direct effects on blockade of microglia/macrophage activation, inhibition of matrix metalloproteinases, and anti-apoptotic effects (Yrjanheikki et al. 1998; Popovic et al. 2002; Elewa et al. 2006; Nikodemova et al. 2007), as well as indirect effects by limitation of CD4+/CD8+ T cell entry into the CNS and peripheral (but not central) immune deviation (Popovic et al. 2002; Nikodemova et al. 2010). Minocycline can also promote the survival and myelination of OPCs transplanted into rodents (Zhang et al. 2003). Clinical trials in MS patients have shown minocycline to beneficial in reducing the number of active scans and minimizing brain volume change (Zhang et al. 2008). However, the neuroprotective properties of minocycline are unclear at the moment given some reports of worsened disease or lack of neuroprotection in animal models (Saganova et al. 2008); improved outcome appears to depend on early initiation of treatment, high dosage, and direct delivery to the CNS (Xue et al. 2010). Of concern are recent clinical trial results showing that minocycline had a harmful effect on patients with amyotrophic lateral sclerosis, an effect not predicted by studies in the SOD1 animal model of this disease (Gordon et al. 2007).

9.3.3 Statins

Statins describe a class of drugs that competitively inhibit the rate-limiting enzyme in the mevalonate pathway, 3-hydroxy-3-methyl-glutaryl-coenzyme A (HMG CoA) reductase (HMGR), primarily reducing the biosynthesis of cholesterol and the post-translational lipid moiety attachments, isoprenoids (Edwards and Ericsson 1999). Their beneficial effects in EAE were attributed to their systemic immune-regulatory actions resulting in changes in protein isoprenylation (Greenwood et al. 2006; Peng et al. 2006; Pahan et al. 1997; Youssef et al. 2002). Based on these studies, simvastatin was selected for a clinical trial in MS and was found to reduce the number of newly emerging lesions by MRI at a dose substantially higher than those typically prescribed for hypercholesterolemia (Vollmer et al. 2004). Statins dampen the expression of proinflammatory cytokines, chemokine receptors, adhesion molecules, and MMPs that would normally facilitate lymphocyte transendothelial migration (Lindberg et al. 2005; Pahan et al. 1997). Additionally, lipophilic statins (i.e., lovastatin, simvastatin) can passively diffuse through cell membranes and directly access the CNS (Saheki et al. 1994). The expression of HMGR in all cell types along with the penetration of this therapy into the brain highlights the potential direct effects of statins on neural cell properties. Accordingly, statins have been attributed neuroprotective properties in neurodegenerative models of stroke, vascular dementia, Alzheimer's disease, and Parkinson's disease, although the mechanisms through which this occurs are not yet fully understood (Wang et al. 2011).

Statins can prevent the release of proinflammatory mediators such as prostaglandin E2, nitric oxide, TNFα, IL-1β, and IL-6 from activated microglia and astrocytes (Pahan et al. 1997; Tringali et al. 2004). Treatment of astrocytes with lovastatin also impairs the function of glutamate synthetase which normally degrades excessive cytotoxic amounts of glutamate sequestered from the extracellular environment (Chou et al. 2003). In vitro studies have also indicated that statins can have variable effects on neuronal morphology and cell survival depending on dose, cell maturity, neuronal subtype, length of treatment, and environmental stress (Kumano et al. 2000; Michikawa and Yanagisawa 1999). Simvastatin has been shown to promote neuritic outgrowth on myelin substrates that would normally prevent regeneration and growth of axons (Holmberg et al. 2006). Statins also have beneficial effects on nerve conduction and generation of neuroblasts in neurodegenerative models (Tarhzaoui et al. 2009; Chen et al. 2009). In vitro studies have demonstrated the ability of lipophilic statins to influence cellular processes in human and rodent OPCs that have been implicated in the remyelination process (Miron et al. 2007; Paintlia et al. 2005). For instance, simvastatin treatment can induce an initial process extension, enhanced differentiation, and impaired spontaneous migration via inhibition of isoprenoid synthesis (Miron et al. 2007). Prolonged treatment induced process retraction via inhibition of cholesterol production and decreased viability and cell death caused by interference with both isoprenoid- and cholesterol-dependent functions in the cells (Miron et al. 2007). Mature oligodendrocytes treated with statins demonstrate similar responses to

OPCs with regard to cytoskeletal dynamics and survival (Miron et al. 2007). Statin treatment may have altered the activity levels of the normally isoprenylated Rho GTPases to regulate effects on process extension, cell survival, differentiation, and migration. In addition, statin exposure may have altered the dynamics of cholesterol-rich lipid rafts important for OL process outgrowth (Decker and Ffrench-Constant 2004; Michikawa and Yanagisawa 1999).

Myelin contains a high content of lipids which consist of the majority of its dry weight, over a quarter of which is composed of cholesterol (Morell and Jurevics 1996). Importantly, brain-derived cholesterol cannot be replaced by dietary or peripheral sources (Jurevics and Morell 1995; Morell and Jurevics 1996; Saher et al. 2005). Short-term daily treatment of EAE-afflicted animals with an immunomodulatory dose of lovastatin was shown to increase myelin recovery in the spinal cord by enhancing survival as well as proliferation and differentiation of OPCs (Paintlia et al. 2004). This was attributed to the creation of an environment favorable to remyelination by altered inflammation and neurotrophic factor production. However, these studies could reflect either a direct impact of the statin on neural cells or a sparing of immune-mediated injury consequent to anti-inflammatory activity. This was addressed by using a relatively non-immune-mediated model of brain-based demyelination (oral cuprizone administration) to assess the direct impact of statins on myelin maintenance and remyelination (Klopfleisch et al. 2008; Miron et al. 2009). Both short-term and long-term administration of an immunomodulatory dose of the more potent simvastatin impaired remyelination when statin was administered during the period of concomitant de- and remyelination, the entire period of OPC responses to demyelination, or the remyelination period alone. Effects were associated with a decrease in OPC numbers when the drug was applied during the period of concomitant de- and remyelination, and an increase in OPCs in an immature state when statin was applied during the remyelination period suggesting a block in differentiation (Miron et al. 2009). Together, these findings suggest the potential of lipophilic statins to have both deleterious and advantageous effects on the function of various neural cell types and repair/regeneration in the CNS.

9.3.4 FTY720

FTY720 (fingolimod/Gilenya) is a recently clinically approved oral agent for the treatment of relapsing MS that induces profound reduction in numbers of lymphocytes circulating in the peripheral blood due to its capacity to enhance ingress and reduce egress of CCR7-expressing naïve and central memory lymphocytes from regional lymph nodes (Compston and Coles 2002). This agent was initially evaluated (and concluded to be ineffective) as a therapy to prevent organ rejection in transplant patients. Based on these immunomodulatory properties, this agent has been successfully used in clinical trials in relapsing–remitting MS and is being evaluated in primary progressive MS clinical trials (Kappos et al. 2006a, b). FTY720 decreases the rate of neurodegeneration in patients compared to an

IFNβ-1a-treated cohort, as measured by brain volume loss (Khatri et al. 2011). The clinically administered native form of FTY720 can readily cross the BBB by virtue of its lipophilicity (Sanchez and Hla 2004). Once it has entered the brain parenchyma, it is rapidly phosphorylated by sphingosine kinase (SphK) and consequently biotransformed into an analogue of the endogenous bioactive lysophospholipid, sphingosine-1-phosphate (S1P) (Billich et al. 2003; Zemann et al. 2006). The active phosphorylated form of the drug can bind to four of five known G protein-coupled receptors which belong to the endothelial differentiation gene-related (EDG) family, termed S1P1, 3, 4, and 5 or EDG 1, 3, 6, and 8, respectively (Davis et al. 2005; Mandala et al. 2002). Generally, S1P3/4/5 activation leads to $G_{12/13}$ signaling and process retraction via the cytoskeletal modulator RhoA GTPase (Jaillard et al. 2005; Toman et al. 2004). S1P1 signaling to $G_{i/o}$ is associated with Rac1, Ras, and extracellular signal-related kinase (ERK) 1/2 activation and subsequent process extension, survival, migration, and proliferation (Goetzl and Rosen 2004; Toman et al. 2004). S1P receptors are ubiquitously expressed throughout the immune and central nervous systems. The biological basis behind its anti-inflammatory effects reflects the ligation of the bound S1P receptor on circulating naïve and central memory lymphocytes, with eventual endocytosis of the bound receptor, downregulation of the receptor at the mRNA level, and subsequent blockade of S1P1-dependent efflux from secondary lymph nodes to target organs such as the brain (Goetzl and Rosen 2004; Liu et al. 1999; Sawicka et al. 2005). Fingolimod has previously been shown to have neuroprotective effects in experimental models of spinal cord injury, stroke, and multiple sclerosis (Zhang et al. 2009; Hasegawa et al. 2010; Foster et al. 2007).

Audoradiographical analysis of rodents that were orally administered C^{14}-labelled FTY720 for 1 week indicate that both the native drug and the phosphorylated metabolite entered the brain parenchyma, steadily increased in concentration over time within this compartment, and was concentrated in myelin sheaths (Foster et al. 2007). Levels of the bioactive form of FTY720 are within the micromolar range in the brain, whereas subnanomolar concentrations are measured in the CSF (Foster et al. 2007).

FTY720 treatment of relapsing EAE animals, either prophylactically or during the chronic phases of the disease, can prevent or reverse clinical disability, respectively (Fujino et al. 2003; Webb et al. 2004; Kataoka et al. 2005; Al-Izki et al. 2011), whereas a recent study has demonstrated that clinical deterioration is not prevented when FTY720 is administered during progressive EAE (Al-Izki et al. 2011). Microarray profiling has revealed that FTY720 treatment of EAE animals also results in a downregulation of proinflammatory cytokine and chemokine-related genes and an increase in transcript levels of myelin-related genes (Foster et al. 2007). Given that the clinically administered native form of FTY720 can readily cross the BBB, such responses in the EAE model could reflect effects in the periphery, at the BBB, and/or on neural cells. The efficacy of FTY720 in relieving clinical disability in the later stages of EAE where inflammation has subsided suggests a neuroprotective effect (Fujino et al. 2003; Webb et al. 2004; Foster et al. 2007).

The combination of FTY720's crossing of the BBB and the expression of S1P1 and S1P3 on human-derived endothelial cells indicates the potential of this agent to impact on endothelial barrier properties in treated patients (Lin et al. 2007). FTY720 has been shown in vivo to promote adherens junction assembly between endothelial cells and decrease BBB permeability via S1P1 signaling (Lin et al. 2007; Foster et al. 2009). Astrocytes and microglia also have the potential to respond to S1P receptor signaling, with the expression of S1P1, 3, and 5 receptor isoforms. FTY720 treatment of astrocytes in vitro promotes their migration and survival via S1P1 signaling (Mullershausen et al. 2007; Osinde et al. 2007). Selective loss of S1P1 expression in astrocytes abrogates the protective effect of FTY720 in EAE (Choi et al. 2011). These findings suggest that S1P receptor signaling on astrocytes can influence cellular events that contribute to the inflammatory response and/or to a nonsupportive microenvironment, i.e., a parallel with the glial scar found in MS lesions. Studies using rat-derived microglia have indicated that their relative S1P receptor levels are modulated based on their activation state and that S1P receptor engagement on these cells alters their cytokine profile (Tham et al. 2003). This agent also increases microglia and astrocyte cell numbers ex vivo and in vivo during remyelination (Miron et al. 2010; Kim et al. 2011). Furthermore, FTY720 inhibits macrophage migration into inflammatory lesions (Zhang et al. 2007). Together these data indicate that FTY720 may influence their roles in phagocytosing debris within lesions and regulating the immune responses within the CNS. FTY720 also normalizes electrophysiological responses in EAE-afflicted animals (Balatoni et al. 2007). Although FTY720 has been shown to induce apoptosis in rat neurons, this only occurs at a concentration which would be supraphysiological in the CNS (Oyama et al. 1998). Although FTY720 does enter the brain parenchyma, it is not primarily localized to neurons suggesting limited potential to impact neuronal properties (Foster et al. 2007).

Rodent-derived mature OLs and OPCs express S1P receptors both in vitro and in vivo, in the relative abundance of S1P5 > S1P1 = S1P2 > S1P3 for OLs and S1P1 = S1P2 = S1P5 > S1P3 for OPCs (Yu et al. 2004; Jaillard et al. 2005; Novgorodov et al. 2007; Miron et al. 2007, 2008b). FTY720 has been shown to induce transient S1P3/5-dependent process retraction with subsequent S1P1-dependent process extension in human OPCs and adult mature OLs but only in rodent OLs matured in vitro when coapplied with neurotrophic factors (Miron et al. 2007, 2008a, b; Coelho et al. 2007). FTY720 induced cyclic down- and upregulation of S1P1 and S1P5 mRNA transcripts over time in cultured oligodendroglial cells, in contrast to the sustained downregulation of S1P1 observed in lymphocytes exposed to the drug in vivo (Miron et al. 2007, 2008a, b). FTY720 inhibits OPC differentiation and spontaneous migration yet does not influence directed migration to the chemoattractant PDGF (Jung et al. 2007). FTY720 was able to enhance OPC and OL survival under death-inducing conditions via S1P1 and S1P5 signaling, respectively (Miron et al. 2007, 2008a, b; Coelho et al. 2007; Jung et al. 2007). Contrasting reports exist on the ability of oral administration of FTY720 to promote OL survival and myelin integrity in experimental models of de- and remyelination (Kim et al. 2011; Hu et al. 2011). Kim et al. reported that 1 mg/kg/day FTY720 orally administered during

6 weeks of cuprizone-induced demyelination attenuated injury to OLs, neurons, and myelin, causing an increase in the percentage of myelinated fibers compared to control and associated with a reduced astro- and microgliotic response. However, when FTY720 was administered during the entire period of OPC responses (6 weeks during cuprizone administration and return to normal diet), an increase in astrogliosis was observed, thereby indicating that FTY720 can induce differential neural responses depending on the time it is administered following injury. Accordingly, Mi et al. observed that FTY720 (1 mg/kg/day) orally administered for 2 weeks at 4 weeks after the onset of cuprizone diet did not have any protective effect on the proportion of myelinated axons or lesion volume (Hu et al. 2011). Neither group observed an increase in remyelination in vivo, perhaps reflecting the difficulty in assessing enhanced remyelination in young animals where this process is rapid and effective. In contrast, application of physiological doses of fingolimod to organotypic mouse cerebellar slice cultures, which maintain the physiological cell–cell interactions and have mature compact myelin, enhanced remyelination following toxin-induced demyelination via S1P3/5 signaling (Miron et al. 2010). This was associated with process extension in both OPCs and mature OLs, as well as an increase in microglial and astrocytic components, which may have contributed to the repair process by production of factors beneficial to OPC and OL function. Differences with in vivo observations may indicate limited bioavailability of the drug in an animal compared to direct application in vitro. Mi et al. (2011) showed that FTY720 caused a decrease in developmental myelination in slices of P17 rat corpus callosum, which may reflect either differential susceptibility of myelin and OLs to FTY720 under myelinating and remyelinating conditions, or distinct responses of specific white matter tracts to the drug. Overall, the net effect of FTY720 on neural cells will reflect the relative levels of the S1P receptor subtypes on cells of interest, the modulation of these receptor levels in response to the environment, and counteracting signals emanating from the inflamed brain. Efforts are underway to assess more selective S1PR agonists, such as BAF312, which specifically binds S1P1 and S1P5, shows encouraging results in decreasing disease severity in EAE (Nuesslein-Hildesheim et al. 2010).

9.3.5 Fumarate Esters

BG00012 (BG12) is a dimethyl fumarate (the active agent of fumarate acid esters; FAE) which has shown beneficial effects in MS clinical trials with regard to relapse rate and neuronal integrity assessed by imaging readouts (Kappos et al. 2008). When applied both in a preventative or therapeutic treatment regimen in MOG-induced EAE, BG12 improved disease course and increased preservation of myelin and axonal integrity (Linker et al. 2011). A marginal increase in remyelination was observed in the cuprizone model of demyelination following treatment with FAE (Moharregh-Khiabani et al. 2010). In vitro studies have elucidated the pro-survival effect of BG12 on neurons and oligodendrocytes to result from nuclear factor-E2-related factor (Nrf)-2 transcriptional pathway-dependent induction of phase II detoxification enzymes and consequent protection from oxidative stress (Linker et al. 2011; van Horssen et al. 2010; Wierinckx et al. 2005). Nrf2 expression was increased in these neural cell types following BG12 treatment of EAE-afflicted mice, as well as in the spinal cord of untreated MS patients (Linker et al. 2011). A further effect may be on decreasing reactive nitrogen species generation from microglia (Moharregh-Khiabani et al. 2010), which may be toxic to neural cells. Other studies have shown the immunomodulatory properties of BG12 by demonstrating decreased peripheral lymphocyte counts and proinflammatory cytokine production, and induction of a Th2 cytokine profile (Hoxtermann et al. 1998), suggesting an active indirect mechanism of action as well.

9.3.6 Lamotrigine

Lamotrigine is a sodium channel blocker most commonly used as an antiepileptic which was conceived to be neuroprotective based on the finding that inflammation (nitric oxide) induces persistent sodium channel activation in axons. This may result in accumulation of intracellular sodium followed by exchange for calcium and resultant injury. This is particularly relevant in demyelinating diseases as axons recovering from acute insults have increased densities of sodium channels (Bostock and Sears 1976; Craner et al. 2005). Blocking sodium channels in EAE preserved conduction in CNS axons and reduced disability scores (Bechtold et al. 2006). Small exploratory clinical trials are being conducted with riluzole and lamotrigine in progressive forms of MS (Kalkers et al. 2002; Miller et al. 1986), showing promising results in decreasing atrophy and reduction in brain volume. Sodium channel blockers would likely have increased efficiency in MS treatment if administered in conjunction with immunomodulatory agents to dampen injury to OLs and myelin.

Fampridine (4-aminopyridine) is a selective potassium channel blocker that has now been approved as a symptomatic therapy for MS, having been shown to enhance speed of ambulation and endurance. This agent has been shown to increase potentiation of synaptic transmission in a model of demyelination (Smith et al. 2000).

9.3.7 Laquinimod

Laquinimod (ABR-215062) is a quinoline-2-carboximide compound with beneficial effects in both acute and chronic EAE by immunomodulation (Yang et al. 2004; Wegner et al. 2010; Brunmark et al. 2002). It shifts T lymphocyte responses to Th2/Th3 profile, reduce T cell migration via a decrease in MMP9 and ICAM-1 expression, and decrease antigen presentation capacities (Yang et al. 2004; Runstrom et al. 2006; Gurevich et al. 2010; Wegner et al. 2010). Although clinical trials in relapsing MS have implicated a neuroprotective effect (Comi et al. 2008), with decreases in brain atrophy as measured by MRI, this mechanism of action is not clearly understood. Laquinimod orally administered in MOG-EAE decreases inflammation, demyelination, and axonal damage (Bruck and Wegner 2011). Its ability to cross the BBB in conjunction with induction of BDNF, NT3, and NT4 expression may contribute to limitation of axonal loss (Thone and Gold 2011; Bruck and Wegner 2011); BDNF levels are increased in the serum of relapsing MS patients at 3 months post initiation of treatment (Linker et al. 2010). Additionally, the increase in TGFβ and IL10 expression following laquinimod treatment of EAE (Yang et al. 2004) may directly impact OPC survival and differentiation.

9.4 Conclusion

Systemic immunomodulators have established their efficacy in reducing the frequency of clinical relapses in MS. MR-based imaging has documented the associated reduction in numbers of inflammatory lesions in the CNS. Studies of peripheral blood cells can be used to assess the mechanisms of action of these agents so as to aid in adjusting dosing regimens to maximize the benefit-to-risk ratio. The observations presented here regarding indirect and direct effects of immunomodulatory agents on the CNS are currently largely derived from in vitro studies and/or experimental disease models. As indicated, such effects could either promote or impede disease-relevant injury and repair processes within the CNS. An ongoing challenge is to develop monitoring techniques, e.g., imaging, biomarkers, and neurophysiological methods, which can be used to assess the neurobiological effects of these agents in the clinical setting.

References

Aharoni R, Teitelbaum D, Leitner O, Meshorer A, Sela M, Arnon R (2000) Specific Th2 cells accumulate in the central nervous system of mice protected against experimental autoimmune encephalomyelitis by copolymer 1. Proc Natl Acad Sci USA 97:11472–11477

Aharoni R, Kayhan B, Eilam R, Sela M, Arnon R (2003) Glatiramer acetate-specific T cells in the brain express T helper 2/3 cytokines and brain-derived neurotrophic factor in situ. Proc Natl Acad Sci USA 100:14157–14162

Aharoni R, Arnon R, Eilam R (2005) Neurogenesis and neuroprotection induced by peripheral immunomodulatory treatment of experimental autoimmune encephalomyelitis. J Neurosci 25: 8217–8228

Aharoni R, Herschkovitz A, Eilam R, Blumberg-Hazan M, Sela M, Bruck W, Arnon R (2008) Demyelination arrest and remyelination induced by glatiramer acetate treatment of experimental autoimmune encephalomyelitis. Proc Natl Acad Sci USA 105:11358–11363

Aharoni R, Vainshtein A, Stock A, Eilam R, From R, Shinder V, Arnon R (2011) Distinct pathological patterns in relapsing-remitting and chronic models of experimental autoimmune encephalomyelitis and the neuroprotective effect of glatiramer acetate. J Autoimmun 37(3):228–241

Aktas O, Schulze-Topphoff U, Zipp F (2007) The role of TRAIL/TRAIL receptors in central nervous system pathology. Front Biosci 12:2912–2921

Al-Izki S, Pryce G, Jackson SJ, Giovannoni G, Baker D (2011) Immunosuppression with FTY720 is insufficient to prevent secondary progressive neurodegeneration in experimental autoimmune encephalomyelitis. Mult Scler 17(8):939–948

Althaus HH (2004) Remyelination in multiple sclerosis: a new role for neurotrophins? Prog Brain Res 146:415–432

Arnett HA, Mason J, Marino M, Suzuki K, Matsushima GK, Ting JP (2001) TNF alpha promotes proliferation of oligodendrocyte progenitors and remyelination. Nat Neurosci 4:1116–1122

Arvin KL, Han BH, Du Y, Lin SZ, Paul SM, Holtzman DM (2002) Minocycline markedly protects the neonatal brain against hypoxic-ischemic injury. Ann Neurol 52:54–61

Axtell RC, de Jong BA, Boniface K, van der Voort LF, Bhat R, De SP, Naves R, Han M, Zhong F, Castellanos JG, Mair R, Christakos A, Kolkowitz I, Katz L, Killestein J, Polman CH, de Waal MR, Steinman L, Raman C (2010) T helper type 1 and 17 cells determine efficacy of interferon-beta in multiple sclerosis and experimental encephalomyelitis. Nat Med 16:406–412

Balatoni B, Storch MK, Swoboda EM, Schonborn V, Koziel A, Lambrou GN, Hiestand PC, Weissert R, Foster CA (2007) FTY720 sustains and restores neuronal function in the DA rat model of MOG-induced experimental autoimmune encephalomyelitis. Brain Res Bull 74:307–316

Barkhof F, Hulst HE, Drulovic J, Uitdehaag BM, Matsuda K, Landin R (2010) Ibudilast in relapsing-remitting multiple sclerosis: a neuroprotectant? Neurology 74:1033–1040

Bar-Or A et al (2007) Induction of antigen-specific tolerance in multiple sclerosis after immunization with DNA encoding myelin basic protein in a randomized, placebo-controlled phase 1/2 trial. Arch Neurol 64:1407–1415

Barouch R, Schwartz M (2002) Autoreactive T cells induce neurotrophin production by immune and neural cells in injured rat optic nerve: implications for protective autoimmunity. FASEB J 16:1304–1306

Bartosik-Psujek H, Belniak E, Mitosek-Szewczyk K, Dobosz B, Stelmasiak Z (2004) Interleukin-8 and RANTES levels in patients with relapsing-remitting multiple sclerosis (RR-MS) treated with cladribine. Acta Neurol Scand 109:390–392

Bechtold DA, Miller SJ, Dawson AC, Sun Y, Kapoor R, Berry D, Smith KJ (2006) Axonal protection achieved in a model of multiple sclerosis using lamotrigine. J Neurol 253:1542–1551

Benner EJ, Mosley RL, Destache CJ, Lewis TB, Jackson-Lewis V, Gorantla S, Nemachek C, Green SR, Przedborski S, Gendelman HE (2004) Therapeutic immunization protects dopaminergic neurons in a mouse model of Parkinson's disease. Proc Natl Acad Sci USA 101:9435–9440

Bieber AJ, Warrington A, Asakura K, Ciric B, Kaveri SV, Pease LR, Rodriguez M (2002) Human antibodies accelerate the rate of remyelination following lysolecithin-induced demyelination in mice. Glia 37:241–249

Bielekova B, Goodwin B, Richert N, Cortese I, Kondo T, Afshar G, Gran B, Eaton J, Antel J, Frank JA, McFarland HF, Martin R (2000) Encephalitogenic potential of the myelin basic protein peptide (amino acids 83-99) in multiple sclerosis: results of a phase II clinical trial with an altered peptide ligand. Nat Med 6:1167–1175

Biernacki K, Prat A, Blain M, Antel JP (2001) Regulation of Th1 and Th2 lymphocyte migration by human adult brain endothelial cells. J Neuropathol Exp Neurol 60:1127–1136

Biernacki K, Antel JP, Blain M, Narayanan S, Arnold DL, Prat A (2005) Interferon beta promotes nerve growth factor secretion early in the course of multiple sclerosis. Arch Neurol 62:563–568

Billich A, Bornancin F, Devay P, Mechtcheriakova D, Urtz N, Baumruker T (2003) Phosphorylation of the immunomodulatory drug FTY720 by sphingosine kinases. J Biol Chem 278: 47408–47415

Bitsch A, Kuhlmann T, Stadelmann C, Lassmann H, Lucchinetti C, Bruck W (2001) A longitudinal MRI study of histopathologically defined hypointense multiple sclerosis lesions. Ann Neurol 49:793–796

Bornstein MB, Miller A, Slagle S, Weitzman M, Crystal H, Drexler E, Keilson M, Merriam A, Wassertheil-Smoller S, Spada V (1987) A pilot trial of Cop 1 in exacerbating-remitting multiple sclerosis. N Engl J Med 317:408–414

Bostock H, Sears TA (1976) Continuous conduction in demyelinated mammalian nerve fibers. Nature 263:786–787

Boutros T, Croze E, Yong VW (1997) Interferon-beta is a potent promoter of nerve growth factor production by astrocytes. J Neurochem 69:939–946

Bruck W, Wegner C (2011) Insight into the mechanism of laquinimod action. J Neurol Sci 306: 173–179

Brundula V, Rewcastle NB, Metz LM, Bernard CC, Yong VW (2002) Targeting leukocyte MMPs and transmigration: minocycline as a potential therapy for multiple sclerosis. Brain 125: 1297–1308

Brunmark C, Runstrom A, Ohlsson L, Sparre B, Brodin T, Astrom M, Hedlund G (2002) The new orally active immunoregulator laquinimod (ABR-215062) effectively inhibits development and relapses of experimental autoimmune encephalomyelitis. J Neuroimmunol 130:163–172

Buttmann M (2010) Treating multiple sclerosis with monoclonal antibodies: a 2010 update. Expert Rev Neurother 10:791–809

Caggiula M, Batocchi AP, Frisullo G, Angelucci F, Patanella AK, Sancricca C, Nociti V, Tonali PA, Mirabella M (2006) Neurotrophic factors in relapsing remitting and secondary progressive multiple sclerosis patients during interferon beta therapy. Clin Immunol 118:77–82

Chari DM, Zhao C, Kotter MR, Blakemore WF, Franklin RJ (2006) Corticosteroids delay remyelination of experimental demyelination in the rodent central nervous system. J Neurosci Res 83: 594–605

Chen J, Cui X, Zacharek A, Chopp M (2009) Increasing Ang1/Tie2 expression by simvastatin treatment induces vascular stabilization and neuroblast migration after stroke. J Cell Mol Med 13:1348–1357

Choi JW, Gardell SE, Herr DR, Rivera R, Lee C-W, Noguchi K, Teo ST, Tung YC, Lu M, Kennedy G, Chun J (2011) FTY720 (fingolimod) efficacy in an animal model of multiple sclerosis requires astrocyte sphingosine1-phosphate receptor 1 (S1P1) modulation. Proc Natl Acad Sci USA 108:751–756

Chou YC, Lin SB, Tsai LH, Tsai HI, Lin CM (2003) Cholesterol deficiency increases the vulnerability of hippocampal glia in primary culture to glutamate-induced excitotoxicity. Neurochem Int 43:197–209

Claussen MC, Korn T (2012) Immune mechanisms of new therapeutic strategies in MS – teriflunomide. Clin Immunol 142(1):49–56

Coelho RP, Payne SG, Bittman R, Spiegel S, Sato-Bigbee C (2007) The immunomodulator FTY720 has a direct cytoprotective effect in oligodendrocyte progenitors. J Pharmacol Exp Ther 323:626–635

Coles AJ, Compston DA, Selmaj KW, Lake SL, Moran S, Margolin DH, Norris K, Tandon PK (2008) Alemtuzumab vs. interferon beta-1a in early multiple sclerosis. N Engl J Med 359: 1786–1801

Compston A, Coles A (2002) Multiple sclerosis. Lancet 359:1221–1231

Comi G, Pulizzi A, Rovaris M, Abramsky O, Arbizu T, Boiko A, Gold R, Havrdova E, Komoly S, Selmaj K, Sharrack B, Filippi M (2008) Effect of laquinimod on MRI-monitored disease activity in patients with relapsing-remitting multiple sclerosis: a multicentre, randomised, double-blind, placebo-controlled phase IIb study. Lancet 371:2085–2092

Craner MJ, Damarjian TG, Liu S, Hains BC, Lo AC, Black JA, Newcombe J, Cuzner ML, Waxman SG (2005) Sodium channels contribute to microglia/macrophage activation and function in EAE and MS. Glia 49:220–229

Davis MD, Clemens JJ, Macdonald TL, Lynch KR (2005) Sphingosine 1-phosphate analogs as receptor antagonists. J Biol Chem 280:9833–9841

Decker L, Ffrench-Constant C (2004) Lipid rafts and integrin activation regulate oligodendrocyte survival. J Neurosci 24:3816–3825

Du S, Sandoval F, Trinh P, Voskuhl RR (2010) Additive effects of combination treatment with anti-inflammatory and neuroprotective agents in experimental autoimmune encephalomyelitis. J Neuroimmunol 219:64–74

Edwards PA, Ericsson J (1999) Sterols and isoprenoids: signaling molecules derived from the cholesterol biosynthetic pathway. Annu Rev Biochem 68:157–185

Elewa HF, Hilali H, Hess DC, Machado LS, Fagan SC (2006) Minocycline for short-term neuroprotection. Pharmacotherapy 26:515–521

Engell T (1989) A clinical patho-anatomical study of clinically silent multiple sclerosis. Acta Neurol Scand 79:428–430

Foster CA, Howard LM, Schweitzer A, Persohn E, Hiestand PC, Balatoni B, Reuschel R, Beerli C, Schwartz M, Billich A (2007) Brain penetration of the oral immunomodulatory drug FTY720 and its phosphorylation in the central nervous system during experimental autoimmune encephalomyelitis: consequences for mode of action in multiple sclerosis. J Pharmacol Exp Ther 323:469–475

Foster CA, Mechtcheriakova D, Storch MK, Balatoni B, Howard LM, Bornancin F, Wlachos A, Sobanov J, Kinnunen A, Baumruker T (2009) FTY720 rescue therapy in the dark agouti rat model of experimental autoimmune encephalomyelitis: expression of central nervous system genes and reversal of blood-brain-barrier damage. Brain Pathol 19:254–266

Fox C, Dingman A, Derugin N, Wendland MF, Manabat C, Ji S, Ferriero DM, Vexler ZS (2005) Minocycline confers early but transient protection in the immature brain following focal cerebral ischemia-reperfusion. J Cereb Blood Flow Metab 25:1138–1149

Freedman MS (2007) Bone marrow transplantation: does it stop MS progression? J Neurol Sci 259:85–89

Fu L, Matthews PM, De SN, Worsley KJ, Narayanan S, Francis GS, Antel JP, Wolfson C, Arnold DL (1998) Imaging axonal damage of normal-appearing white matter in multiple sclerosis. Brain 121(Pt 1):103–113

Fujino M, Funeshima N, Kitazawa Y, Kimura H, Amemiya H, Suzuki S, Li XK (2003) Amelioration of experimental autoimmune encephalomyelitis in Lewis rats by FTY720 treatment. J Pharmacol Exp Ther 305:70–77

Giovannoni G, Comi G, Cook S, Rammohan K, Rieckmann P, Soelberg SP, Vermersch P, Chang P, Hamlett A, Musch B, Greenberg SJ (2010) A placebo-controlled trial of oral cladribine for relapsing multiple sclerosis. N Engl J Med 362:416–426

Glezer I, Lapointe A, Rivest S (2006) Innate immunity triggers oligodendrocyte progenitor reactivity and confines damages to brain injuries. FASEB J 20:750–752

Goetzl EJ, Rosen H (2004) Regulation of immunity by lysosphingolipids and their G protein-coupled receptors. J Clin Invest 114:1531–1537

Gomez-Gallego M, Meca-Lallana J, Fernandez-Barreiro A (2008) Multiple sclerosis onset during etanercept treatment. Eur Neurol 59:91–93

Gordon PH, Moore DH, Miller RG, Florence JM, Verheijde JL, Doorish C, Hilton JF, Spitalny GM, MacArthur RB, Mitsumoto H, Neville HE, Boylan K, Mozaffar T, Belsh JM, Ravits J, Bedlack RS, Graves MC, McCluskey LF, Barohn RJ, Tandan R (2007) Efficacy of minocycline in patients with amyotrophic lateral sclerosis: a phase III randomised trial. Lancet Neurol 6:1045–1053

Greenwood J, Steinman L, Zamvil SS (2006) Statin therapy and autoimmune disease: from protein prenylation to immunomodulation. Nat Rev Immunol 6:358–370

Greil R, Anether G, Johrer K, Tinhofer I (2003) Tracking death dealing by Fas and TRAIL in lymphatic neoplastic disorders: pathways, targets, and therapeutic tools. J Leukoc Biol 74: 311–330

Gronseth GS, Ashman EJ (2000) Practice parameter: the usefulness of evoked potentials in identifying clinically silent lesions in patients with suspected multiple sclerosis (an evidence-based review): report of the Quality Standards Subcommittee of the American Academy of Neurology. Neurology 54:1720–1725

Gurevich M, Gritzman T, Orbach R, Tuller T, Feldman A, Achiron A (2010) Laquinimod suppress antigen presentation in relapsing-remitting multiple sclerosis: in-vitro high-throughput gene expression study. J Neuroimmunol 221:87–94

Hasegawa Y, Suzuki H, Sozen T, Rolland W, Zhang JH (2010) Activation of sphingosine 1-phosphate receptor-1 by FTY720 is neuroprotective after ischemic stroke in rats. Stroke 41:368–374

Hauser SL, Waubant E, Arnold DL, Vollmer T, Antel J, Fox RJ, Bar-Or A, Panzara M, Sarkar N, Agarwal S, Langer-Gould A, Smith CH (2008) B-cell depletion with rituximab in relapsing-remitting multiple sclerosis. N Engl J Med 358:676–688

Hempstead BL, Salzer JL (2002) Neurobiology. A glial spin on neurotrophins. Science 298: 1184–1186

Hendrix S, Nitsch R (2007) The role of T helper cells in neuroprotection and regeneration. J Neuroimmunol 184:100–112

Hirsch M, Knight J, Tobita M, Soltys J, Panitch H, Mao-Draayer Y (2009) The effect of interferon-beta on mouse neural progenitor cell survival and differentiation. Biochem Biophys Res Commun 388:181–186

Hohlfeld R, Barkhof F, Polman C (2011) Future clinical challenges in multiple sclerosis: relevance to sphingosine 1-phosphate receptor modulator therapy. Neurology 76:S28–S37

Holmberg E, Nordstrom T, Gross M, Kluge B, Zhang SX, Doolen S (2006) Simvastatin promotes neurite outgrowth in the presence of inhibitory molecules found in central nervous system injury. J Neurotrauma 23:1366–1378

Hoxtermann S, Nuchel C, Altmeyer P (1998) Fumaric acid esters suppress peripheral CD4- and CD8-positive lymphocytes in psoriasis. Dermatology 196:223–230

Hu Y, Lee X, Ji B, Guckian K, Apicco D, Pepinsky RB, Miller RH, Mi S (2011) Sphingosine 1-phosphate receptor modulator fingolimod (FTY720) does not promote remyelination in vivo. Mol Cell Neurosci 48:72–81

Jacobs LD, Cookfair DL, Rudick RA, Herndon RM, Richert JR, Salazar AM, Fischer JS, Goodkin DE, Granger CV, Simon JH (1995) A phase III trial of intramuscular recombinant interferon beta as treatment for exacerbating-remitting multiple sclerosis: design and conduct of study and baseline characteristics of patients. Multiple Sclerosis Collaborative Research Group (MSCRG). Mult Scler 1:118–135

Jaillard C, Harrison S, Stankoff B, Aigrot MS, Calver AR, Duddy G, Walsh FS, Pangalos MN, Arimura N, Kaibuchi K, Zalc B, Lubetzki C (2005) Edg8/S1P5: an oligodendroglial receptor with dual function on process retraction and cell survival. J Neurosci 25:1459–1469

Javed A, Reder AT (2006) Therapeutic role of beta-interferons in multiple sclerosis. Pharmacol Ther 110:35–56

Jung CG, Kim HJ, Miron VE, Cook S, Kennedy TE, Foster CA, Antel JP, Soliven B (2007) Functional consequences of S1P receptor modulation in rat oligodendroglial lineage cells. Glia 55:1656–1667

Jurevics H, Morell P (1995) Cholesterol for synthesis of myelin is made locally, not imported into brain. J Neurochem 64:895–901

Kalkers NF, Barkhof F, Bergers E, van Schijndel R, Polman CH (2002) The effect of the neuroprotective agent riluzole on MRI parameters in primary progressive multiple sclerosis: a pilot study. Mult Scler 8:532–533

Kappos L, Traboulsee A, Constantinescu C, Eralinna JP, Forrestal F, Jongen P, Pollard J, Sandberg-Wollheim M, Sindic C, Stubinski B, Uitdehaag B, Li D (2006a) Long-term subcutaneous interferon beta-1a therapy in patients with relapsing-remitting MS. Neurology 67:944–953

Kappos L, Antel J, Comi G, Montalban X, O'Connor P, Polman CH, Haas T, Korn AA, Karlsson G, Radue EW (2006b) Oral fingolimod (FTY720) for relapsing multiple sclerosis. N Engl J Med 355:1124–1140

Kappos L, Gold R, Miller DH, Macmanus DG, Havrdova E, Limmroth V, Polman CH, Schmierer K, Yousry TA, Yang M, Eraksoy M, Meluzinova E, Rektor I, Dawson KT, Sandrock AW, O'Neill GN (2008) Efficacy and safety of oral fumarate in patients with relapsing-remitting multiple sclerosis: a multicentre, randomised, double-blind, placebo-controlled phase IIb study. Lancet 372:1463–1472

Kataoka H, Sugahara K, Shimano K, Teshima K, Koyama M, Fukunari A, Chiba K (2005) FTY720, sphingosine 1-phosphate receptor modulator, ameliorates experimental autoimmune encephalomyelitis by inhibition of T cell infiltration. Cell Mol Immunol 2:439–448

Kerschensteiner M, Gallmeier E, Behrens L, Leal VV, Misgeld T, Klinkert WE, Kolbeck R, Hoppe E, Oropeza-Wekerle RL, Bartke I, Stadelmann C, Lassmann H, Wekerle H, Hohlfeld R (1999)

Activated human T cells, B cells, and monocytes produce brain-derived neurotrophic factor in vitro and in inflammatory brain lesions: a neuroprotective role of inflammation? J Exp Med 189:865–870

Keszthelyi E, Karlik S, Hyduk S, Rice GP, Gordon G, Yednock T, Horner H (1996) Evidence for a prolonged role of alpha 4 integrin throughout active experimental allergic encephalomyelitis. Neurology 47:1053–1059

Khatri B, Barkhof F, Comi G, Hartung HP, Kappos L, Montalban X, Pelletier J, Stites T, Wu S, Holdbrook F, Zhang-Auberson L, Francis G, Cohen JA (2011) Comparison of fingolimod with interferon beta-1a in relapsing-remitting multiple sclerosis: a randomised extension of the TRANSFORMS study. Lancet Neurol 10:520–529

Kidd D, Barkhof F, McConnell R, Algra PR, Allen IV, Revesz T (1999) Cortical lesions in multiple sclerosis. Brain 122(Pt 1):17–26

Kim HJ, Ifergan I, Antel JP, Seguin R, Duddy M, Lapierre Y, Jalili F, Bar-Or A (2004) Type 2 monocyte and microglia differentiation mediated by glatiramer acetate therapy in patients with multiple sclerosis. J Immunol 172:7144–7153

Kim HJ, Miron VE, Dukala D, Proia RL, Ludwin SK, Traka M, Antel JP, Soliven B (2011) Neurobiological effects of sphingosine 1-phosphate receptor modulation in the cuprizone model. FASEB J 25:1509–1518

Kipnis J, Schwartz M (2002) Dual action of glatiramer acetate (Cop-1) in the treatment of CNS autoimmune and neurodegenerative disorders. Trends Mol Med 8:319–323

Kipnis J, Yoles E, Porat Z, Cohen A, Mor F, Sela M, Cohen IR, Schwartz M (2000) T cell immunity to copolymer 1 confers neuroprotection on the damaged optic nerve: possible therapy for optic neuropathies. Proc Natl Acad Sci USA 97:7446–7451

Kivisakk P, Mahad DJ, Callahan MK, Trebst C, Tucky B, Wei T, Wu L, Baekkevold ES, Lassmann H, Staugaitis SM, Campbell JJ, Ransohoff RM (2003) Human cerebrospinal fluid central memory CD4+ T cells: evidence for trafficking through choroid plexus and meninges via P-selectin. Proc Natl Acad Sci USA 100:8389–8394

Klopfleisch S, Merkler D, Schmitz M, Kloppner S, Schedensack M, Jeserich G, Althaus HH, Bruck W (2008) Negative impact of statins on oligodendrocytes and myelin formation in vitro and in vivo. J Neurosci 28:13609–13614

Kotter MR, Li WW, Zhao C, Franklin RJ (2006) Myelin impairs CNS remyelination by inhibiting oligodendrocyte precursor cell differentiation. J Neurosci 26:328–332

Kumano T, Mutoh T, Nakagawa H, Kuriyama M (2000) HMG-CoA reductase inhibitor induces a transient activation of high affinity nerve growth factor receptor, trk, and morphological differentiation with fatal outcome in PC12 cells. Brain Res 859:169–172

Lee SM, Yune TY, Kim SJ, Park DW, Lee YK, Kim YC, Oh YJ, Markelonis GJ, Oh TH (2003) Minocycline reduces cell death and improves functional recovery after traumatic spinal cord injury in the rat. J Neurotrauma 20:1017–1027

Lin CI, Chen CN, Lin PW, Chang KJ, Hsieh FJ, Lee H (2007) Lysophosphatidic acid regulates inflammation-related genes in human endothelial cells through LPA1 and LPA3. Biochem Biophys Res Commun 363:1001–1008

Lin W, Kunkler PE, Harding HP, Ron D, Kraig RP, Popko B (2008) Enhanced integrated stress response promotes myelinating oligodendrocyte survival in response to interferon-gamma. Am J Pathol 173:1508–1517

Lindberg C, Crisby M, Winblad B, Schultzberg M (2005) Effects of statins on microglia. J Neurosci Res 82:10–19

Linker RA, Lee DH, Demir S, Wiese S, Kruse N, Siglienti I, Gerhardt E, Neumann H, Sendtner M, Luhder F, Gold R (2010) Functional role of brain-derived neurotrophic factor in neuroprotective autoimmunity: therapeutic implications in a model of multiple sclerosis. Brain 133:2248–2263

Linker RA, Lee DH, Ryan S, van Dam AM, Conrad R, Bista P, Zeng W, Hronowsky X, Buko A, Chollate S, Ellrichmann G, Bruck W, Dawson K, Goelz S, Wiese S, Scannevin RH, Lukashev M, Gold R (2011) Fumaric acid esters exert neuroprotective effects in neuroinflammation via activation of the Nrf2 antioxidant pathway. Brain 134:678–692

Liu CH, Thangada S, Lee MJ, Van Brocklyn JR, Spiegel S, Hla T (1999) Ligand-induced trafficking of the sphingosine-1-phosphate receptor EDG-1. Mol Biol Cell 10:1179–1190

Liu J, Johnson TV, Lin J, Ramirez SH, Bronich TK, Caplan S, Persidsky Y, Gendelman HE, Kipnis J (2007) T cell independent mechanism for copolymer-1-induced neuroprotection. Eur J Immunol 37:3143–3154

Luger TA, Bhardwaj RS, Grabbe S, Schwarz T (1996) Regulation of the immune response by epidermal cytokines and neurohormones. J Dermatol Sci 13:5–10

Lum M, Croze E, Wagner C, McLenachan S, Mitrovic B, Turnley AM (2009) Inhibition of neurosphere proliferation by IFNgamma but not IFNbeta is coupled to neuronal differentiation. J Neuroimmunol 206:32–38

Mandala S, Hajdu R, Bergstrom J, Quackenbush E, Xie J, Milligan J, Thornton R, Shei GJ, Card D, Keohane C, Rosenbach M, Hale J, Lynch CL, Rupprecht K, Parsons W, Rosen H (2002) Alteration of lymphocyte trafficking by sphingosine-1-phosphate receptor agonists. Science 296:346–349

Maysami S, Nguyen D, Zobel F, Pitz C, Heine S, Hopfner M, Stangel M (2006) Modulation of rat oligodendrocyte precursor cells by the chemokine CXCL12. Neuroreport 17:1187–1190

Menge T, Hartung HP, Stuve O (2005) Statins – a cure-all for the brain? Nat Rev Neurosci 6:325–331

Mews I, Bergmann M, Bunkowski S, Gullotta F, Bruck W (1998) Oligodendrocyte and axon pathology in clinically silent multiple sclerosis lesions. Mult Scler 4:55–62

Michikawa M, Yanagisawa K (1999) Inhibition of cholesterol production but not of nonsterol isoprenoid products induces neuronal cell death. J Neurochem 72:2278–2285

Miller AA, Wheatley P, Sawyer DA, Baxter MG, Roth B (1986) Pharmacological studies on lamotrigine, a novel potential antiepileptic drug: I. Anticonvulsant profile in mice and rats. Epilepsia 27:483–489

Miller DH, Khan OA, Sheremata WA, Blumhardt LD, Rice GP, Libonati MA, Willmer-Hulme AJ, Dalton CM, Miszkiel KA, O'Connor PW (2003) A controlled trial of natalizumab for relapsing multiple sclerosis. N Engl J Med 348:15–23

Miller A, Spada V, Beerkircher D, Kreitman RR (2008) Long-term (up to 22 years), open-label, compassionate-use study of glatiramer acetate in relapsing-remitting multiple sclerosis. Mult Scler 14:494–499

Miron VE, Rajasekharan S, Jarjour AA, Zamvil SS, Kennedy TE, Antel JP (2007) Simvastatin regulates oligodendroglial process dynamics and survival. Glia 55:130–143

Miron VE, Hall JA, Kennedy TE, Soliven B, Antel JP (2008a) Cyclical and dose-dependent responses of adult human mature oligodendrocytes to fingolimod. Am J Pathol 173:1143–1152

Miron VE, Jung CG, Kim HJ, Kennedy TE, Soliven B, Antel JP (2008b) FTY720 modulates human oligodendrocyte progenitor process extension and survival. Ann Neurol 63:61–71

Miron VE, Zehntner SP, Kuhlmann T, Ludwin SK, Owens T, Kennedy TE, Bedell BJ, Antel JP (2009) Statin therapy inhibits remyelination in the central nervous system. Am J Pathol 174:1880–1890

Miron VE, Ludwin SK, Darlington PJ, Jarjour AA, Soliven B, Kennedy TE, Antel JP (2010) Fingolimod (FTY720) enhances remyelination following demyelination of organotypic cerebellar slices. Am J Pathol 176:2682–2694

Moharregh-Khiabani D, Blank A, Skripuletz T, Miller E, Kotsiari A, Gudi V, Stangel M (2010) Effects of fumaric acids on cuprizone induced central nervous system de- and remyelination in the mouse. PLoS One 5:e11769

Morell P, Jurevics H (1996) Origin of cholesterol in myelin. Neurochem Res 21:463–470

Mullershausen F, Craveiro LM, Shin Y, Cortes-Cros M, Bassilana F, Osinde M, Wishart WL, Guerini D, Thallmair M, Schwab ME, Sivasankaran R, Seuwen K, Dev KK (2007) Phosphorylated FTY720 promotes astrocyte migration through sphingosine-1-phosphate receptors. J Neurochem 102:1151–1161

Narayanan S, De SN, Francis GS, Arnaoutelis R, Caramanos Z, Collins DL, Pelletier D, Arnason BGW, Antel JP, Arnold DL (2001) Axonal metabolic recovery in multiple sclerosis patients treated with interferon beta-1b. J Neurol 248:979–986

Nikodemova M, Watters JJ, Jackson SJ, Yang SK, Duncan ID (2007) Minocycline down-regulates MHC II expression in microglia and macrophages through inhibition of IRF-1 and protein kinase C (PKC)alpha/betaII. J Biol Chem 282:15208–15216

Nikodemova M, Lee J, Fabry Z, Duncan ID (2010) Minocycline attenuates experimental autoimmune encephalomyelitis in rats by reducing T cell infiltration into the spinal cord. J Neuroimmunol 219:33–37

Novgorodov AS, El Alwani M, Bielawski J, Obeid LM, Gudz TI (2007) Activation of sphingosine-1-phosphate receptor S1P5 inhibits oligodendrocyte progenitor migration. FASEB J 21: 1503–1514

Nuesslein-Hildesheim B, Zecri FJ, Bruns C, Cooke N, Seabrook T, Smith PA (2010) BAF312, a potent and selective S1P1/5 receptor modulator reverses ongoing chronic EAE in mice. In: American Academy of Neurology 62nd Annual Meeting, Toronto, Canada

O'Connor PW, Goodman A, Willmer-Hulme AJ, Libonati MA, Metz L, Murray RS, Sheremata WA, Vollmer TL, Stone LA (2004) Randomized multicenter trial of natalizumab in acute MS relapses: clinical and MRI effects. Neurology 62:2038–2043

Osinde M, Mullershausen F, Dev KK (2007) Phosphorylated FTY720 stimulates ERK phosphorylation in astrocytes via S1P receptors. Neuropharmacology 52:1210–1218

Oyama Y, Chikahisa L, Kanemaru K, Nakata M, Noguchi S, Nagano T, Okazaki E, Hirata A (1998) Cytotoxic actions of FTY720, a novel immunosuppressant, on thymocytes and brain neurons dissociated from the rat. Jpn J Pharmacol 76:377–385

Pahan K, Sheikh FG, Namboodiri AM, Singh I (1997) Lovastatin and phenylacetate inhibit the induction of nitric oxide synthase and cytokines in rat primary astrocytes, microglia, and macrophages. J Clin Invest 100:2671–2679

Paintlia AS, Paintlia MK, Singh AK, Stanislaus R, Gilg AG, Barbosa E, Singh I (2004) Regulation of gene expression associated with acute experimental autoimmune encephalomyelitis by Lovastatin. J Neurosci Res 77:63–81

Paintlia AS, Paintlia MK, Khan M, Vollmer T, Singh AK, Singh I (2005) HMG-CoA reductase inhibitor augments survival and differentiation of oligodendrocyte progenitors in animal model of multiple sclerosis. FASEB J 19:1407–1421

Panitch HS, Hirsch RL, Schindler J, Johnson KP (1987) Treatment of multiple sclerosis with gamma interferon: exacerbations associated with activation of the immune system. Neurology 37:1097–1102

Peng X, Jin J, Giri S, Montes M, Sujkowski D, Tang Y, Smrtka J, Vollmer T, Singh I, Markovic-Plese S (2006) Immunomodulatory effects of 3-hydroxy-3-methylglutaryl coenzyme-A reductase inhibitors, potential therapy for relapsing remitting multiple sclerosis. J Neuroimmunol 178:130–139

Phadke JG, Best PV (1983) Atypical and clinically silent multiple sclerosis: a report of 12 cases discovered unexpectedly at necropsy. J Neurol Neurosurg Psychiatry 46:414–420

Popovic N, Schubart A, Goetz BD, Zhang SC, Linington C, Duncan ID (2002) Inhibition of autoimmune encephalomyelitis by a tetracycline. Ann Neurol 51:215–223

Prineas JW, Connell F (1979) Remyelination in multiple sclerosis. Ann Neurol 5:22–31

Raine CS, Wu E (1993) Multiple sclerosis: remyelination in acute lesions. J Neuropathol Exp Neurol 52:199–204

Revesz T, Kidd D, Thompson AJ, Barnard RO, McDonald WI (1994) A comparison of the pathology of primary and secondary progressive multiple sclerosis. Brain 117(Pt 4):759–765

Riley CP, Cope TC, Buck CR (2004) CNS neurotrophins are biologically active and expressed by multiple cell types. J Mol Histol 35:771–783

Rodriguez M, Zoecklein LJ, Howe CL, Pavelko KD, Gamez JD, Nakane S, Papke LM (2003) Gamma interferon is critical for neuronal viral clearance and protection in a susceptible mouse strain following early intracranial Theiler's murine encephalomyelitis virus infection. J Virol 77:12252–12265

Rose JW, Burns JB, Bjorklund J, Klein J, Watt HE, Carlson NG (2007) Daclizumab phase II trial in relapsing and remitting multiple sclerosis: MRI and clinical results. Neurology 69: 785–789

Runstrom A, Leanderson T, Ohlsson L, Axelsson B (2006) Inhibition of the development of chronic experimental autoimmune encephalomyelitis by laquinimod (ABR-215062) in IFN-beta k.o. and wild type mice. J Neuroimmunol 173:69–78

Saganova K, Orendacova J, Cizkova D, Vanicky I (2008) Limited minocycline neuroprotection after balloon-compression spinal cord injury in the rat. Neurosci Lett 433:246–249

Saheki A, Terasaki T, Tamai I, Tsuji A (1994) In vivo and in vitro blood-brain barrier transport of 3-hydroxy-3-methylglutaryl coenzyme A (HMG-CoA) reductase inhibitors. Pharm Res 11:305–311

Saher G, Brugger B, Lappe-Siefke C, Mobius W, Tozawa R, Wehr MC, Wieland F, Ishibashi S, Nave KA (2005) High cholesterol level is essential for myelin membrane growth. Nat Neurosci 8:468–475

Sanchez T, Hla T (2004) Structural and functional characteristics of S1P receptors. J Cell Biochem 92:913–922

Sawicka E, Dubois G, Jarai G, Edwards M, Thomas M, Nicholls A, Albert R, Newson C, Brinkmann V, Walker C (2005) The sphingosine 1-phosphate receptor agonist FTY720 differentially affects the sequestration of CD4+/CD25+ T-regulatory cells and enhances their functional activity. J Immunol 175:7973–7980

Schori H, Yoles E, Schwartz M (2001) T-cell-based immunity counteracts the potential toxicity of glutamate in the central nervous system. J Neuroimmunol 119:199–204

Shapiro AM, Jack CS, Lapierre Y, Arbour N, Bar-Or A, Antel JP (2006) Potential for interferon beta-induced serum antibodies in multiple sclerosis to inhibit endogenous interferon-regulated chemokine/cytokine responses within the central nervous system. Arch Neurol 63:1296–1299

Skihar V, Silva C, Chojnacki A, Doring A, Stallcup WB, Weiss S, Yong VW (2009) Promoting oligodendrogenesis and myelin repair using the multiple sclerosis medication glatiramer acetate. Proc Natl Acad Sci USA 106:17992–17997

Smith KJ, Felts PA, John GR (2000) Effects of 4-aminopyridine on demyelinated axons, synapses and muscle tension. Brain 123(Pt 1):171–184

Stadelmann C, Bruck W (2008) Interplay between mechanisms of damage and repair in multiple sclerosis. J Neurol 255(Suppl 1):12–18

Tarhzaoui K, Valensi P, Leger G, Cohen-Boulakia F, Lestrade R, Behar A (2009) Rosuvastatin positively changes nerve electrophysiology in diabetic rats. Diabetes Metab Res Rev 25:272–278

Tham CS, Lin FF, Rao TS, Yu N, Webb M (2003) Microglial activation state and lysophospholipid acid receptor expression. Int J Dev Neurosci 21:431–443

Thone J, Gold R (2011) Laquinimod: a promising oral medication for the treatment of relapsing-remitting multiple sclerosis. Expert Opin Drug Metab Toxicol 7:365–370

Toman RE, Payne SG, Watterson KR, Maceyka M, Lee NH, Milstien S, Bigbee JW, Spiegel S (2004) Differential transactivation of sphingosine-1-phosphate receptors modulates NGF-induced neurite extension. J Cell Biol 166:381–392

Trapp BD, Peterson J, Ransohoff RM, Rudick R, Mork S, Bo L (1998) Axonal transection in the lesions of multiple sclerosis. N Engl J Med 338:278–285

Tringali G, Vairano M, Dello RC, Preziosi P, Navarra P (2004) Lovastatin and mevastatin reduce basal and cytokine-stimulated production of prostaglandins from rat microglial cells in vitro: evidence for a mechanism unrelated to the inhibition of hydroxy-methyl-glutaryl CoA reductase. Neurosci Lett 354:107–110

van HJ, Drexhage JA, Flor T, Gerritsen W, van der Valk P, de Vries HE (2010) Nrf2 and DJ1 are consistently upregulated in inflammatory multiple sclerosis lesions. Free Radic Biol Med 49:1283–1289

van Oosten BW, Barkhof F, Truyen L, Boringa JB, Bertelsmann FW, von Blomberg BM, Woody JN, Hartung HP, Polman CH (1996) Increased MRI activity and immune activation in two multiple sclerosis patients treated with the monoclonal anti-tumor necrosis factor antibody cA2. Neurology 47:1531–1534

Vela JM, Yanez A, Gonzalez B, Castellano B (2002) Time course of proliferation and elimination of microglia/macrophages in different neurodegenerative conditions. J Neurotrauma 19: 1503–1520

Vollmer T, Key L, Durkalski V, Tyor W, Corboy J, Markovic-Plese S, Preiningerova J, Rizzo M, Singh I (2004) Oral simvastatin treatment in relapsing-remitting multiple sclerosis. Lancet 363:1607–1608

Wandinger KP, Lunemann JD, Wengert O, Bellmann-Strobl J, Aktas O, Weber A, Grundstrom E, Ehrlich S, Wernecke KD, Volk HD, Zipp F (2003) TNF-related apoptosis inducing ligand (TRAIL) as a potential response marker for interferon-beta treatment in multiple sclerosis. Lancet 361:2036–2043

Wang Q, Yan J, Chen X, Li J, Yang Y, Weng J, Deng C, Yenari MA (2011) Statins: multiple neuroprotective mechanisms in neurodegenerative diseases. Exp Neurol 230(1):27–34

Webb M, Tham CS, Lin FF, Lariosa-Willingham K, Yu N, Hale J, Mandala S, Chun J, Rao TS (2004) Sphingosine 1-phosphate receptor agonists attenuate relapsing-remitting experimental autoimmune encephalitis in SJL mice. J Neuroimmunol 153:108–121

Wegner C, Stadelmann C, Pfortner R, Raymond E, Feigelson S, Alon R, Timan B, Hayardeny L, Bruck W (2010) Laquinimod interferes with migratory capacity of T cells and reduces IL-17 levels, inflammatory demyelination and acute axonal damage in mice with experimental autoimmune encephalomyelitis. J Neuroimmunol 227:133–143

Wierinckx A, Breve J, Mercier D, Schultzberg M, Drukarch B, van Dam AM (2005) Detoxication enzyme inducers modify cytokine production in rat mixed glial cells. J Neuroimmunol 166:132–143

Wosik K, Antel J, Kuhlmann T, Bruck W, Massie B, Nalbantoglu J (2003) Oligodendrocyte injury in multiple sclerosis: a role for p53. J Neurochem 85:635–644

Wu DC, Jackson-Lewis V, Vila M, Tieu K, Teismann P, Vadseth C, Choi DK, Ischiropoulos H, Przedborski S (2002) Blockade of microglial activation is neuroprotective in the 1-methyl-4-phenyl-1,2,3,6-tetrahydropyridine mouse model of Parkinson disease. J Neurosci 22:1763–1771

Xue M, Mikliaeva EI, Casha S, Zygun D, Demchuk A, Yong VW (2010) Improving outcomes of neuroprotection by minocycline: guides from cell culture and intracerebral hemorrhage in mice. Am J Pathol 176:1193–1202

Yang JS, Xu LY, Xiao BG, Hedlund G, Link H (2004) Laquinimod (ABR-215062) suppresses the development of experimental autoimmune encephalomyelitis, modulates the Th1/Th2 balance and induces the Th3 cytokine TGF-beta in Lewis rats. J Neuroimmunol 156:3–9

Yednock TA, Cannon C, Fritz LC, Sanchez-Madrid F, Steinman L, Karin N (1992) Prevention of experimental autoimmune encephalomyelitis by antibodies against alpha 4 beta 1 integrin. Nature 356:63–66

Yoshimura S, Ochi H, Isobe N, Matsushita T, Motomura K, Matsuoka T, Minohara M, Kira J (2010) Altered production of brain-derived neurotrophic factor by peripheral blood immune cells in multiple sclerosis. Mult Scler 16:1178–1788

Youssef S, Stuve O, Patarroyo JC, Ruiz PJ, Radosevich JL, Hur EM, Bravo M, Mitchell DJ, Sobel RA, Steinman L, Zamvil SS (2002) The HMG-CoA reductase inhibitor, atorvastatin, promotes a Th2 bias and reverses paralysis in central nervous system autoimmune disease. Nature 420:78–84

Yrjanheikki J, Keinanen R, Pellikka M, Hokfelt T, Koistinaho J (1998) Tetracyclines inhibit microglial activation and are neuroprotective in global brain ischemia. Proc Natl Acad Sci USA 95:15769–15774

Yu N, Lariosa-Willingham KD, Lin FF, Webb M, Rao TS (2004) Characterization of lysophosphatidic acid and sphingosine-1-phosphate-mediated signal transduction in rat cortical oligodendrocytes. Glia 45:17–27

Zanon RG, Cartarozzi LP, Victório SCS, Moraes JC, Morari J, Velloso LA, Oliveira ALR (2010) Interferon (IFN) beta treatment induces major histocompatibility complex (MHC) class I expression in the spinal cord and enhances axonal growth and motor function recovery following sciatic nerve crush in mice. Neuropathol Appl Neurobiol 36:515–534

Zarkou S, Carter JL, Wellik KE, Demaerschalk BM, Wingerchuk DM (2010) Are corticosteroids efficacious for preventing or treating neutralizing antibodies in multiple sclerosis patients treated with beta-interferons? A critically appraised topic. Neurologist 16:212–214

Zemann B, Kinzel B, Muller M, Reuschel R, Mechtcheriakova D, Urtz N, Bornancin F, Baumruker T, Billich A (2006) Sphingosine kinase type 2 is essential for lymphopenia induced by the immunomodulatory drug FTY720. Blood 107:1454–1458

Zhang SC, Goetz BD, Duncan ID (2003) Suppression of activated microglia promotes survival and function of transplanted oligodendroglial progenitors. Glia 41:191–198

Zhang Y, Taveggia C, Melendez-Vasquez C, Einheber S, Raine CS, Salzer JL, Brosnan CF, John GR (2006) Interleukin-11 potentiates oligodendrocyte survival and maturation, and myelin formation. J Neurosci 26:12174–12185

Zhang Z, Zhang Z, Fauser U, Artelt M, Burnet M, Schluesener HJ (2007) FTY720 attenuates accumulation of EMAP-II+ and MHC-II+ monocytes in early lesions of rat traumatic brain injury. J Cell Mol Med 11:307–314

Zhang Y, Metz LM, Yong VW, Bell RB, Yeung M, Patry DG, Mitchell JR (2008) Pilot study of minocycline in relapsing-remitting multiple sclerosis. Can J Neurol Sci 35:185–191

Zhang J, Zhang A, Sun Y, Cao X, Zhang N (2009) Treatment with immunosuppressants FTY720 and tacrolimus promotes functional recovery after spinal cord injury in rats. Tohoku J Exp Med 219:295–302

Zhang Y, Jalili F, Ouamara N, Zameer A, Cosentino G, Mayne M, Hayardeny L, Antel JP, Bar-Or A, John GR (2010) Glatiramer acetate-reactive T lymphocytes regulate oligodendrocyte progenitor cell number in vitro: role of IGF-2. J Neuroimmunol 227:71–79

Ziemssen T, Kumpfel T, Klinkert WE, Neuhaus O, Hohlfeld R (2002) Glatiramer acetate-specific T-helper 1- and 2-type cell lines produce BDNF: implications for multiple sclerosis therapy. Brain-derived neurotrophic factor. Brain 125:2381–2391

Chapter 10
Imaging of Demyelination and Remyelination in Multiple Sclerosis

Douglas L. Arnold, Catherine M. Dalton, Klaus Schmierer,
G. Bruce Pike, and David H. Miller

10.1 Introduction

The ability to estimate myelin content in MS brain is of particular interest given the primary role of demyelination in this disease and the significance of remyelination for axonal preservation and functional recovery. As treatment strategies potentially facilitating remyelination are developed, accurate non-invasive techniques are needed to monitor remyelination in patients with MS. This chapter reviews techniques potentially able to provide this information: multi-component T2 relaxometry, magnetization transfer imaging, diffusion tensor imaging and positron emission tomography, starting with a brief discussion of how they do this, and then reviewing published data on post-mortem MRI correlations and clinical applications.

D.L. Arnold (✉) • G.B. Pike
McConnell Brain Imaging Centre, WB323, Montreal Neurological Institute,
3801 University Street, Montreal, QC, Canada H3A 2B4
e-mail: doug@mrs.mni.mcgill.ca

C.M. Dalton • D.H. Miller
Department of Neuroinflammation, The Institute of Neurology (Queen Square),
University College London, London WC1N 3BG, UK

K. Schmierer
Centre for Neuroscience & Trauma, Blizard Institute, Barts and The London School
of Medicine & Dentistry, 4 Newark Street, London E1 2AT, UK

Centre for Neuroscience & Trauma, Blizard Institute, Barts and The London School
of Medicine & Dentistry, 4 Newark Street,,
London E1 2AT, UK

10.2 Technical Considerations

10.2.1 Conventional MRI and Myelin

Conventional T2w and T1w MRI both are sensitive to myelin, as evidenced by the age-related changes that occur in association with myelination during normal development (for review, see Barkovich 2005). T1w signal intensity increases with myelination due to shortening of T1, and T2w signal intensity decreases due to shortening of T2. However, changes in signal intensity on T2w and T1w MRI reflect a complex combination of changes in the micro-structural and chemical environment of bulk water and, as such, cannot be used as a pathologically specific marker of myelin. For example, in multiple sclerosis, only about 50% of lesions that are abnormal on T2w MRI alone show evidence of demyelination on post-mortem examination (Fisher et al. 2007). More pathologically specific techniques are required to image myelin in vivo.

10.2.2 Potential Myelin-Specific MRI Techniques

Increased pathological specificity for myelin can be achieved by the use of non-conventional MRI acquisition techniques that take advantage of the particular chemistry and structure of myelin.

10.2.2.1 Multi-component T2 Relaxometry

Multi-component T2 relaxometry can image myelin by exploiting the intrinsic difference in T2 relaxation time between water trapped within the bilayers of myelin and water in other central nervous system tissue compartments in order to estimate myelin water content (MacKay et al. 1994).

T2 relaxation describes the random loss of phase coherence of transverse magnetization. In biological tissue, water will lose phase coherence faster (shorter T2) when its motion is restricted. Water in large open spaces, such as CSF, has a long T2 relaxation time (on the order of seconds) whereas water bound to large macromolecules, which have much more restricted motion, has a very short T2 (on the order of microseconds). Water trapped between the tightly wrapped layers of myelin has a T2 of 10–50 ms. It is therefore believed that the percentage of the total water signal with relaxation time in this range, known as the myelin water fraction (MWF), is a measure of the amount of myelin present. To measure the so-called MWF, it is necessary to perform multi-component T2 relaxometry.

Multi-component T2 relaxometry is performed by acquiring a series of spin-echo images (typically 32 echo times) and estimating the distribution of T2 relaxation times present in each voxel by fitting a large number of exponential curves to

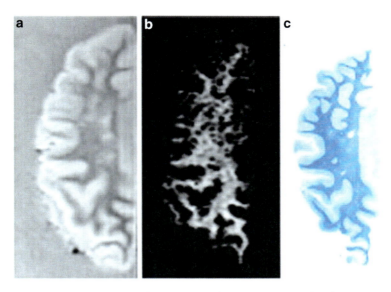

Fig. 10.1 MRI (*left*), MWF (*middle*) and Luxol fast blue myelin stain of post-mortem tissue (*right*) (figure from Laule et al. 2006)

the observed T2 signal decay. In white and grey matter, the T2 distributions are characterized by only a few peaks, which have been experimentally assigned to compartmentalized water populations: a short T2 peak around 20 ms (between 10 and 50 ms) representing the water trapped between the layers of myelin, a second peak around 70–90 ms assigned to intra- and extracellular water and a third peak with T2 > 2 s assigned to either CSF signal or the result of noise (Menon and Allen 1991; Stewart et al. 1993). The myelin water fraction is computed as the fraction of signal below 40–50 ms in the T2 distribution. This can be done voxel by voxel to yield distribution maps (MacKay et al. 1994), a technique that has been referred to as "myelin water imaging" (Fig. 10.1).

The number of myelin water imaging studies of patients is small, as the specialized acquisition and analysis required to calculate myelin water maps are not routinely available on commercial scanners and must be implemented locally.

In cross-sectional studies of patients with MS, the MWF and the mean of the T2 relaxation time distribution (<T2>) are significantly altered in lesions and the normal-appearing white matter (Laule et al. 2004; Vavasour et al. 1998; Whittall et al. 2002). The MWF changes are mainly due to myelin loss, as confirmed by pathology studies using LFB (Moore et al. 2000), whereas increases in T2 of the intermediate component (and by extension, <T2>) have been proposed as a marker of inflammation and oedema (Stanisz et al. 2004; Webb et al. 2003). In a study of MS lesions, myelin water imaging did not show the expected differences in myelin water content between T1-hypointense and T1-isointense lesions (Vavasour et al. 2007).

Serial studies of the evolution of changes in the MWF of MS lesions and NAWM are extremely limited. Levesque et al. (2008) found the MWF in lesions to be abnormal but too variable to show the trends over time expected in association with the acute demyelination and remyelination in Gd-enhancing lesions. In two patients with MS, Vavasour et al. (2009) found changes in MWF consistent with remyelination in two lesions, one of which showed evidence of recovery over time and one of which did not.

10.2.2.2 MWF Advantages and Disadvantages

An important advantage of MW imaging is that it provides theoretical specificity for myelin. Quantitatively, unmyelinated tissue shows a logical MWF of zero, and in post-mortem MS brains, the correlation between MWF and LFB was moderately strong ($R2=0.67$) (Laule et al. 2006).

Disadvantages of MW imaging include the fact that the acquisitions are long, having a typical scan time of approximately 25 min for a single slice, and require custom acquisition and analysis software not provided on commercial scanners. Another general issue with MW imaging is that the signal to noise ratio of the data is low and the conversion of the multi-echo data to produce a spectrum of T2 values, usually performed using a non-negative least squares (NNLS) technique, is underdetermined. Typically, 32 data points (echoes) are processed to produce estimates of the relative amount of water at 120 T2 times. This can result in unstable fits, and therefore data smoothing (regularization) is required. This, coupled with the relatively course sampling of echo times (TE spacing of 10 ms) in the range of MW, makes the reliable estimation of the MWF in vivo difficult (Laule et al. 2004). In fact, much of the ex vivo experimental work that established the multi-exponential nature of white matter relaxation utilized thousands of very closely spaced echoes (Menon et al. 1992). Other potential complications are that MW imaging may not distinguish intact myelin and myelin debris (Webb et al. 2003) and that changes in the rate of exchange of water between compartments can alter the apparent MWF (Levesque et al. 2006). The latter phenomenon may underlie the fact that MWF can vary by a factor of two in regions with the same myelin content but different microanatomy of myelin (Dula et al. 2010).

10.2.2.3 Magnetization Transfer

The signal acquired in conventional MRI originates primarily from the hydrogen nuclei of water that are freely tumbling. Hydrogen nuclei bound to macromolecules have highly restricted motion and have T2 relaxation times that are orders of magnitude too short to contribute directly to the MRI signal. However, these bound or semi-solid protons can affect the observed signal indirectly by magnetization transfer (for a review, see Henkelman et al. 2001 and van Buchem and Tofts 2000).

Magnetization transfer (MT) refers to the MR phenomenon in which spins in two or more distinct environments exchange their magnetization via cross relaxation

Fig. 10.2 MTR image. Note the high MTR in white matter, low MTR in grey matter and absent MTR in CSF. Lesions show variable MTR depending on their myelin content

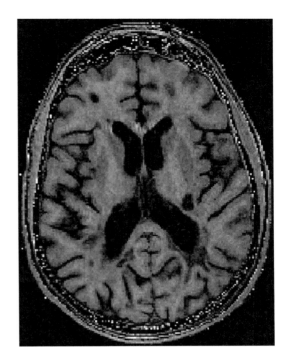

and/or chemical exchange. The simplest and most extensively used MT tissue model is a two-pool system in which protons are considered to be either in a highly mobile liquid state (water) or in a semi-solid macromolecular state of relatively restricted motion, such as large lipids (Wolff and Balaban 1989). The difference in mobility of the water and the semi-solids results in T2 relaxation times of >10 ms for the water and <100 μs for the semi-solids. Experimentally, selectively saturating the semi-solid component and measuring the reduction in the liquid component, undergoing exchange, enables observation of the MT effect. To isolate the MT effect, acquisitions are often performed with and without semi-solid saturation pulses and ratio or percent difference images (so-called MTR images) are calculated (Fig. 10.2). The resulting semi-quantitative images reflect a combination of sequence and relaxation parameters in addition to MT (Pike 1996; Pike et al. 1992). Detailed information about these spin populations can also be calculated using a variety of off-resonance saturation pulses and an appropriate model of MT in tissue, so-called quantitative magnetization transfer imaging (QMTI) (Henkelman et al. 1993; Morrison et al. 1995). The technique of Sled and Pike yields the relative size of the restricted proton pool (F), the first-order forward magnetization exchange rate (kf), as well as most relaxation parameters of the free and restricted pools (R1f and T2f, and T2r) (Sled and Pike 2001).

MT occurs in most biological tissues that have significant interacting liquid and semi-solid constituents. However, in brain white matter, magnetization transfer is dominated by interactions between water and the macromolecular components of myelin, in particular the hydroxyl and amino moieties of cholesterol and glycocerebrosides on the myelin surface (Fralix et al. 1991; Kucharczyk et al. 1994). Thus, while MT measurements cannot provide absolute specificity to the molecular constituents of tissue, there is strong and convergent evidence to support the interpretation of MT changes in white matter as reflecting changes in myelin content.

The onset of T1 shortening seen on T1-weighted images during the early phases of myelination corresponds temporally and spatially to changes in magnetization transfer. Thus, it appears that the T1 shortening on spin echo images with myelin maturation results, in part, from a strong magnetization transfer interaction with myelin. The amount of magnetization transfer in the brain increases during myelination and parallels myelination as determined by histology (Engelbrecht et al. 1998). Destruction of myelin results in decreased magnetization transfer (Dousset et al. 1992). In a study of fixed post-mortem brain samples, Schmierer et al. (2007) showed a strong correlation between the relative size of the restricted pool and the myelin lipid content observed with the Luxol fast blue (LFB) stain. In a follow-up study, the same authors showed there was also a correlation in unfixed post-mortem tissue (Schmierer et al. 2008).

In vivo imaging studies have also contributed to the validation of MT as a marker for myelin. Logical variations have been reported in the QMTI parameters of white matter of healthy controls (Sled et al. 2004), and densely myelinated white matter fibre tracts are distinguishable in maps of the restricted pool fraction (Yarnykh and Yuan 2004).

In multiple sclerosis (MS), of the restricted pool size has been observed to decrease substantially in chronic lesions (Davies et al. 2004; Levesque et al. 2005; Sled and Pike 2001; Tofts et al. 2005), while small but significant decreases in F in the normal-appearing white matter (NAWM) have been reported (Davies et al. 2004; Narayanan et al. 2006). The impact of MS pathology on the T2 of the restricted pool (T2r) is still a matter for discussion, as both decreases (Tofts et al. 2005) and increases (Davies et al. 2004) have been reported in the lesions of patients with MS. In a study combining QMTI and myelin water imaging in 9 controls and 19 MS patients (Tofts et al. 2005), Tofts et al. reported significantly decreased restricted pool and myelin water fractions in lesions and significantly decreased restricted pool fraction in NAWM. These authors concluded that, while pathology affects each measure differently (potentially explained by the presence of inflammation and axonal loss), the restricted pool fraction and the myelin water fraction both reflect demyelination to some extent.

Despite the growing number of studies of MS employing these two quantitative techniques, there has been limited work on one of the most dynamic phases of the disease, acute gadolinium-enhancing lesions. The MT ratio (MTR), which combines the entire MT effect into a single parameter, has been shown to decrease with acute demyelination and to increase with remyelination in an animal model of demyelination (Deloire-Grassin et al. 2000). In acute gadolinium-enhancing (Gd+) lesions of MS patients, the MTR has been observed to decrease acutely, followed by variable recovery over subsequent months (Chen et al. 2007; Dousset et al. 1998; Lai et al. 1997; Richert et al. 2001). Preliminary data show changes in

MTR in acute lesions, in fact, correlate extremely well with quantitative measurements of macromolecular content (Levesque et al. 2008), and the correlation of increases in the MTR of acute Gd-enhancing lesions has been validated by post-mortem examination of an MS patient who died 6 months after MRI examination (Chen et al. 2007).

10.2.2.4 MT Imaging Advantages and Disadvantages

MT contrast is now a well-established mainstream imaging method with most manufacturers providing basic MT imaging capabilities on their scanners. MTR imaging can be performed in under 15 min with whole brain coverage, high spatial resolution and excellent within-site reproducibility. QMTI measurements, while more complex than MTR and still in the realm of custom acquisition and analysis, provide much more detailed information about the size and characteristics of the liquid and semi-solid constituents of tissue. Although the MT effect is not specific to myelin, post-mortem studies of MS brains have shown a correlation between semi-solid pool size and myelin content similar to that reported for myelin water imaging, and animal studies have indicated that the correlation of MTR with myelin content may be more reliable than that of myelin water imaging (Dula et al. 2010). With regard to the longer scan times and limited coverage associated with first generation QMTI, new approaches are now emerging that enable whole brain, high-resolution, QMTI to be performed in under 30 min (Gloor et al. 2008).

10.2.2.5 Diffusion Tensor Imaging

Diffusion tensor imaging (DTI) uses magnetic field gradients applied in different directions to measure the extent to which the diffusion of water has a directional preference. Water molecules in pure water diffuse randomly and have an equal probability of motion in all directions. The probability distribution function describing this is spherical, and the diffusion is described as isotropic. Water molecules in tissue that has an oriented microstructure diffuse preferentially along the axis of the oriented structure and may be restricted in the axis perpendicular to the structure. In white matter tracts, for example, water is hindered in its diffusion perpendicular to the fibres and has a greater probability of motion along the tracts. The probability distribution function corresponding to this is ellipsoidal and can be characterized by a vector describing the long axis (apparent diffusion coefficient, ADC, parallel) and two orthogonal vectors describing the short axes (ADC perpendicular or radial). The more restricted the motion is, the more elongated the ellipse becomes, and the more anisotropic the diffusion. The relative size of the long axis of the ellipse to its short axis can be used to quantify the anisotropy in terms of fractional anisotropy (FA).

DTI is most commonly acquired using fast imaging techniques, such as single-shot echo planar imaging (EPI), to minimize problems associated with the technique's inherent sensitivity to motion. Diffusion encoding is typically achieved using magnetic field gradient pulses to dephase diffusing spins while leaving stationary ones unaffected. To calculate a diffusion tensor, a minimum of six diffusion encodings must be performed with gradients encoding different diffusion directions. DTI acquisition and analysis is now available on most commercial MRI scanners, and basic whole brain data sets can be acquired with isotropic resolution, on the order of a few millimetres, in a few minutes.

Myelin, the axonal membrane, microtubules and microfilaments within axons all are highly oriented structures that could hinder water diffusion and contribute to fractional anisotropy. It has been shown that microtubules and microfilaments are not significant contributors (Beaulieu and Allen 1994). Myelin contributes to some extent. However, myelin is not essential for diffusion anisotropy, as nerves which are normally unmyelinated show anisotropy that is, in some cases, comparable to that of myelinated nerves (Beaulieu and Allen 1994), and diffusion is only decreased by ~20% in the myelin-deficient rat (Gulani et al. 2001; Tyszka et al. 2006). Thus, it appears that the axonal membrane is the major source of anisotropy.

In MS lesions, the perpendicular component of the ellipsoid describing diffusion anisotropy increases and fractional anisotropy decreases. The extent to which this reflects demyelination of intact axons is, however, unclear, as MS lesions are known to be associated with oedema and axonal injury, which also could be associated with increased diffusion perpendicular to fibre orientation and decreased FA.

10.2.2.6 DTI Advantages and Disadvantages

The use of DTI to assess myelination also has other disadvantages. Aside from the fact that the axonal membrane (and not myelin) is the primary determinant of anisotropy, crossing, branching and bending of fibres within a voxel reduce the anisotropy independent of myelination. Thus, it is possible that loss of axons in one of two tracts that have crossing fibres within a voxel could result in an increase rather than a decrease of anisotropy. This explanation has been proposed to explain paradoxical behaviour of FA in certain pathological situations (Oouchi et al. 2007). Since EPI is typically employed in DTI, geometric distortions and signal loss in areas of magnetic field inhomogeneities are another problem.

10.2.3 PET Markers of Myelin

It is also possible to assess myelin in vivo using positron emission tomography (PET) and tracers such as 1,4-bis(p-aminostyryl)-2-methoxy benzene (BMB) and PIB (Stankoff et al. 2006, 2011) (Fig. 10.3). These tracers crosses the blood–brain barrier and bind to myelin in a dose-dependent and reversible manner, enabling the detection of demyelinating lesions in experimental models of demyelination and dysmyelination and post-mortem in the brains of patients with multiple sclerosis. PIB is

10 Imaging of Demyelination and Remyelination in Multiple Sclerosis

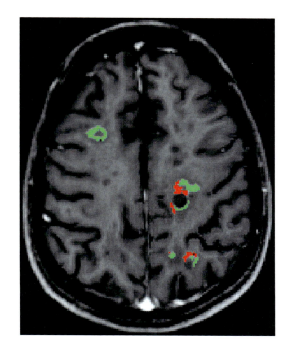

Fig. 10.3 Voxel-based analysis of changes in MTR between two time points. Significant increases in MTR consistent with remyelination are shown in *green*. Significant decreases in MTR consistent with demyelination are shown in *red* (after Chen et al. 2007)

already being used in humans for the study of Alzheimer's disease and is under development as a potential brain myelin imaging technique for use in humans.

10.2.4 Conclusions

A variety of techniques are available that offer relative specificity for myelin and the possibility to assess demyelination and remyelination in vivo. Multi-component T2 relaxometry may be the most specific but has low signal to noise ratio and is not widely available. MTR imaging is a practical alternative, albeit less specific than fully quantitative MT imaging. Diffusion tensor imaging is theoretically less specific but FA is generally modulated by myelination. Positron emission tomography is still in the very early stages of development.

10.3 Clinical Applications

Off all the techniques used for assessing myelin, by far the greatest experience has been with MTR. In addition to brain, a number of studies have focused on the optic nerve because of its unique ability to provide pathophysiological correlates of demyelination and remyelination.

10.3.1 Imaging Demyelination and Remyelination in the Optic Nerve

Optic neuritis occurs when an acute inflammatory-demyelinating lesions affects the optic nerve and is a frequent presenting symptom or relapse event that occurs at the onset or during the relapsing–remitting phase. It is possible to characterize the functional effects of optic neuritis using quantitative measures of vision: standard and low contrast acuity, visual fields and colour vision. Furthermore, the electrophysiological consequences of the lesion can be evaluated using the visual evoked potential (VEP), and abnormalities of latency can be used to infer the presence of demyelination or remyelination. Finally, imaging of the retinal nerve fibre layer (RNFL) using optic coherence tomography (OCT) provides a measure of axonal loss that occurs as a consequence of the inflammatory-demyelinating episode (Trip et al. 2005).

In such a context, imaging the symptomatic optic nerve lesion directly enables investigation of the relationship between the lesion morphology per se and function (quantitative vision measures), nerve conduction (VEP) and axonal damage (OCT). Specifically, identifying a relationship between a quantitative imaging measure from the lesion and the latency of the VEP would provide supportive evidence that the imaging measure reflects myelination: a prolonged well-formed P100 response suggests demyelination, and subsequent shortening of the latency towards normal suggests remyelination.

10.3.1.1 MRI of the Optic Nerve in Optic Neuritis

MRI of the optic nerve is technically challenging because of its small size, motion during image acquisition, chemical shift artefacts from orbital fat and CSF signal in the nerve sheath. Nevertheless, the symptomatic lesion is detectable in over 95% of people with acute optic neuritis as a region of high signal on T2-weighted fat-suppressed images (Fisher et al. 2006). The acute lesion displays gadolinium enhancement reflecting inflammation, and its presence is associated with more severe visual loss and conduction block in comparison with the post-inflammatory non-enhancing lesion state (Gass et al. 1996).

Other MR imaging measures that have been obtained in the optic nerve include cross-sectional area (a measure of atrophy), diffusion and magnetisation transfer ratio (MTR). Of these, MTR has emerged as the most promising potential marker of demyelination and remyelination in optic nerve.

10.3.1.2 Optic Nerve MTR Studies in Optic Neuritis

Thorpe et al. (1995) investigated the optic nerves of 20 patients with optic neuritis using MT imaging and correlated the imaging findings with clinical and VEP measures. MTR was measured from a 2D gradient echo sequence. There was a significant correlation between MTR reduction and prolonged whole ($r=-0.554$, $p<0.01$) and central field ($p<0.05$) VEP latency.

Inglese et al. (2002) obtained conventional and MTR images from the optic nerves in 30 patients with MS as well as 18 controls and 10 patients with Lebers Hereditary Optic Neuropathy (LHON). In MS subject eyes, three subgroups were investigated: clinically unaffected, affected with good recovery of vision and affected with incomplete or no visual recovery. Reduced optic nerve volume and MTR was found in MS eyes with incomplete or no visual recovery and in LHON eyes, whereas such reductions were not observed in MS-unaffected or good recovery eyes. There was no significant correlation between optic nerve MTR and VEP latency.

Hickman et al. (2004) performed a serial clinical and imaging follow-up study of 21 of 29 patients who presented with acute optic neuritis. The follow-up period was 1 year, and 3D gradient echo MTR scans were obtained upon each visit along with conventional MRI, VEPs and visual function assessments. Whole nerve and lesion MTR measures were investigated. The initial MTR (during the acute episode) was not significantly different between affected and control nerves (47.3 pu vs. 47.9 pu), but in the affected nerve it decreased to a nadir of 44.2 pu after ~240 days before partially recovering (45.1 pu at 1 year). For each 0.01 increase in the time-averaged lesion to normal-appearing nerve MTR ratio, the logMAR visual acuity recovery improved by 0.03 (95% CI, 0.002, 0.08, $p=0.02$). The time-averaged VEP central field latency was shorter by 6.1 ms (95% CI 1.5, 10.7, $p=0.012$) per 1 pu rise in time-averaged diseased optic nerve MTR. The investigators proposed that the early fall in diseased optic nerve MTR is consistent with demyelination and Wallerian degeneration of transected axons. A later nadir of MTR was evident when compared with studies of multiple sclerosis brain lesions (see elsewhere in this chapter), and this may have been due to slow clearance of myelin debris. Remyelination may have contributed to the non-significant increase in MTR seen between nadir and 1 year of follow-up.

Trip et al. (2005) studied optic nerve MTR in a cohort of 25 patients with previous optic neuritis; in many of whom, there was incomplete recovery of vision. The patients were also studied with OCT and VEPs and also had detailed assessments of visual function. Lesion MTR was significantly correlated with both central ($r=-0.53$, $p=0.008$) and whole field ($r=-0.40$, $p=0.05$) VEP latency. Whole nerve MTR was correlated with central VEP latency ($r=0.45$, $p=0.03$). No significant correlations were observed between optic nerve MTR measures and visual function, VEP amplitude or lesion length. However, OCT-measured RNFL thickness was significantly correlated with both whole nerve ($r=0.50$, $p=0.01$) and lesion ($r=0.44$, $p=0.03$) MTR. In this poorer visual outcome cohort, there was considerable axonal—as well as myelin—loss (Trip et al. 2005).

10.3.2 Imaging Demyelination and Remyelination in Brain

With the evidence from MR-pathology studies and optic nerve imaging suggesting that reduced MTR and T1 hypointensity are associated with demyelination and/or axonal loss, the next section reviews studies that have investigated these measures serially in new and evolving MS brain lesions and considers their potential as measures of demyelination and remyelination in such lesions

10.3.2.1 MTR Changes in NAWM Before a Lesion Appears

In a longitudinal study of newly appearing T2-weighted lesions, Pike et al. (2000) found that MTR decreased steadily in pre-lesional NAWM for months to years before lesion appearance on MRI. In order to explore the time course of pre-lesional MTR changes, Filippi performed a 3-month longitudinal serial study where ten patients were scanned every 4 weeks; 48 new enhancing lesions that arose out of NAWM were identified (Filippi et al. 1998). Three months before enhancement, the mean MTR in the corresponding NAWM was significantly lower than the mean MTR in NAWM outside enhancing areas; the MTR thenceforth continued to decrease prior to the appearance of the enhancing lesion. In similar studies of lesion evolution, Fazekas found a significant decrease of the MTR 4 months before lesion enhancement in 12 patients with 44 enhancing lesions (Fazekas et al. 2002). Goodkin, in a serial study of 22 patients who developed 129 enhancing lesions over 12 months, found differences in pre-lesional contralateral NAWM several months prior to the lesion appearing (Goodkin et al. 1998). Richert et al. in 2001 co-registered the regions of interest of 185 new contrast-enhancing lesions to the MTR image set to determine the mean lesion MTR on each monthly exam. The lesion MTR was reduced compared with MTR of NAWM as early as 12 months prior to enhancement (Richert et al. 2001).

However, not all groups have found a prolonged or steady decrease in MTR prior to the lesion appearing with breakdown of the blood–brain barrier and lesion enhancement. Dousset studied 15 new enhancing lesions in four MS patients for 9–12 months and found no change in MTR prior to enhancement (Dousset et al. 1998). Also, Silver et al. (1998) studied new gadolinium-enhancing lesions and MTR in three patients who were scanned weekly for 3 months and found no reduction in MTR prior to enhancement.

Overall, there does appear to be a pre-lesional change occurring in NAWM in some but not all instances before a lesion appears. The small magnitude of the MTR changes observed is potentially explained by a number of pathological processes and does not necessarily indicate demyelination.

10.3.2.2 MTR Patterns During Lesion Enhancement

Silver et al. (1999) performed a weekly enhanced MRI and MT scanning in three patients with MS and evaluated 25 lesions greater than 20 mm^2 which first enhanced after the baseline and found significant reductions in MTR in the first week of enhancement (28.2 pu) compared with prior to enhancement (29.6 pu) with a nadir reached at week three following enhancement. Large lesions greater than 45 mm^2 behaved differently showing a greater decrease and more significant recovery in MTR compared with smaller lesions. Silver went on to study a single large lesion measuring 101 mm^2, which started to enhance at week 2 and continued to throughout the 12-week study period. The decrease in MTR occurred initially at the lesion centre 1–3 weeks following initial enhancement (21.5 pu week 2 and 12.9 pu week 3), followed by slower decreases at the lesion rim, despite cessation of enhancement. The MTR nadir reached a minimum of 7.4 pu (24% of the pre-enhancement value) with positive correlations between duration of gadolinium enhancement and the MTR lesion nadir.

Dousset also found marked decreases in MTR in 13/15 lesions with the nadir MTR values occurring in the initial enhancing scan or a month later, whereas the MTR declined progressively in the remaining two lesions which reinforces the idea that each lesion had an individual MTR profile (Dousset et al. 1998). Richert showed the MTR nadir occurred between months 0 and 1 following enhancement (Richert et al. 2001). Goodkin et al. (1998) noted the median paired difference between MTR values in lesions and contralateral NAWM was most marked with new focal enhancement. Fazekas found a highly significant difference in MTR between active lesions and control NAWM at first enhancement (month 0: 21.4%) (Fazekas et al. 2002). In summary there is a body of evidence confirming that the MTR nadir in new inflammatory (enhancing) brain lesions occurs during or immediately following the early enhancing phase of lesion evolution.

10.3.2.3 Baseline Lesion Enhancement Characteristics and MTR Outcome

The baseline pattern of enhancement has been used to determine MTR outcome: Rovira et al. (1999) performed a longitudinal study of 11 patients over 1 year including 47 lesions. Lesions classified as baseline ring-enhancing lesions with a T1-hypointense centre had initially low baseline MTR (mean 19.7%) compared with nodular-enhancing lesions (mean 32.6%). At 12 months, the MTR recovered to 34.3 in the ring-enhancing lesions and 38.3 in the nodular-enhancing group. These differences may reflect lesion age and/or size (not reported) or may indicate a different pathological basis for ring- and nodular-enhancing lesions.

10.3.2.4 Enhancement Duration and Contrast Dosage: Relationship with Lesion MTR

Acute inflammatory MS lesions appear heterogeneous with regard to the duration of their enhancing phase, which varies from less than 2 weeks to more than 3 months, though in most instances will be between 2 weeks and 2 months. Filippi evaluated MTR in a serial study of ten patients scanned every 4 weeks for

4 months. A total of 54 new enhancing lesions were co-registered, and the mean MTR during enhancement was 33.6% (Filippi et al. 1998). The mean MTR values were higher at 35.3% for lesions enhancing on only one scan compared with 29.3% for those enhancing on at least two consecutive scans suggesting that there is more severe acute tissue damage with more prolonged inflammation (the latter inferred from the persistence of enhancement). In a further study of ten patients who were scanned twice every 4 weeks for 3 months and given either single- or triple-dose gadolinium chelate at each 4-week visit, Filippi noted higher MTR values in newly enhancing lesions that enhanced only with triple-dose contrast (38.2% vs. 31.4%) perhaps reflecting less severe inflammation (Filippi et al. 1998).

10.3.2.5 Spectroscopy, Lesion Enhancement and MTR

Narayana et al., in a spectroscopy study of lesions in 14 MS patients using a short TE found prominent resonances in the 0.5–2.0-ppm region in the spectra of six of nine enhancing lesions in seven patients. These resonances were attributed to products of lipid and myelin breakdown. Similar resonances were detected in only seven of 21 non-enhancing plaques. The authors speculate whether active inflammation (inferred by the presence of enhancement) might not be a necessary precursor of myelin breakdown (Levesque et al. 2005).

Pike also performed a combined MTR and spectroscopy study in 30 patients with relapsing–remitting, primary progressive or secondary progressive MS. The average lesion MTR and NAA were correlated in patients with relapsing–remitting MS (greater in patients with new lesions), implying that axonal damage occurs with new demyelination (Levesque et al. 2005).

10.3.2.6 Serial Evolution of Enhancing Lesions on T1-Weighted Images

A serial study of 126 MS enhancing lesions that were observed in 11 patients found that in the acute enhancing phase, 25 (20%) appeared T1 isointense and 101 (80%) appeared T1 hypointense compared to NAWM (Nijeholt et al. 2001). The investigators observed four evolutionary MR patterns during the 6-month follow-up period: (1) initially T1-isointense lesions which remained isointense (15%), (2) initially T1-isointense lesions which became hypointense (5%; mostly these were lesions that re-enhanced), (3) initially T1 hypointense which became isointense (44%) and (4) initially T1 hypointense which remained hypointense (36%). Decreased MTR at baseline enhancement, the duration of enhancement and to a lesser extent ring enhancement were predictive of persistent T1 hypointensity—suggesting irreversible structural damage to myelin and axons—at 6 months.

10.3.2.7 MTR Recovery Within Lesions

Another important part of the pathological evolution of new MS lesions is the occurrence of repair. This can be inferred in a broad sense by observing reduction or resolution of MRI abnormalities on serial imaging. Filippi noted significant increases in the mean lesion MTR values (more so for lesions seen only using triple-dose contrast) at each of the monthly scans during the 3-month follow-up period (Filippi et al. 1998). In serial studies that had a longer period of follow-up, Richert et al. (2001) and Van Waesberghe et al. (Nijeholt et al. 2001) found MTR recovery continuing to occur within lesions over 4–6 months following enhancement.

Rovira et al., in their 12-month study, observed that MTR recovery is dependent on lesion type. Initially ring-enhancing lesions with a T1-hypointense centre had a low baseline MTR (mean 19.7%) with most recovery apparent at 3 months but continued recovery seen over 12 months (at which time, mean MTR was 34.3%). In contrast, initial non-enhancing T1-hypointense lesions had a mean MTR of 26.3 and showed only a slight recovery of MTR within lesions at month 12 (mean 28.8%) (Rovira et al. 1999).

10.3.2.8 Lesion Recovery and Voxel Inhomogeneity

Chen et al. (2007) studied 14 subjects in a 2-month follow-up study to determine whether the change in MTR between baseline and follow-up scans might be related to the local MTR signal inhomogeneity (MTRinhomog) at baseline. The initial step was to create a baseline-blurred MTR volume, which was subtracted from the baseline MTR image. The absolute difference of MTR of the voxel and the average of its neighbours was interpreted as a measure of local signal inhomogeneity. Using segmented lesions and NAWM, Chen calculated MTR signal inhomogeneity maps and MTR lesion difference maps between baseline and follow-up. Lesion voxels were grouped to calculate their MTR and MTR inhomogeneity according to 25th and 75th percentiles. Lesions with low MTR inhomogeneity at baseline (supposed inactive plaques) experienced little further change in MTR on follow-up. In contrast, regions with high MTR inhomogeneity at baseline (supposed active plaques) experienced large longitudinal MTR changes. These high inhomogeneity voxels were divided into two groups. The voxels with high MTR in regions of high MTR inhomogeneity at baseline showed a decrease in MTR between baseline and follow-up (interpreted as demyelination and oedema). Voxels with low MTR in regions of high MTR inhomogeneity at baseline showed a mean increase in MTR between baseline and follow-up (interpreted as remyelination or resolution of oedema). The investigators concluded that the level of magnetization transfer together with local MTR inhomogeneity was predictive of subsequent changes in MTR in both the lesions and NAWM and was potentially suitable as a biomarker predictive of changes in myelin content for evaluating therapies targeting remyelination.

10.3.2.9 Voxel-Based Analysis of Lesion MTR

To deal more appropriately with lesion inhomogeneity, Chen et al. developed an advanced processing technique to quantify MTR changes within individual lesion voxels and infer which areas were undergoing demyelination and remyelination. This technique was validated in a lesion that was followed for 6.5 months after enhancement in a patient who had participated in a trial of immunoablation and autologous hematopoietic stem cell transplantation and died of hepatic insufficiency (Chen et al. 2007). Detailed post-mortem histological analysis revealed that the central area of the lesion which showed stable low MTR was associated with demyelination, macrophages/microglia and loss of oligodendrocytes, whereas areas of MTR increase at the periphery of the lesion adjacent to NAWM were associated with partial remyelination, macrophages/microglia and oligodendrocytes (Fig. 10.4).

Chen et al. (2007) applied this technique to four other patients participating in the same trial. The average mean normalized lesion MTR over all lesions exhibited partial recovery over 2–4 months after Gd enhancement. Voxel-based MTR analysis showed ongoing significant increases and decreases in MTR consistent with ongoing demyelination and remyelination after the mean MTR of all voxels had stabilized (Fig. 10.4). MTR remained low in 70% of the initially enhancing lesion volume. The MTR evolution differed between lesions within a given patient and also differed between patients. This voxel-level, registration-based evaluation of lesion MTR evolution has promise for meaningful investigation of remyelination in clinical trials.

10.3.2.10 Summary

Serial MTR of new enhancing brain lesions in MS has revealed that MTR values decline subtly in pre-lesional NAWM months to years prior to the development of enhancing lesions and that there is a sharp and much larger decline in MTR with the onset of enhancement that quickly reaches a nadir (<1 month). The pattern (ring or nodular) and duration of enhancement and the formation of a T1-hypointense lesions in serial studies all have some relationship with the likelihood of a lesion having a higher or lower MTR that may reflect demyelination and remyelination. MTR recovery occurs over 4–6 months. Analysis of significant changes in MTR at the level of individual voxels and of the level of MTR inhomogeneity may help to discriminate between areas within lesions that are demyelinating and remyelinating.

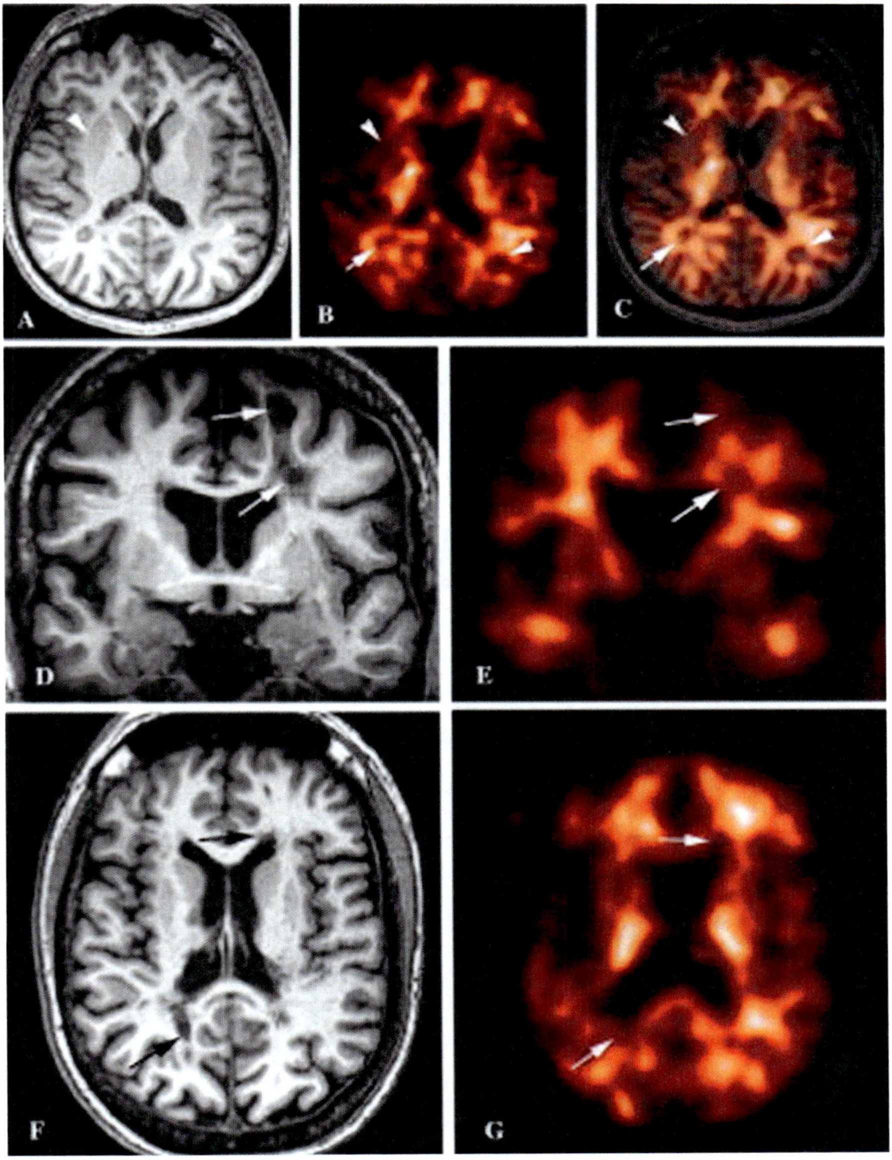

Fig. 10.4 Positron emission tomography (PET) imaging using [methyl-^{11}C]-2-(4′-methylaminophenyl)-6-hydroxybenzothiazole ([^{11}C]PIB) in multiple sclerosis patients. PET images (Patient 1, **a–c**: 327 MBq [^{11}C]PIB injected; Patient 2, **d–g**: 320 MBq [^{11}C]PIB injected) were co-registered onto the corresponding slices of the brain magnetization-prepared rapid acquisition gradient echo (MP-RAGE) 3T images and corresponded to the activity between 30 and 70 min. *Arrows* point to MS plaques, and *arrowheads* point to grey matter structures, both appearing as a loss of uptake on PET images. (**a, d, f**) MP-RAGE images, (**b, e, g**) corresponding [^{11}C]PIB PET images (which were co-registered with MP-RAGE images) and (**c**) superposition of an MP-RAGE and the corresponding [^{11}C]PIB PET images for Patient 1 (from Stankoff et al. 2011)

References

Barkovich AJ (2005) Magnetic resonance techniques in the assessment of myelin and myelination. J Inherit Metab Dis 28:311–343

Beaulieu C, Allen PS (1994) Water diffusion in the giant axon of the squid: implications for diffusion-weighted MRI of the nervous system. Magn Reson Med 32:579–583

Chen JT, Kuhlmann T, Jansen GH, Collins DL, Atkins HL, Freedman MS, O'Connor PW, Arnold DL (2007) Voxel-based analysis of the evolution of magnetization transfer ratio to quantify remyelination and demyelination with histopathological validation in a multiple sclerosis lesion. Neuroimage 36:1152–1158

Davies GR, Tozer DJ, Cercignani M, Ramani A, Dalton CM, Thompson AJ, Barker GJ, Tofts PS, Miller DH (2004) Estimation of the macromolecular proton fraction and bound pool T2 in multiple sclerosis. Mult Scler 10:607–613

Deloire-Grassin MS, Brochet B, Quesson B, Delalande C, Dousset V, Canioni P, Petry KG (2000) In vivo evaluation of remyelination in rat brain by magnetization transfer imaging. J Neurol Sci 178:10–16

Dousset V, Grossman RI, Ramer KN, Schnall MD, Young LH, Gonzalez-Scarano F, et al. (1992) Experimental allergic encephalomyelitis and multiple sclerosis: lesion characterization with magnetization transfer imaging. Radiology. 182(2):483–91. Epub 1992/02/01.

Dousset V, Gayou A, Brochet B, Caille JM (1998) Early structural changes in acute MS lesions assessed by serial magnetization transfer studies. Neurology 51:1150–1155

Dula AN, Gochberg DF, Valentine HL, Valentine WM, Does MD (2010) Multiexponential T2, magnetization transfer, and quantitative histology in white matter tracts of rat spinal cord. Magn Reson Med 63:902–909

Engelbrecht V, Rassek M, Preiss S, Wald C, Modder U (1998) Age-dependent changes in magnetization transfer contrast of white matter in the pediatric brain. AJNR Am J Neuroradiol 19(10):1923–9. Epub 1999/01/05.

Fazekas F, Ropele S, Enzinger C, Seifert T, Strasser-Fuchs S (2002) Quantitative magnetization transfer imaging of pre-lesional white-matter changes in multiple sclerosis. Mult Scler 8:479–484

Filippi M, Rocca MA, Comi G (1998) Magnetization transfer ratios of multiple sclerosis lesions with variable durations of enhancement. J Neurol Sci 159:162–165

Fisher JB, Jacobs DA, Markowitz CE, Galetta SL, Volpe NJ, Nano-Schiavi ML, Baier ML, Frohman EM, Winslow H, Frohman TC, Calabresi PA, Maguire MG, Cutter GR, Balcer LJ (2007) Relation of visual function to retinal nerve fiber layer thickness in multiple sclerosis. Ophthalmology 113:324–332. doi:10.1002/ana.21113

Fisher E, Chang A, Fox RJ, Tkach JA, Svarovsky T, Nakamura K, et al. (2006) Imaging correlates of axonal swelling in chronic multiple sclerosis brains. Ann Neurol 62(3):219–28. Epub 2007/04/13.

Fralix TA, Ceckler TL, Wolff SD, Simon SA, Balaban RS (1991) Lipid bilayer and water proton magnetization transfer: effect of cholesterol. Magnetic Resonance in Medicine 18(1):214–23.

Gass A, Moseley IF, Barker GJ, Jones S, MacManus D, McDonald WI, Miller DH (1996) Lesion discrimination in optic neuritis using high-resolution fat-suppressed fast spin-echo MRI. Neuroradiology 38:317–321

Gloor M, Scheffler K, Bieri O (2008) Quantitative magnetization transfer imaging using balanced SSFP. Magn Reson Med 60:691–700. doi:10.1002/mrm.21705

Goodkin DE, Rooney WD, Sloan R, Bacchetti P, Gee L, Vermathen M, Waubant E, Abundo M, Majumdar S, Nelson S, Weiner MW (1998) A serial study of new MS lesions and the white matter from which they arise. Neurology 51:1689–1697

Gulani V, Webb AG, Duncan ID, Lauterbur PC (2001) Apparent diffusion tensor measurements in myelin-deficient rat spinal cords. Magn Reson Med 45:191–195

Henkelman RM, Huang X, Xiang QS, Stanisz GJ, Swanson SD, Bronskill MJ (1993) Quantitative interpretation of magnetization transfer. Magn Reson Res 29:759–766

Henkelman RM, Stanisz GJ, Graham SJ (2001) Magnetization transfer in MRI: a review. NMR Biomed 14:57–64. doi:10.1002/nbm.683 [pii]

Hickman SJ, Toosy AT, Jones SJ, Altmann DR, Miszkiel KA, MacManus DG, Barker GJ, Plant GT, Thompson AJ, Miller DH (2004) Serial magnetization transfer imaging in acute optic neuritis. Brain 127:692–700. doi:10.1093/brain/awh076

Inglese M, Salvi F, Iannucci G, Mancardi GL, Mascalchi M, Filippi M (2002) Magnetization transfer and diffusion tensor MR imaging of acute disseminated encephalomyelitis. AJNR Am J Neuroradiol 23:267–272

Kucharczyk W, Macdonald PM, Stanisz GJ, Henkelman RM (1994) Relaxivity and magnetization transfer of white matter lipids at MR imaging: importance of cerebrosides and pH. Radiology 192:521–529

Lai HM, Davie CA, Gass A, Barker GJ, Webb S, Tofts PS, Thompson AJ, McDonald WI, Miller DH (1997) Serial magnetisation transfer ratios in gadolinium-enhancing lesions in multiple sclerosis. J Neurol 244:308–311

Laule C, Vavasour IM, Moore GR, Oger J, Li DK, Paty DW, MacKay AL (2004) Water content and myelin water fraction in multiple sclerosis. A T2 relaxation study. J Neurol 251:284–293

Laule C, Leung E, Lis DK, Traboulsee AL, Paty DW, MacKay AL, Moore GR (2006) Myelin water imaging in multiple sclerosis: quantitative correlations with histopathology. Mult Scler 12:747–753

Levesque I, Sled JG, Narayanan S, Santos AC, Brass SD, Francis SJ, Arnold DL, Pike GB (2005) The role of edema and demyelination in chronic T(1) black holes: a quantitative magnetization transfer study. J Magn Reson Imaging 21:103–110

Levesque I, Chia CL, Pike GB (2006) The impact of compartmental exchange on estimates of the myelin water fraction. In: ISMRM workshop on imaging myelin: formation, destruction and repair, Vancouver

Levesque I, Giacomini PS, Narayanan S, Ribeiro L, Sled JG, Arnold DL, Pike GB (2008) Evolution of quantitative magnetization transfer imaging parameters in acute lesions of multiple sclerosis. In: Proceedings of the 16th scientific meeting of the International Society for Magnetic Resonance in Medicine, Toronto

MacKay A, Whittall K, Adler J, Li D, Paty D, Graeb D (1994) In vivo visualization of myelin water in brain by magnetic resonance. Magn Reson Med 31:673–677

Menon RS, Allen PS (1991) Application of continuous relaxation time distributions to the fitting of data from model systems and excised tissue. Magn Reson Med 20:214–227

Menon RS, Rusinko MS, Allen PS (1992) Proton relaxation studies of water compartmentalization in a model neurological system. Magn Reson Med 28:264–274

Moore GR, Leung E, MacKay AL, Vavasour IM, Whittall KP, Cover KS, Li DK, Hashimoto SA, Oger J, Sprinkle TJ, Paty DW (2000) A pathology-MRI study of the short-T2 component in formalin-fixed multiple sclerosis brain. Neurology 55:1506–1510

Morrison C, Stanisz G, Henkelman RM (1995) Modeling magnetization transfer for biological-like systems using a semi-solid pool with a super-Lorentzian lineshape and dipolar reservoir. J Magn Reson B 108:103–113

Narayanan S, Francis SJ, Sled JG, Santos AC, Antel S, Levesque I, Brass S, Lapierre Y, Sappey-Marinier D, Pike GB, Arnold DL (2006) Axonal injury in the cerebral normal-appearing white matter of patients with multiple sclerosis is related to concurrent demyelination in lesions but not to concurrent demyelination in normal-appearing white matter. Neuroimage 29:637–642

Nijeholt GJ, Bergers E, Kamphorst W, Bot J, Nicolay K, Castelijns JA, van Waesberghe JH, Ravid R, Polman CH, Barkhof F (2001) Post-mortem high-resolution MRI of the spinal cord in multiple sclerosis: a correlative study with conventional MRI, histopathology and clinical phenotype. Brain 124:154–166

Oouchi H, Yamada K, Sakai K, Kizu O, Kubota T, Ito H, Nishimura T (2007) Diffusion anisotropy measurement of brain white matter is affected by voxel size: underestimation occurs in areas with crossing fibers. AJNR Am J Neuroradiol 28:1102–1106. doi:10.3174/ajnr.A0488

Pike GB (1996) Pulsed magnetization transfer contrast in gradient echo imaging: a two-pool analytic description of signal response. Magn Reson Med 36:95–103

Pike GB, Glover GH, Hu BS, Enzmann DR (1992) Pulsed magnetization transfer spin-echo MR imaging. J Magn Reson 3:531–539

Pike GB, De Stefano N, Narayanan S, Worsley KJ, Pelletier D, Francis GS, Antel JP, Arnold DL (2000) Multiple sclerosis: magnetization transfer MR imaging of white matter before lesion appearance on T2-weighted images. Radiology 215:824–830

Richert ND, Ostuni JL, Bash CN, Leist TP, McFarland HF, Frank JA (2001) Interferon beta-1b and intravenous methylprednisolone promote lesion recovery in multiple sclerosis. Mult Scler 7:49–58

Rovira A, Alonso J, Cucurella G, Nos C, Tintore M, Pedraza S, Rio J, Montalban X (1999) Evolution of multiple sclerosis lesions on serial contrast-enhanced T1-weighted and magnetization-transfer MR images. AJNR Am J Neuroradiol 20:1939–1945

Schmierer K, Tozer DJ, Scaravilli F, Altmann DR, Barker GJ, Tofts PS, Miller DH (2007) Quantitative magnetization transfer imaging in postmortem multiple sclerosis brain. J Magn Reson Imaging 26:41–51

Schmierer K, Wheeler-Kingshott CA, Tozer DJ, Boulby PA, Parkes HG, Yousry TA, Scaravilli F, Barker GJ, Tofts PS, Miller DH (2008) Quantitative magnetic resonance of postmortem multiple sclerosis brain before and after fixation. Magn Reson Med 59:268–277. doi:10.1002/mrm.21487

Silver NC, Lai M, Symms MR, Barker GJ, McDonald WI, Miller DH (1998) Serial magnetization transfer imaging to characterize the early evolution of new MS lesions. Neurology 51:758–764

Silver N, Lai M, Symms M, Barker G, McDonald I, Miller D (1999) Serial gadolinium-enhanced and magnetization transfer imaging to investigate the relationship between the duration of blood-brain barrier disruption and extent of demyelination in new multiple sclerosis lesions. J Neurol 246:728–730. doi:92460728.415 [pii]

Sled JG, Pike GB (2001) Quantitative imaging of magnetization transfer exchange and relaxation properties in vivo using MRI. Magn Reson Med 46:923–931

Sled JG, Levesque I, Santos AC, Francis SJ, Narayanan S, Brass SD, Arnold DL, Pike GB (2004) Regional variations in normal brain shown by quantitative magnetization transfer imaging. Magn Reson Med 51:299–303

Stanisz GJ, Webb S, Munro CA, Pun T, Midha R (2004) MR properties of excised neural tissue following experimentally induced inflammation. Magn Reson Med 51:473–479

Stankoff B, Wang Y, Bottlaender M, Aigrot MS, Dolle F, Wu C, Feinstein D, Huang GF, Semah F, Mathis CA, Klunk W, Gould RM, Lubetzki C, Zalc B (2006) Imaging of CNS myelin by positron-emission tomography. Proc Natl Acad Sci USA 103:9304–9309

Stankoff B, Freeman L, Aigrot MS, Chardain A, Dolle F, Williams A, Galanaud D, Armand L, Lehericy S, Lubetzki C, Zalc B, Bottlaender M (2011) Imaging central nervous system myelin by positron emission tomography in multiple sclerosis using [methyl-(1)(1)C]-2-(4'-methylaminophenyl)-6-hydroxybenzothiazole. Ann Neurol 69:673–680

Stewart WA, MacKay AL, Whittall KP, Moore GR, Paty DW (1993) Spin-spin relaxation in experimental allergic encephalomyelitis. Analysis of CPMG data using a non-linear least squares method and linear inverse theory. Magn Reson Med 29:767–775

Thorpe JW, Barker GJ, Jones SJ, Moseley I, Losseff N, MacManus DG, Webb S, Mortimer C, Plummer DL, Tofts PS et al (1995) Magnetisation transfer ratios and transverse magnetisation decay curves in optic neuritis: correlation with clinical findings and electrophysiology. J Neurol Neurosurg Psychiatry 59:487–492

Tofts PS, Cercignani M, Tozer DJ, Symms MR, Davies GR, Ramani A, Barker GJ (2005) Tozer et al. Quantitative magnetization transfer mapping of bound protons in multiple sclerosis, Magn Reson Med 2003;50:83-91. Magn Reson Med 53:492–493

Trip SA, Schlottmann PG, Jones SJ, Altmann DR, Garway-Heath DF, Thompson AJ, Plant GT, Miller DH (2005) Retinal nerve fiber layer axonal loss and visual dysfunction in optic neuritis. Ann Neurol 58:383–391. doi:10.1002/ana.20575

Tyszka JM, Readhead C, Bearer EL, Pautler RG, Jacobs RE (2006) Statistical diffusion tensor histology reveals regional dysmyelination effects in the shiverer mouse mutant. Neuroimage 29:1058–1065

van Buchem MA, Tofts PS (2000) Magnetization transfer imaging. Neuroimaging Clin N Am 10:771–788

Vavasour IM, Whittall KP, MacKay AL, Li DK, Vorobeychik G, Paty DW (1998) A comparison between magnetization transfer ratios and myelin water percentages in normals and multiple sclerosis patients. Magn Reson Med 40:763–768

Vavasour IM, Li DK, Laule C, Traboulsee AL, Moore GR, Mackay AL (2007) Multi-parametric MR assessment of T(1) black holes in multiple sclerosis: evidence that myelin loss is not greater in hypointense versus isointense T(1) lesions. J Neurol 254:1653–1659. doi:10.1007/s00415-007-0604-x

Vavasour IM, Laule C, Li DK, Oger J, Moore GR, Traboulsee A, MacKay AL (2009) Longitudinal changes in myelin water fraction in two MS patients with active disease. J Neurol Sci 276:49–53

Webb S, Munro CA, Midha R, Stanisz GJ (2003) Is multicomponent T2 a good measure of myelin content in peripheral nerve? Magn Reson Med 49:638–645

Whittall KP, MacKay AL, Li DK, Vavasour IM, Jones CK, Paty DW (2002) Normal-appearing white matter in multiple sclerosis has heterogeneous, diffusely prolonged T(2). Magn Reson Med 47:403–408. doi:10.1002/mrm.10076 [pii]

Wolff SD, Balaban RS (1989) Magnetization transfer contrast (MTC) and tissue water proton relaxation in vivo. Magn Reson Med 10:135–144

Yarnykh VL, Yuan C (2004) Cross-relaxation imaging reveals detailed anatomy of white matter fiber tracts in the human brain. Neuroimage 23:409–424

Chapter 11
Designing Clinical Trials to Test Neuroprotective Therapies in Multiple Sclerosis

P. Connick, M. Kolappan, A. Compston, and S. Chandran

11.1 Introduction

11.1.1 The Unmet Need for Neuroprotective Therapy in MS

Multiple sclerosis (MS) is the commonest cause of acquired neurological disability in young adults. It is a multi-focal and multi-phasic immune-mediated disorder characterised pathologically by inflammatory demyelination, axonal injury and partial remyelination (Compston and Coles 2008). Treatments for MS thus have two aims: to prevent and to repair damage that has already occurred. Although important advances in treatment to reduce relapse rate have been made in the last two decades, more limited progress has been achieved in terms of definitive treatments for relapses and the slowing, cessation or reversal of disease progression (see Chap. 10). The lack of such therapies represents a substantial gap in the treatment of MS.

P. Connick (✉) • A. Compston
Department of Clinical Neurosciences, School of Clinical Medicine,
Addenbrookes Hospital, University of Cambridge, Cambridge CB2 0SP, UK
e-mail: pc349@cam.ac.uk

M. Kolappan
NMR Research Unit, Department of Neuroinflammation, The Institute of Neurology
(Queen Square), University College London, London WC1N 3BG, UK

S. Chandran
Centre for Neuroregeneration, University of Edinburgh,
Chancellor's Building, 49 Little France Crescent, Edinburgh EH16 4SB, UK
e-mail: siddharthan.chandran@ed.ac.uk

11.1.2 The Challenge of Detecting and Measuring Neuroprotection in MS

Multiple sclerosis has two distinct clinical phases, reflecting interrelated pathological processes: focal inflammation drives activity during the relapse-remitting stage, and axonal degeneration represents the principal substrate of progressive disability. Observations from clinical trials of existing disease-modifying therapies demonstrate that although successful reduction of focal inflammatory activity can be achieved throughout the disease, impact on progression is dependent upon the phase of disease at which the intervention is administered (Coles et al. 2008; Kappos et al. 2007). These insights from clinical trials illustrate the complex pathogenic relationship between inflammation, degeneration and timing of intervention, and the limitations of therapeutic strategies for established or progressive MS that are exclusively anti-inflammatory. Here the need is for strategies that promote myelin repair and thus prevent ongoing axonal loss. Pathological and experimental studies have shown that these two processes are tightly linked and that remyelination is neuroprotective.

Against this background, the challenge for future neuroprotective clinical trials is to identify those patients who will both benefit from putative neuroprotective treatments as well as prove informative in understanding mode of action of any intervention. This requires clinical trial protocols that differ from those used in the evaluation of disease-modifying therapies in terms of both participant selection and measurement(s) of efficacy. Current methods to detect repair are generally limited to comparatively insensitive assessments of overall response. These composite assessments risk failure to detect subtle but nonetheless meaningful effects of treatment intervention on components of the disease process that drive cumulative disability. Future neuroprotective trial design must therefore aim to select outcomes that are informative with respect to stages of the disease other than the relapsing–remitting phase, and which demonstrate the success and advantages of targeting both immunological and neurobiological components of the complex pathogenesis. Meeting this aim will likely require a combination of clinical outcome measures to register improved function in addition to the use of paraclinical observations and novel biomarkers in order to inform on the mechanisms of any therapeutic effect and detect benefits that lie below the clinical threshold. In view of the evidence for ongoing inflammation, demyelination, gliosis, neuronal injury and remyelination, the need is for mechanistic-based measures that will inform on one or more of these interdependent pathological processes. Recent technical innovations and methodological advances now offer a range of direct and indirect quantitative measures of inflammation, myelination status and neurodegeneration that can begin to dissect pathological components of the disease and register change in response to intervention.

Outcome measures in neuroprotective trials must therefore address two principle tiers of enquiry: demonstration of therapeutic effects on function (clinical response) and definition of therapeutic effects on neurobiology (mechanistic response). Within each tier, no single outcome is capable of addressing all of the relevant dimensions, and groups of outcomes targeting specific components are therefore required. A "three-step" model represents a potentially useful framework for outcome measure selection: (1) clinical outcomes to establish clinical efficacy, (2) paraclinical *anatomical* outcomes to exclude symptomatic benefit or CNS plasticity as the mechanism of effect and (3) paraclinical outcomes addressing inflammation, neuronal/axonal integrity and myelination status to define the biological mechanism(s) of effect.

This chapter summarises the key principles of early phase neuroprotective trial design in MS and highlights the need to develop approaches that are distinct to studies that have focussed on relapse-remitting disease. A complementary approach of studying highly selected patient cohorts with novel paraclinical measures to demonstrate proof of concept is illustrated through a current neuroprotective study that adopts a sentinel lesion approach to inform on both putative mechanism and functional efficacy.

11.1.3 *Clinical Outcome Measures*

Despite poor specificity for therapeutic effects on disease pathogenesis, clinical outcome measures remain the key "bottom line" for clinical trials. However, selection of suitable clinical outcomes for evaluation of neuroprotective therapies from the large range available is difficult and no consensus exists. Current approaches frequently involve a combination of historically established "global benchmark" scales such as Expanded Disability Status Scale (EDSS) with those providing improved metric performance such as the MS Functional Composite (MSFC). Although necessary, these functional multidimensional outcome scales are insensitive and when deployed alone risk failure to register subtle but potentially important effects of treatment. In order to reduce dependence on these outcome scales, an alternative strategy uses complementary selection of patients and outcomes with targeted anatomical specificity. Here, cohorts in whom detailed protocols can be evaluated are prospectively identified based on a clinically articulate reference lesion(s) such as those involving the optic nerve, cerebellum or spinal cord. The experience of the "sentinel lesion"—selected on the basis that this provides measurable clinical and paraclinical outcomes that are objective and useful to the recipient—can then be extrapolated to the disease as a whole. The anterior optic pathway is perhaps the best example of this approach and benefits from readily available quantitative measures of physiology, structure and function. Assuming that the demonstration of protection in the anterior optic pathway is representative of the wider CNS, such a strategy offers potential to minimise phenotypic heterogeneity within trial cohorts, increases statistical power to detect efficacy and offers a range of measures to inform on putative mechanism of action.

Clinical outcomes can be classified as physician/observer-based or patient/subject-based. Whilst the former remains of central importance, there is increasing recognition of the independent value of patient-based measures that address quality of life

Table 11.1 Multidimensional clinical outcome measures

Scale	Description	Reference
(A) Physician measured		
Disease specific		
Expanded (Kurtzke) Disability Status Scale (EDSS)	20-Step ordinal scale. Benchmark position despite poor metric properties	Hobart et al. (2000)
Multiple Sclerosis Functional Composite (MSFC)	3 Functional domains with good metric performance	Cutter et al. (1999)
UK (Guys) Neurological Disability Scale (UKNDS)	12 Functional domains with good metric performance	Sharrack et al. (1999)
Generic		
UK FIM+FAM	Generic scale with wide application in neurorehabilitation	Law et al. (2009)
(B) Patient measured		
Disease specific		
Multiple Sclerosis Impact Scale 29 (MSIS-29)	29-Item scale. Robust metric development and performance	Hobart et al. (2001)
Functional Assessment of MS (FAMS)	59-Item scale	Ritvo et al. (1997)
MS Quality of Life 54 (MSQOL-54)	54-Item scale incorporates the SF-36	Vickrey et al. (1995)
Leeds MSQoL	8-Item scale. Metrics suggest unidimensional performance (well-being)	Ford et al. (2001)
Generic		
Short Form 36 Health Survey Questionnaire (SF-36)	Established generic instrument, metric performance in MS questioned	Hobart et al. (2004)

outcomes (Rothwell et al. 1997). Further subdivision of clinical outcomes into unidimensional (where the concept can be measured directly—e.g. relapse rate) vs. multidimensional (where the concept cannot be measured directly, requiring a series of component measures—e.g. "quality of life" or "disability" scales) is also possible and provides a useful framework for further discussion.

11.1.3.1 Multidimensional Measures

In a disease notable for its wide-ranging phenotypic complexity, multidimensional outcomes aim to capture a global representation of function that allows comparison between individual patients, groups of patients and across trial populations (Table 11.1). The physician-based ordinal EDSS exhibits formidable resilience and remains the industry benchmark multidimensional outcome measure notwithstanding well-recognised limitations (Thompson and Hobart 1998). Patient-based multidimensional instruments are a diverse group with no equivalent benchmark, although the Multiple Sclerosis Impact Scale (MSIS-29) is increasingly adopted. Performance of the main contenders and the challenge of selecting from the available options are comprehensively reviewed by Riazi (2006).

11.1.3.2 Unidimensional (Function Specific) Measures

The primary value of function-specific measures is as an adjunct to paraclinical "site-specific" measures in trials employing a "sentinel lesion" approach where patients are selected based on disease at a clinically eloquent CNS site such as the optic nerve, cerebellum or spinal cord. *Symptom-specific* assessment instruments (of which many are available—spasticity, fatigue scales, *etc.*) other than measures of affective function are not discussed here as their relevance lies with trials for symptomatic therapies rather than neuroprotective and repair therapies. Assessment of affective function is discussed due to its potential for confounding effect on patient/subject-based outcomes.

Visual Function

Visual impairment is a common clinical feature of MS with a significant impact on quality of life (Mowry et al. 2009). Pathological evidence of optic nerve involvement is found in 94–99% at autopsy (Toussaint et al. 1983). Normal visual function involves the complex integration of multiple systems, of which acuity, contrast sensitivity, colour vision and field-of-vision are amenable to detailed quantitative non-invasive clinical assessment. Acuity has traditionally been measured using the Snellen chart; however, logMAR scoring using a retro-illuminated Early Treatment Diabetic Retinopathy Study (ETDRS) chart is preferred for research application. ETRDS generates continuous data scores that are more amenable to statistical analysis. Sloan charts are of similar style but combine decrementing levels of contrast. The resultant contrast visual acuity provides a sensitive and reproducible measure of acuity deficits in MS (Kolappan et al. 2009). Colour vision can be comprehensively and quantitatively assessed by the Farnsworth-Munsell 100 hue test, requiring the subject to place 85 coloured tiles in order of hue. This test provides both quantification and spectral localisation of dyschromatopsia. The pivotal optic neuritis treatment trial confirmed that low contrast visual acuity and contrast sensitivity are sensitive and reliable measures for the detection and monitoring of visual dysfunction in MS (Balcer 2001; Optic Neuritis Study Group 2008). Visual fields can also be quantitatively assessed with static targets using a Humphrey perimeter or with kinetic targets using the Goldmann method. Taken together, the high quality of clinical outcomes marks the visual system as an attractive candidate pathway for detailed assessment in clinical trials

Cerebellar Function

Cerebellar dysfunction is common in multiple sclerosis with lifetime prevalence >70%. Assessment of tremor and/or dexterity forms the mainstay of available outcomes. The nine-hole peg test and finger-tapping test both provide objective and valid quantitative assessments of upper limb function but lack specificity to cerebellar dysfunction. Observer-dependent rating scales such as the *Scale for the Assessment and Rating of Ataxia* (SARA) or the *International Cooperative Ataxia Rating Scale* (ICARS) are either unvalidated in MS or shown to exhibit significant metric limitations (Schmitz-Hübsch et al. 2006). The Composite Cerebellar Functional Score (CCFS) comprising a dominant hand "nine-hole peg test" and "click test" (based on kinematic data regarding optimum assessment for upper limb goal directed multi-joint movement) represents a promising option for quantitative cerebellar assessment (du Montcel et al. 2008). However, validation in MS is awaited. A further challenge to cerebellar assessment (in any disease) is the avoidance of floor effect. Severe dysfunction renders assessment impossible with all currently available outcomes. These limitations make the use of cerebellar disease unattractive to trials adopting a "sentinel lesion" approach.

Spinal Function

The potential value of a spinal-specific scale is highlighted by the observation that a large component of disability due to MS reflects spinal pathology (Filippi et al. 1996). Accepting there are currently no high performing clinical scales for the assessment of spinal function in MS, commonly used outcomes are timed walk (as a component of MSFC) and/or inference from the EDSS. The challenge of measuring spinal function will be familiar to clinicians dealing with spinal cord injury (SCI) where only the detailed American Spinal Injury Association (ASIA) motor and sensory scoring systems have been demonstrated to exhibit favourable psychometric properties for their longitudinal use in clinical trials. However, formal metric evaluation of these scales in MS patients is awaited.

Affective Function

Mental health is a key determinant of patient perceptions regarding quality of life and can therefore be viewed as an independent clinically significant outcome (Rothwell et al. 1997). Furthermore, assessment of affective function is essential to interpreting outcomes based on global function and patient-reported measures. The most widely used assessment scale for depression in MS is the 21-item Beck Depression Inventory (BDI)—recently revised as the BDI-II. This generic instrument offers strengths of clinical familiarity, the ability to make cross-disease comparisons and an extensive existing literature. The BDI has also been endorsed by international consensus guidelines for the treatment of depression in MS (Goldmann Consensus Group 2005). However, the challenge of attributing physical symptoms to the syndrome of depression rather than MS per se has led to a debate regarding the specificity of physical symptoms of depression in MS. This complicates the use of generic instruments such as the BDI and the Hospital Anxiety and Depression Scale (HADS). Two disease-specific instruments have therefore been developed to address these issues. The comprehensive 42-item Chicago Multiscale Depression Inventory (CMDI) has been psychometrically validated in MS but lacks familiarity amongst clinicians (Chang et al. 2003). The rival 7-item Beck Fast Screen for Medically Ill Patients (B-FS) has also been validated in MS (Benedict et al. 2003). Comparative analysis of these two instruments, particularly with regard to responsiveness, is not yet available. Those designing a clinical trial at present therefore face a difficult choice between the generic BDI-II, and the two psychometrically favourable but less familiar disease-specific scales available.

11.1.4 Paraclinical Outcome Measures

Paraclinical outcome measures can be used as surrogates for clinical outcomes with the potential to avoid observer bias and increase statistical power to detect efficacy. However, they can also be used as direct measures of a biological process—required to demonstrate mechanism of putative therapeutic effect. They are therefore of particular importance in neuroprotective trials. Paraclinical parameters may have advantages over clinical parameters in their pathological specificity but crucially may also be more responsive (to changes in underlying pathology) and reproducible (with attendant benefit in detecting efficacy "signal" from test–retest "noise").

11.1.5 Magnetic Resonance Imaging Measures

Magnetic resonance imaging (MRI) measures have consistently proven to be the most useful surrogate outcomes in MS clinical trials. Conventional MRI techniques (T2, T1 ± gadolinium (Gd), and fast fluid-attenuated inverse recovery [FLAIR]) have long been used as biomarkers for disease activity in clinical trials. Together with the emergence of methodologies that can begin to discriminate between distinct pathological processes such as inflammation, myelination status and neuronal loss, imaging-based measures represent a powerful tool for mechanistic evaluation of therapeutic intervention (Fig. 11.1). A summary is given below of the main techniques and how they may be best deployed in the setting of an experimental trial seeking to demonstrate efficacy as well as inform on mechanism.

11.1.5.1 T2-Weighted Sequences

T2-weighted sequences are sensitive for the detection of focal white matter lesions with approximately 5–10 new lesions on imaging per clinical relapse—resulting in significant improvement of statistical power to detect efficacy (Miller and Thompson 1999).

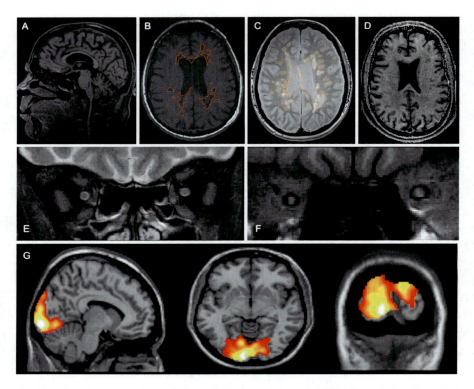

Fig. 11.1 Imaging approaches to define pathological substrate. (**a**) 3D high-resolution T1 sagittal brain image for volume and atrophy measurements; (**b**) T1 axial brain image suitable for T1 hypointense (*black hole*) lesion volume analysis (*ringed*); (**c**) proton density axial brain image suitable for T2-PD lesion volume analysis (*ringed*); (**d**) magnetisation transfer ratio map suitable for whole and segmented brain and lesional MTR analysis; (**e**) fat-suppressed T2 fast spin echo coronal image of optic nerves suitable for detection of optic nerve lesion and lesion length measurement (*left* optic nerve showing hyperintense lesion); (**f**) fat-suppressed coronal T1 fast FLAIR image of optic nerves suitable for analysis of ON area (*right* optic nerve atrophy shown—consistent with previous optic neuritis); (**g**) fMRI brain images showing activation patterns following visual stimulation

Gadolinium administration doubles sensitivity but is relatively expensive and may not be cost-effective in neuroprotective/repair trials as new T2 lesions in progressive disease can develop without gadolinium enhancement. Approaches that improve the sensitivity of non-enhanced images include fast fluid-attenuated inverse recovery (FLAIR) sequences and higher field-strength scanners. Despite impressive sensitivity (responsiveness), poor pathological specificity (validity) limits the utility of T2 lesions as an outcome measure in neuroprotective trials. *Total lesion number (load)* or *total lesion volume* can be used as a marker of cumulative *focal* inflammatory disease; however, neither measure correlates with current or future clinical disability nor informs about more diffuse inflammatory processes (Simon and Miller 2007).

11.1.5.2 T1-Weighted Sequences

Approximately 20–30% of T2 hyperintense lesions appear hypointense ("black holes") on T1-weighted MRI. The natural history of these lesions over several weeks is either resolution to isointensity—thought to reflect resolution of oedema and remyelination, or persistent focal hypointensity reflecting axonal loss—such lesions being descriptively labelled as "chronic black holes". Potential outcome measures therefore include total chronic black hole number or volume—reflecting a marker of cumulative axonal loss due to focal inflammation. These measures correlate modestly with clinical disability and can be used as surrogates, although the resultant gain in statistical power is marginal. A more sophisticated approach involves combining T1 data with gadolinium enhancement (or new T2 lesions) in order that single lesion natural history outcomes can be assessed, e.g. the proportion of new Gd/T2 lesions evolving to chronic black holes. This uncoupling of inflammation from axonal loss offers the potential to define intervention as anti-inflammatory *vs.* neuroprotective/reparative. Further complexity can be added by quantitative assessment of lesion T1 intensity recovery (therefore axonal loss) through the normalised contrast/hypointensity ratio (Zivadinov 2007). This approach allows a quantitative assessment of neuroprotective/reparative intervention.

11.1.5.3 Gadolinium Enhancement

Gadolinium enhancement is the most sensitive measure of new focal white matter lesions with pathological specificity for focal inflammation (Simon and Miller 2007). In RR-MS, monthly Gd-MRI reveals about ten new enhancing lesions for every clinical relapse. Gd-MRI can therefore be used as a surrogate for clinical relapses in order to improve statistical power. Sensitivity can be further increased by more frequent scanning (weekly), triple-dose Gd, off-resonance magnetisation transfer (MT) pulse, delayed scanning (post-Gd administration) and thinner slices. However, these approaches do not have a significant impact on sample size requirements because patient variability also increases (Silver et al. 2001). Aside from providing a measure of focal inflammatory disease activity, the combination of Gd-MRI and T1 in order to define the pathological evolution of individual lesions (see above) is of particular benefit to neuroprotective studies.

11.1.5.4 Atrophy Measures

CNS atrophy (measurement of volume loss) has been widely used as an outcome in MS clinical trials. Global brain atrophy is a moderate but significant predictor of neurological impairment that is independent of conventional MRI lesions (Zivadinov and Bakshi 2004). However, responsiveness is limited (particularly in early RR-MS—possibly reflecting greater functional reserve, and/or more effective endogenous repair and plasticity), and the gain in statistical power compared to direct clinical assessment is therefore low. Regional atrophy measures may be preferable in this context, notably (brain) grey matter atrophy occurs at nearly twice the rate of whole brain or white matter atrophy and is a better predictor of cognitive deficits, and cervical spine atrophy is a better predictor of change in EDSS (Fisher 2007). Accepting that atrophy measures are non-specific for axonal/neuronal loss, with myelin, glia and vascular components, it represents an important outcome for longitudinal assessment of axonal/neuronal loss and can be further segregated into global (brain), grey matter (brain) and regional scores.

11.1.5.5 Magnetisation Transfer Ratio Sequences

Magnetisation transfer ratio (MTR) forms an indirect measure of macromolecular structure (such as myelin) and can be used to infer myelination status globally (whole brain) or within specific regions of interest. Pathological specificity of MTR is imperfect and asymmetric; a decrease in MTR reflects demyelination and axonal loss, whereas an elevation of MTR is reflective of possible remyelination or resolution of oedema (Schmierer et al. 2004). Consequently, a number of potential roles have been proposed for MTR as an outcome measure in clinical trials. Firstly, low MTR at baseline is predictive of progression, and there may therefore be a role for MTR-based eligibility criteria in neuroprotective trials looking to identify an enriched cohort of patients likely to exhibit clinical progression during the trial period. Secondly, there is some evidence that the decline of MTR in evolving lesions (before Gd enhancement) is a marker of lesion severity—with more marked decline being predictive of persistent T1 hypointensity (axonal loss). MTR decrement in lesions might therefore be used as a marker for the severity of myelin and axonal damage secondary to focal inflammatory insult—i.e. a quantitative assessment of the neuroprotective consequence of anti-inflammatory intervention(s). Finally an increase in MTR may be used to quantitatively assess the effect of repair therapies on myelin.

11.1.5.6 Diffusion Tensor Imaging Sequences

Diffusion tensor imaging (DTI) is a sensitive technique for the assessment of tissue microstructure based on water molecule diffusion properties in tissues. In white matter, diffusion is facilitated along fibre tracts and is slower in the direction perpendicular to the main axis of the tract. This physiological restriction is disturbed by pathological processes like demyelination, resulting in increased total diffusivity and disruption of the directional selectivity (anisotropy) of the fibres to allow diffusion. *Fractional anisotropy* (FA) indicates the orientation of diffusion and is high along well-defined pathways such as the corpus callosum, pyramidal tracts and optic radiations. A reduction in FA and an increase in the *mean diffusivity* (MD)/apparent diffusion coefficient (ADC) are potential markers for the structural integrity of myelinated axons. A modest correlation between clinical progression and DTI parameters of single well-defined tracts (but not global diffusion parameters) means that DTI is a poor surrogate for clinical progression. However, potential application to any white matter fibre bundle with high orientational coherence and associated functional outcomes (e.g. optic nerve DTI and visual function) makes it an attractive measure of axonal integrity for trials employing a sentinel lesion approach.

11.1.5.7 Magnetic Resonance Spectroscopy

Proton MR spectroscopy (MRS) enables quantitative assessment of the molecular composition in CNS tissues. The main molecular peak in adult human CNS is from the amino acid *N*-acetyl aspartate (NAA) which is found almost exclusively in neurons and axons—a reduction in NAA providing evidence of axonal dysfunction or loss. Therefore, NAA quantification potentially represents a measure of axonal function and number in neuroprotective/repair trials. Other metabolites offer potential to assess non-neuronal CNS structures and processes; e.g. choline-containing compounds can be used as a marker of cell membrane integrity, myo-inositol as a glial cell marker, lipids as products of brain destruction, lactate as a product of anaerobic glycolysis, and creatine/phosphocreatine as a marker of energy metabolism (Narayana 2005). However, low signal-to-noise ratio results in modest reproducibility, and the use of proton MRS has to date been limited mainly to single-centre trials. Clearly, this evolving technique represents a rich resource for further development of pathologically specific outcomes relevant to neuroprotective/repair trials.

11.1.5.8 Functional Magnetic Resonance Imaging

Neuronal activity increases local levels of deoxyhaemoglobin; this is reflected by changes in functional magnetic resonance imaging (fMRI) signal intensity. Consequently, fMRI can potentially be used as a technique to measure dynamic changes or plasticity in the brain in response to disease per se or to therapeutic intervention. In the setting of an experimental neuroprotective or repair therapeutic trial, fMRI is most powerfully harnessed in the context of structural imaging changes. For example, fMRI in a "sentinel lesion" trial allows examination of a relationship between lesion(s) in a single white matter tract (e.g. optic nerve) and the clinical function (visual acuity) not entirely accounted by the structural recovery (e.g. remyelination as indicated by increase in MTR values or decrease in the visual evoked potential (VEP) latency).

In summary, MRI measures currently represent the most powerful and well-validated paraclinical outcomes for use in neuroprotective trials that can begin to provide non-invasive and quantitative assessments of key pathological and repair processes. Future imaging approaches will benefit from increased sensitivity through technical advances such as higher field-strength scanners, use of surface coils to improve both resolution and signal-to-noise ratio and faster acquisition methods.

Methodological advances also offer the prospect for application to challenging CNS sites such as the spinal cord and optic nerve. The spinal cord is a more difficult structure to image than brain due to its smaller size, mobility and proximity to the heart and great vessels. However, these difficulties can be largely overcome by approaches involving cardiac gating, spatial pre-saturation slabs and the development of phased array coils enabling rapid imaging of the whole spinal cord. The challenges of optic nerve MRI reflect its small size, mobility, surrounding fat and CSF, and the bony optic canal. MR sequences have been developed to overcome these challenges through suppression of fat and CSF signal, fast sequences, use of surface coils and high field MR systems to allow high-resolution imaging.

11.1.6 Quantitative Assessment of the Retinal Nerve Fibre Layer

The retinal nerve fibre layer (RNFL) consists of unmyelinated axons within the retina. Consequently, measurements of RNFL thickness in MS are not confounded by loss of myelin. The RNFL is therefore an attractive structure to visualise processes of neurodegeneration and potentially neural repair. Optical coherence tomography (OCT) uses the echo time delay of low-coherence light to delineate the RNFL. Layers of the retina have different reflectivity and can thus be distinguished and measured: RNFL thickness (giving an estimate of axonal number) and macular volume (giving an estimate of ganglion cell number). Reductions in RNFL thickness and macular volume are significantly correlated with reductions in visual function, and correlation with brain atrophy has also been described (Fisher et al. 2006; Gordon-Lipkin et al. 2007). OCT measurements may therefore be useful in clinical trials to detect and monitor neuroprotection. High-speed OCT using a "Fourier" or "spectral" detection technique is also now becoming widely available. This technique is approximately 50 times faster—resulting in reduced eye movement artefact, and has a superior sensitivity compared to standard OCT.

11.1.7 Neurophysiological Measures

11.1.7.1 Visual Evoked Potential

Conventional VEP measures the cortical response to monocular stimulation in the central 30° of the visual field (known as the P100). In MS, the waveform is characteristically delayed with well-preserved amplitude. Response latency can be used as a measure of myelination in the afferent visual pathway (increased with demyelination), and amplitude can be used as a measure of axonal conduction (reduced with axon loss or conduction block due to demyelination) (Diem et al. 2003). The multifocal VEP (mVEP) has been developed to examine conduction in the parts of the visual field not covered by full field VEP. The mVEP uses a paradigm of sectoral stimulation with pseudo-stimulation at other sites, using the fellow eye and normal controls for comparison at each point. mVEP has the advantage that a particular sector of the visual field can be examined for abnormality and compared with the results of other retinotopic tests (e.g. standard automated perimetry). mfVEP may also be more sensitive than conventional VEP as a marker of clinical progression (Fraser et al. 2006).

Somatosensory evoked potential (SSEP); SSEPs measure central responses to electrical stimulation at sites in the peripheral nervous system—they therefore offer potential for use as markers of CNS function. However, their application is currently limited by poor metric performance in both reliability and validity, and practical issues such as the requirement for equipment and skilled operators.

11.1.8 Outcomes Based on CSF and Blood Measures

A number of CSF/blood biomarkers are available as potential outcomes. These have particular relevance to measuring the dynamic pathobiology of MS and are discussed further in Chap. 13.

11.2 Trial Design

In the face of a multi-focal and multi-phasic disease with variable natural history, even carefully selected outcomes that are highly specific for neuroprotective processes will be inadequate unless they are supported by appropriate trial design. This requires the definition of endpoints (clinical and paraclinical) that assess meaningful effect(s) of the proposed neuroprotective intervention; identification and recruitment of an informative and pathogenically relevant group of patients; and the application of design elements such as appropriate control groups, blinding and randomisation that together maximise the likelihood of meaningful conclusions regarding the success or failure of any given putative neuroprotective intervention. As with all therapeutic development, the evaluation of putative neuroprotective agents represents a stepwise process with specific objectives and design characteristics at each phase (box 11.1).

Box 11.1 The Clinical Trials Process

Phase	Participants	Typical number	Objective
I	Healthy volunteers	5–20	Establish initial safety profile in man
IIA	Patients	10–20	Establish initial safety profile in patients
IIB	Patients	20–300	Establish initial efficacy in patients, clarify safety profile
III	Patients	300+	Confirm efficacy and safety profile
IV	Patients	Post-marketing	Safety surveillance, post-marketing efficacy monitoring

11.2.1 Trial Endpoints

A trial "endpoint" is the predefined change in an outcome considered to be worth detecting. This can be expressed in terms of clinical measures for safety or efficacy, or through paraclinical measures of mechanistic response. Decisions on endpoints are required *prior* to trial commencement as they inform power calculations and frame interpretation of the final result, particularly when the result is negative.

11.2.1.1 Safety Endpoints

Guidance is available for safety endpoints in phase I/IIA trials required by regulatory authorities. These represent a relatively standard combination of clinical adverse events, laboratory testing and other measurements collected from specific testing (such as electrocardiogram [ECG]). Comprehensive details can be found in standard texts of clinical trial design (Sahajwalla 2004; Gad 2002).

11.2.1.2 Clinical Efficacy Endpoints

Clinical endpoints based on preventing or reversing fixed disability form the key "bottom line" for a neuroprotective or repair therapy. Furthermore, they represent the only relevant endpoint for patients who do not experience relapses—i.e. all patients with primary progressive MS and the majority of patients in the secondary progressive phase. Two broad strategies are available: comparing group *differences in disability* (measured by an appropriate scale) or comparing group differences in the *time taken to reach disability milestones*.

Approaches Based on Differences in Disability Across Comparator Groups

Investigators may wish to define an endpoint based on differences in a disability score between treatment and comparator groups. The challenges of this approach are insensitivity of clinical outcomes for measuring disability, variability in the natural history of MS (necessitating large group sizes to achieve adequate statistical power) and the contribution of inflammatory activity (relapses) to *reversible* (although not necessarily brief) changes in disability. The latter being distinct from a more insidious development of fixed disability that reflects pathological neurodegeneration (Confavreux and Vukusic 2004).

Approaches Based on Time to Disability Milestones

Alternatively, endpoints based on *time to the development of disability* milestones can be employed. The first potential milestone is *onset* of the progressive phase [where progression is defined as a steady worsening of symptoms or signs over 12 months (Polman et al. 2005)]. A delay in the time to the onset of progression is amenable to quantitative assessment with reasonable inter-examiner reliability (at the level of determining year of onset)—kappa values of 0.76–0.92. Although attractive, this paradigm is ultimately impractical in clinical trials due to the prolonged duration of observation required for all members of a cohort to develop confirmed progression. Comparing proportions with confirmed progression across groups at the end of a defined observation period is a potential solution; however, large numbers of participants are required in order to achieve adequate statistical power within the timescale of a feasible clinical trial [see commentary on the BENEFIT Trial (Coles 2007)]. Furthermore, these approaches are only applicable to the testing of putative neuroprotective therapies in patients with relapsing–remitting disease. Endpoints based on the time to attainment of specific disability milestones such as EDSS scores defining use of walking aids allow the inclusion of patients with progressive disease who have accumulated fixed disability less than the landmark chosen; however, they also require large cohort sizes.

11.2.1.3 Paraclinical Endpoints Based on Measures of Neuroprotection and Repair

Endpoints based on putative mechanisms of action may be suitable for proof of principle trials bridging phases IIA and IIB. For example, anti-inflammatory imaging-based endpoints (such as a reduction in the rate of acquiring gadolinium-enhancing lesions) are familiar from the development of existing disease-modifying therapies. Equivalent endpoints for neuroprotective trials might include changes in brain atrophy measures and tissue-specific measures for myelination/axonal integrity, *etc*. The merits and limitations of the *techniques* available to measure relevant pathogenic processes are discussed above (paraclinical outcome measures); however, defining a suitable *magnitude of change* deemed to be worth detecting is more problematic. Endpoints based on modest mechanistic thresholds may be justifiable based on the a priori view that any single intervention is unlikely to exert a dramatic net effect on neuronal or glial loss even if highly effective on one aspect of disease pathogenesis. Such interventions would therefore be suitable for inclusion in future combinatorial paradigms.

11.2.2 *Identifying the Patient Cohort for Neuroprotection Studies*

Selecting the correct patient group is key to the success of neuroprotective trials and a comparatively overlooked area of clinical trial design. The challenge facing investigators is to identify a phenotypically informative cohort who share pathogenic activity relevant to the intervention whilst reconciling desirability of this "ideal" (to maximise the probability of demonstrating efficacy) patient population with a need to recruit and retain sufficient numbers of patients to power and complete a trial in a timely manner. Refining the patient cohort for study is achieved by application of routine and novel eligibility criteria that include age, disease duration, disease course, disease activity, level of disability, paraclinical parameters and evidence of disease involvement at specific anatomical sites—a "sentinel lesion" approach.

11.2.2.1 Age

Defining an upper age limit is of particular relevance to neuroprotective studies, given that age appears to be negatively correlated with regenerative capacity and some paraclinical measures can be confounded by age-related change (Blakemore et al. 2000).

11.2.2.2 Disease Duration

Available evidence suggests that early treatment is advantageous to achieve both neuroprotection and facilitation of endogenous repair (Goldschmidt et al. 2009). However, in practice, disease duration can only be vaguely defined, and trials involving patients with progression (by definition) require disease maturation. Calculation of disease duration based on the *time from diagnosis* is arbitrary, depending on when patients present and when investigation is undertaken. Furthermore, a systematic bias may be introduced by earlier diagnosis in those with more severe disease, as demonstrated in the large Ontario natural history cohort (Weinshenker et al. 1989). Alternatively, calculation of disease duration based on *time from the first symptom attributable to MS* is liable to imprecision due to recall bias/failure and reporting bias (involving the potential participant and/or the screening physician), in addition to error resulting from uncertainty regarding the relevance of remote minor symptoms. Consequently, a useful role for disease duration eligibility criteria is largely restricted to trials involving patients with clinically severe from onset relapsing–remitting disease.

11.2.2.3 Disease Course

Classification of MS subgroups by clinical course effectively segregates disease into active focal inflammatory (relapse-remitting) or neurodegeneration (progressive) dominant cohorts (Lublin et al. 1996; Confavreaux and Compston 2006). Thus, a priori predictions of putative mode of action determine the correct cohort: dependent on whether the primary target is inflammation, or mechanisms independent of inflammation such as remyelination. In practice, this results in a decision on whether to recruit a cohort of patients who experience relapses (RRMS and PRMS) or those who do not (SPMS and PPMS). For studies concentrating on patients without relapses, a second decision centres on the inclusion or exclusion of patients with PPMS on grounds that disease pathogenesis and thus response to treatment may be fundamentally different to SPMS (Smith et al. 2006).

11.2.2.4 Disease Activity

Unlike relapse-remitting disease where gadolinium enhancement provides a reliable barometer of activity, there is no comparable measure for cell/axonal loss relevant to neuroprotective and repair therapy trials. Poor specificity and responsiveness limits current measures. Furthermore, threshold selection can be problematic. Thresholds defining high levels of disease activity are initially attractive in the context of neuroprotective trials as they maximise statistical power to detect efficacy; however, there is no guarantee that high levels of activity at randomisation will be sustained throughout the duration of the clinical trial. Indeed placebo arms of the pivotal disease-modifying therapy trials clearly demonstrate the contribution of post-randomisation *regression to the mean* (Martínez-Yélamos et al. 2006). Consequently, setting a high disease activity eligibility threshold results in a dynamic (time-dependent) reduction of the *intra-trial* statistical power to detect efficacy (Davis 1976). As a result, *pretrial* power calculations are liable to overestimate true power and the possibility of type II error increases. Despite this caveat, such an approach may be useful in phase II trials of neuroprotective agents aiming to establish "proof of principle". In contrast, trials involving reparative therapies require thresholds defining a low rate of cell/axonal loss in order to maximise statistical power to detect efficacy.

11.2.2.5 Disability Thresholds

A lower limit of disability may be necessary in two settings. Firstly, trials of repair therapies require the presence of fixed disability in order to avoid "ceiling effect". However, a single assessment is inadequate to define *fixed* disability. Interval assessment at no less than 12 months is necessary but impractical in a prospective setting. Retrospective assessment based on the patient's account is subject to recall bias/failure and reporting bias, and assessments based on clinical records may be inadequate and/or incomparable to the contemporary assessment. An optimal solution would be recruitment from an extant cohort of patients with *pretrial* monitoring through the application of standardised patient assessments. Secondly, setting a lower limit of disability may be ethically relevant where therapy-associated risk is potentially high. An upper limit of disability may also be relevant in order to avoid "floor effect" in neuroprotective trials and useful in order to minimise dropout due to intolerance of the trial schedule.

11.2.2.6 Paraclinical Parameters

Similar to the use of clinical disability thresholds to avoid clinical "floor effect" (described above), paraclinical eligibility criteria can be used in order to identify subjects (or sentinel lesions) with evidence of both neuronal loss/injury and adequate neuronal reserve to enable testing of neuroprotective efficacy.

11.2.2.7 The Sentinel Lesion

Targeting a clinically articulate anatomical site is attractive in a multifocal disease. This approach has been pioneered in Cambridge and London (UK) and is currently being applied in a neuroprotective clinical trial (NCT00395200). An ideal "sentinel lesion" would offer a combination of clinically articulate deficits and measurable outcomes to delineate structure and function. Candidate CNS "sentinel lesion" sites include the optic nerve, spinal cord and cerebellum. Pilot data favours the optic nerve as a feasible choice with attractive outcome measures (Brierley 2002); however, paraclinical outcome measures relating to assessment of the spinal cord are improving, and this may emerge as an alternative (Bot and Barkhof 2009). The potential role for this approach is in small phase II "proof of principle" trials where dedicated endpoints for the sentinel site increase statistical power to detect efficacy and potentially allow the independent assessment of effects on specific pathological processes.

11.3 Comparison of Current Neuroprotective Trial Protocols

Clinical trial design to evaluate neuroprotective therapies in MS is in evolution, and early experiences are likely to inform future choices (Kapoor 2006). Comparison of three contemporary protocols is illustrative of the challenges and opportunities throughout the development process (Table 11.2). The Mesenchymal Stem Cells in Multiple Sclerosis (MSCIMS) trial is a phase IIA trial embedded in the sentinel lesion approach discussed above. This allows a small (safety) study to concentrate outcomes on pathogenically relevant tissue and therefore maximises the potential to gather mechanistic information on efficacy. Given the reality of finite resources, evidence of some impact on mechanistically relevant processes becomes crucial at phase IIA where isolated demonstration of a benign safety profile (although essential) represents a missed opportunity to inform a key decision point regarding the investment required to proceed with therapeutic development. The larger lamotrigine (phase IIB) and CUPID (phase III) studies demonstrate a convergence towards clinical endpoints as development proceeds. Eligibility criteria based on disease activity are common to the lamotrigine and CUPID studies. Both protocols require defined changes in disability over a time period preceding recruitment. The lamotrigine study threshold was set as *clinical documentation of steadily increasing disability*, or *an increase of at least one EDSS point* over the preceding 2 years. The CUPID threshold was defined as *disease progression (increase in physical disability not due to major relapse) in the preceding year*. Phase III studies represent a key licensing decision point requiring investigators to make early contact with the relevant regulatory authorities during the planning stage. As a result, design flexibility is limited; a multi-centre parallel two-arm double-blind randomised controlled trial is standard, and endpoints may be dictated by established norms. Resources are invariably stretched to meet these requirements, and the potential to add complex

11 Clinical Trials – Measuring Neuroprotection and Repair in Multiple Sclerosis

Table 11.2 Examples of neuroprotective trial designs in MS

	MSCIMS	Lamotrigine in secondary progressive MS	CUPID
Phase	IIA	IIB	III
Trial registration number	NCT00395200	NCT00257855	ISRCTN62942668
Intervention	Autologous mesenchymal stem cells	Lamotrigine	Δ^9-Tetrahydrocannabinol (Δ^9-THC)
Route of delivery	Intravenous	Oral	Oral
Dose	$\leq 2 \times 10^6$/kg (single administration)	≤ 200 mg bd	3.5–14 mg/day
Design	Single arm (pretest–post-test)	Double-arm RCT	Double-arm RCT
Comparator	None	Placebo	Placebo
Total number of participants (Treatment/comparator)	10 1:0	120 1:1	492 2:1
Trial duration	12 Months	24 Months	36 Months
Blinding	None	Double	Double
Primary endpoint	Safety profile over 6 months	Brain atrophy over 2 years	Time to progression on EDSS and mean change in MSIS-29
Main eligibility criteria	18–65 years EDSS 4.0–6.5 Evidence of disease at optic nerve	18–55 years EDSS 4.0–6.5 Active progression in preceding 2 years	18–65 years EDSS 4.0–6.5 Active progression in preceding year
Key outcome measures	Adverse events Visual function Optic nerve imaging Optical coherence tomography Brain imaging Clinical disability scales	Brain imaging Cervical cord imaging Clinical disability scales	Clinical disability scales Adverse events Brain imaging

multilevel endpoints capable of determining the *mechanism* of efficacy is therefore limited. This highlights the importance of phase II studies, the focus of this chapter, as an opportunity for innovation.

11.4 Conclusions

The challenge to detect neuroprotection and repair in clinical trials for MS is formidable. Only a comprehensive approach involving consideration of all elements in trial design is likely to achieve success. Those designing such trials should ideally be involved with trial design across the spectrum of neurodegenerative disease, as many of the issues are generic. A traditional randomised parallel two-arm design remains the gold standard, although multi-interventional (adaptive and factorial) designs may become increasingly relevant in the near future. In all trials, patient selection is crucial to maximising statistical power to detect efficacy, with the "sentinel lesion" approach representing a novel and efficient paradigm for early phase II. Careful selection of outcome measures is also required to achieve the objectives of neuroprotective/repair therapy trials. A "three-step" approach represents a useful model in this regard: clinical outcomes to establish clinical efficacy, anatomical surrogates to exclude symptomatic benefit or CNS plasticity as the mechanism of effect, and further paraclinical outcomes specific to aspects of disease pathogenesis. The latter currently represents the weakest point in the chain—measures of inflammation are available, but there is a need to improve outcome specificity with regard to the quantitative assessment of structure and function for neuronal and glial cell populations.

References

Balcer LJ (2001) Clinical outcome measures for research in multiple sclerosis. J Neuroophthalmol 21:296–301
Benedict RH, Fishman I, McClellan MM, Bakshi R, Weinstock-Guttman B (2003) Validity of the Beck Depression Inventory-Fast Screen in multiple sclerosis. Mult Scler 9:393–396
Blakemore WF, Gilson JM, Crang AJ (2000) Transplanted glial cells migrate over a greater distance and remyelinate demyelinated lesions more rapidly than endogenous remyelinating cells. J Neurosci Res 61:288–294
Bot JC, Barkhof F (2009) Spinal-cord MRI in multiple sclerosis: conventional and nonconventional MR techniques. Neuroimaging Clin N Am 19:81–99
Brierley C (2002) Remyelination therapy in multiple sclerosis: assessment of three target sites and the cell implantation potential of human Schwann cells. PhD Thesis, University of Cambridge
Chang CH, Nyenhuis DL, Cella D, Luchetta T, Dineen K et al (2003) Psychometric evaluation of the. Chicago Multiscale Depression Inventory in multiple sclerosis patients. Mult Scler 9:160–170
Coles AJ (2007) The fragile benefit of BENEFIT. Lancet Neurol 6:753–754

Coles AJ, Compston DA, Selmaj KW, Lake SL, Moran S et al (2008) Alemtuzumab vs. interferon beta-1a in early multiple sclerosis. N Engl J Med 359:1786–1801

Compston A, Coles A (2008) Multiple sclerosis. Lancet 372:1502–1517

Confavreaux C, Compston DAS (2006) The natural history of multiple sclerosis. In: Compston DAS (ed) McAlpine's multiple sclerosis. Churchill Livingstone, New York, pp 183–272

Confavreux C, Vukusic S (2004) Accumulation of irreversible disability in multiple sclerosis – lessons from natural history studies and therapeutic trials. J R Coll Physicians Edinb 34:268–273

Cutter GR, Baier ML, Rudick RA, Cookfair DL, Fischer JS et al (1999) Development of a multiple sclerosis functional composite as a clinical trial outcome measure. Brain 122:871–882

Davis CE (1976) The effect of regression to the mean in epidemiologic and clinical studies. Am J Epidemiol 104:493–498

Diem R, Tschirne A, Bähr M (2003) Decreased amplitudes in multiple sclerosis patients with normal visual acuity: a VEP study. J Clin Neurosci 10:67–70

du Montcel ST, Charles P, Ribai P, Goizet C, Le Bayon A et al (2008) Composite cerebellar functional severity score: validation of a quantitative score of cerebellar impairment. Brain 131:1352–1361

Filippi M, Campi A, Colombo B, Pereira C, Martinelli V, Baratti C, Comi G (1996) A spinal cord MRI study of benign and secondary progressive multiple sclerosis. J Neurol 243:502–505

Fisher E (2007) Measurement of central nervous system atrophy in multiple sclerosis. In: Cohen JA, Ruddick R (eds) Multiple sclerosis therapeutics, 3rd edn. Informa Healthcare, London, pp 173–200

Fisher JB, Jacobs DA, Markowitz CE, Galetta SL, Volpe NJ et al (2006) Relation of visual function to retinal nerve fiber layer thickness in multiple sclerosis. Ophthalmology 113:324–332

Ford HL, Gerry E, Tennant A, Whalley D, Haigh R, Johnson MH (2001) Developing a disease specific quality of life measure for people with multiple sclerosis. Clin Rehabil 15:247–258

Fraser C, Klistorner A, Graham S, Garrick R, Billson F, Grigg J (2006) Multifocal visual evoked potential latency analysis: predicting progression to multiple sclerosis. Arch Neurol 63:847–850

Gad SC (2002) Drug safety evaluation. Wiley, New York

Goldmann Consensus Group (2005) The Goldman Consensus statement on depression in multiple sclerosis. Mult Scler 11:328–337

Goldschmidt T, Antel J, König FB, Brück W, Kuhlmann T (2009) Remyelination capacity of the MS brain decreases with disease chronicity. Neurology 72:1914–1921

Gordon-Lipkin E, Chodkowski B, Reich DS, Smith SA, Puliicken M, Balcer LJ, Frohman EM, Cutter G, Calabresi PA (2007) Retinal nerve fiber layer is associated with brain atrophy in multiple sclerosis. Neurology 69:1603–1609

Hobart JC, Freeman JA, Thompson AJ (2000) Kurtzke scales revisited: the application of psychometric methods to clinical intuition. Brain 123:1027–1040

Hobart J, Lamping D, Fitzpatrick R, Riazi A, Thompson A (2001) The Multiple Sclerosis Impact Scale (MSIS-29): a new patient-based outcome measure. Brain 124:962–973

Hobart JC, Riazi A, Lamping DL, Fitzpatrick R, Thompson AJ (2004) Improving the evaluation of therapeutic interventions in multiple sclerosis: development of a patient-based measure of outcome. Health Technol Assess 8:1–48

Kapoor R (2006) Neuroprotective strategies in multiple sclerosis: therapeutic strategies and clinical trial design. Curr Opin Neurol 19:255–259

Kappos L, Freedman MS, Polman CH, Edan G, Hartung HP et al (2007) Effect of early versus delayed interferon beta-1b treatment on disability after a first clinical event suggestive of multiple sclerosis: a 3-year follow-up analysis of the BENEFIT study. Lancet 370:389–397

Kolappan M, Henderson APD, Jenkins TM, Wheeler-Kingshott CAM, Plant GT, Thompson AJ, Miller DH (2009) Assessing structure and function of the afferent visual pathway in multiple sclerosis and associated optic neuritis. J Neurol 256:305–319

Law J, Fielding B, Jackson D, Turner-Stokes L (2009) The UK FIM+FAM Extended Activities of Daily Living module: evaluation of scoring accuracy and reliability. Disabil Rehabil 31:825–830

Lublin FD, Reingold SC, the National Multiple Sclerosis Society (USA) Advisory Committee on Clinical Trials of New Agents in Multiple Sclerosis (1996) Defining the clinical course of multiple sclerosis. Results of an international survey. Neurology 46:907–911

Martínez-Yélamos S, Martínez-Yélamos A, Martín Ozaeta G, Casado V, Carmona O et al (2006) Regression to the mean in multiple sclerosis. Mult Scler 12:826–829

Miller DH, Thompson AJ (1999) Nuclear magnetic resonance monitoring of treatment and prediction of outcome in multiple sclerosis. Philos Trans R Soc Lond B 334:1687–1695

Mowry EM, Loguidice MJ, Daniels AB, Jacobs DA, Markowitz CE et al (2009) Vision related quality of life in multiple sclerosis: correlation with new measures of low and high contrast letter acuity. J Neurol Neurosurg Psychiatry 80:767–772

Narayana PA (2005) Magnetic resonance spectroscopy in the monitoring of multiple sclerosis. J Neuroimaging 15:46S–57S

Polman CH, Reingold SC, Edan G, Filippi M, Hartung HP et al (2005) Diagnostic criteria for multiple sclerosis: 2005 revisions to the "McDonald Criteria". Ann Neurol 58:840–846

Riazi A (2006) Patient-reported outcome measures in multiple sclerosis. Int MS J 13:92–99

Ritvo PG, Fisher JS, Miller DM, Andrews H, Paty DW, LaRocca NG (1997) Multiple sclerosis quality of life inventory: a user's manual. National Multiple Sclerosis Society, New York

Rothwell PM, McDowell Z, Wong CK, Dorman PJ (1997) Doctors and patients don't agree: cross sectional study of patients' and doctors' perceptions and assessments of disability in multiple sclerosis. BMJ 315:1305–1306

Sahajwalla CG (2004) New drug development. Marcel Decker, New York

Schmierer K, Scaravilli F, Altmann DR, Barker GJ, Miller DH (2004) Magnetization transfer ratio and myelin in postmortem multiple sclerosis brain. Ann Neurol 56:407–415

Schmitz-Hübsch T, Giunti P, Stephenson DA, Globas C, Baliko L et al (2006) SCA Functional Index: a useful compound performance measure for spinocerebellar ataxia. Neurology 66:1717–1720

Sharrack B, Hughes RA, Soudain S, Dunn G (1999) The psychometric properties of clinical rating scales used in multiple sclerosis. Brain 122:141–159

Silver NC, Good CD, Sormani MP, MacManus DG, Thompson AJ, Filippi M, Miller DH (2001) A modified protocol to improve the detection of enhancing brain and spinal cord lesions in multiple sclerosis. J Neurol 248:215–224

Simon JH, Miller DHM (2007) Measures of gadolinium enhancement, T1 black holes and T2-hyperintense lesions on magnetic resonance imaging. In: Cohen JA, Ruddick R (eds) Multiple sclerosis therapeutics, 3rd edn. Informa, London, pp 113–142

Smith K, McDonald I, Miller DHM, Lassmann H (2006) The pathophysiology of multiple sclerosis. In: Compston DAS (ed) McAlpine's multiple sclerosis. Churchill Livingstone, New York, pp 601–659

The Optic Neuritis Study Group (2008) Multiple sclerosis risk after optic neuritis: final optic neuritis treatment trial follow-up. Arch Neurol 65:727–732

Thompson AJ, Hobart JC (1998) Multiple sclerosis: assessment of disability and disability scales. J Neurol 254:189–196

Toussaint D, Perrier O, Verstappen A, Bervoets S (1983) Clinicopathological study of the visual pathways, eyes and cerebral hemispheres in 32 cases of disseminated sclerosis. J Clin Neuroopthalmol 3:211–220

Vickrey BG, Hays RD, Harooni R, Myers LW, Ellison GW (1995) A health-related quality of life measure for multiple sclerosis. Qual Life Res 4:187–206

Weinshenker BG, Bass B, Rice GP, Noseworthy J, Carriere W et al (1989) The natural history of multiple sclerosis: a geographically based study. I. Clinical course and disability. Brain 112:133–146

Zivadinov R (2007) Can imaging techniques measure neuroprotection and remyelination in multiple sclerosis? Neurology 68:S72–S82

Zivadinov R, Bakshi R (2004) Central nervous system atrophy and clinical status in multiple sclerosis. J Neuroimaging 14:27S–35S

Index

A
A2B5, 2, 3, 107, 110
Active indirect mechanisms, 206
Adaptive immune response, 49
A disintegrin and metalloprotease (ADAM) secretase, 134
Adult PNS stem/precursor cells, 140
Affective function, mental health, 260–261
Ageing, remyelination failure, 80–81
Akt pathway, 73, 74
Alemtuzumab, 206
Allografts, 104
4-Aminopyridine, 220
Amoeboid microglia, 51
AMPA receptors, 11
Anoxia, 180, 194
Apoptosis, oligodendrocytes, 27, 28
Apoptotic cells, 56
Astrocytes
 CNS, 4
 CXCL1, 13
 demyelination, cellular components, 29
 myelination development, 103
 OEC, 135
 type 2, 7
Astrogliosis, 29
Autologous cell transplantation therapy, 142, 143
Avonex. *See* Interferon beta (IFNβ)
Axl receptors, 56
Axonal protection. *See also* Sodium channel blocking agents
 endogenous remyelination, CNS, 75–76
 functional protection, 185–186
 inflammatory infiltrates
  iNOS expression, 189, 190
  microglial migration, 188, 189
  microglial phagocytosis activity, 186, 188
  phenytoin, 188, 189
Axonal regeneration
 schwann cells (SC), 130, 131, 133
 tissue trophic support, 163
Axons
 demyelination, cellular components, 31–32
 functional protection, sodium channel blocking agents, 185–187
 phagocytic activity targets, 55
 remyelination, 35
 sodium channel blocking agents, 194–195

B
B cell follicles, 34
BDNF. *See* Brain-derived neurotrophic factor (BDNF)
Beck Depression Inventory (BDI), 260–261
Betaseron. *See* Interferon beta (IFNβ)
BG00012 (BG12), 219–220
Blood-brain barrier (BBB)
 CNS effects, 209–210
 neural cells, immunomodulatory therapies, 206
Bone morphogenetic proteins (BMPs)
 immune modulation, NPCs, 165
 OPCs, 6–8, 142
 signaling pathway, 6–7
Boundary cap cells, 138–139
Brain-derived neurotrophic factor (BDNF), 131, 132
Brain imaging
 baseline lesion enhancement characteristics, 246
 description, 244

Brain imaging (cont.)
  enhancement duration and contrast dosage, 246
  lesion enhancement, 246–247
  lesion recovery and voxel inhomogeneity, 248
  MTR
    patterns, 245–246
    recovery, 247
  NAWM, 244–245
  spectroscopy, 246–247
  T1-weighted images, 247
  voxel-based analysis, 248–249

C
cAMP-response-element-binding (CREB) protein, 57
Carbamazepine
  inflammatory rebound, 192–193
  sodium channel blockade, 182
CCR2 receptors, 159, 160
CCR7 receptors, 216
CD11b$^+$ cells, 59, 188, 194
CD133+, bone marrow, 141
CD4 cells, 30, 59
CD8 cells, 29–31
CD45$^+$ cells, 189, 193
CD4+Th1 cells, 58
Cell replacement, NPCs, 161–162
Central nervous system (CNS) myelin repair
  committed PNS glia
    OEC, 134–136
    SC, 131–134
    SCIP, 135
    VEGF, 134
  ectopic sources
    ESCs, 143
    iPS cells, 144
    MSC, 141
    niche, 141
    OPCs, 142
    SKP, 141–142
    Sox10, 144
  PNS progenitors
    adult PNS stem/precursor cells, 140
    boundary cap cells, 138–139
    OEp, 140
    SCp, 138
Cerebellar dysfunction, 260
Cerebellar peduncles, 54, 85, 95, 99, 116
Cerebellar slice cultures, 219
CG4 cells, 101, 102, 110
Chemokine receptors

CCR5, 54, 160
CXCL1, 13, 64
CXCL12, 11, 159
CXCR2, 11, 13, 64
CXCR3, 54, 160
CX3CR1, 52
CXCR4, 11, 159
pro-inflammatory cytokines, 64
Chicago Multiscale Depression Inventory (CMDI), 261
Chondroitin sulfate proteoglycans (CSPG), 133, 135–137
Ciliary neurotrophic factor (CNTF), 60, 131, 163
9-cis-retinoic acid (9cRA), 85
Cladribine, 206, 214
Clathrin, 55
Clodronate, 61
Cluster designation (CD) 20, 206
Composite cerebellar functional score (CCFS), 260
Compound action potential (CAP), 185–187
Copaxone. See Glatiramer acetate (GA)
Cre-lox labelling strategy, 76, 77
CSPG. See Chondroitin sulfate proteoglycans (CSPG)
Cuprizone-induced demyelination
  Gas6, 60
  MHCII role, 63
  near-infrared imaging, 111
  TNFα, 62
Cyclophosphamide, 210, 214
Cyclosporin A, 117
Cytokines, pro-inflammatory
  chemokine receptor, 64
  IGF-1, 63–64
  IL-12, 62
  IL-1β, 63
  osteopontin, 63
  TLR-mediated inflammation, 61–62
  TNFα receptor, 62–63

D
Daclizumab, 206
Damage-associated molecular patterns (DAMPs), 58
Dawson's finger, 30
Demyelination
  cellular components, MS lesion
    astrocytes, 29
    axons, 31–32
    blood vessels, 30
    lymphocytes, 30–31
    microglia/macrophages, 29–30

oligodendrocytes and myelin, 28–29
perivascular macrophages, 30
wallerian degeneration, 31
diseases
  acquired, 156–157
  genetically transmitted, 157
  imaging (see Imaging demyelination and remyelination)
  inflammation, 25–26
  lesion
    classification, 24
    location, 24–25
    myelin/myelin degeneration products, 26–27
    pathological heterogeneity, 27–28
Dentate gyrus (DG), 154–155
Diffusion anisotropy, 241
Diffusion tensor imaging (DTI)
  advantages and disadvantages, 241
  description, 240
  diffusion encoding, 241
  EPI, 241
  MRI measures, 264–265
  myelin, 241
Dimethyl fumarate, 219–220
Direct cell-replacement therapy, 164
DNAX-activating protein, 56
Dorsal corticospinal tract, 183, 184
Dorsal funiculus (DF), 98, 183, 184
DTI. See Diffusion tensor imaging (DTI)
Dysregulation hypothesis, 83

E

EAE. See Experimental autoimmune encephalomyelitis (EAE)
Echo planar imaging (EPI), 241
EDSS. See Expanded Disability Status Scale (EDSS)
eGFP. See Enhanced green fluorescent protein (eGFP)
Electron microscopy (EM), 29, 35, 97, 139
Embryoid bodies (EBs), 106, 107
Embryonic stem cells (ESCs)
  central nervous system myelin repair, 143
  exogenous cell-based remyelination, 106–108
  neural rosettes, 143
  s100, 143
  SMAD, 143
Endogenous remyelination, CNS
  Akt pathway, 73, 74
  axonal protection, 75–76
  demyelinated CNS axons, 79–80
  demyelination, normal response, 73–75
  enhancement therapies
    caveats regarding models, 84
    9cRA, 85
    description, 83–84
    drug-based remyelination-enhancing medicines, 84–85
    tankyrase, 84
    Theiler's virus-induced demyelination model, 84
  failure cause
    action timings, 83
    ageing effect, 80–81
    differentiation and maturation, 82
    efficiency, 80
    generic factors, 81
    MS lesions, 81–82
    non-disease-related factors, 80
    potential inhibitory factors, 82–83
    putative inhibitory signals, 83
  identification, 72–73
  mechanisms
    adult OPCs, 76–79
    and inflammation, 79
    oligodendrocytes, 76, 77
    RXR, 79
    transcriptional factors, 77
  myelinated axons, 72
  PMD, 75
  restores function, 75
  Schwann cells, 79–80
Endoplasmic reticulum (ER), 56, 99, 209
Enhanced green fluorescent protein (eGFP), 51, 52
EPI. See Echo planar imaging (EPI)
Epidermal growth factor (EGF), 107, 108, 110
Epstein–Barr virus, 34
Estrogen receptor ligand β, 211–212
Ethidium bromide, 54, 72, 85, 95, 115, 142
Exogenous cell-based remyelination
  choice of stages, 115–116
  EAE, 95–96
  ESCs, 106–108
  focal demyelination model, 95
  functional recovery after transplantation
    β-galactosidase, 115
    impulse conduction restoration, 114
    marmosets, 114
    myelin sheath thickness, 113–114
    nodes of Ranvier, 113–114
    phenotype improvement, 114–115
  human oligodendroglia differentiation, 105, 106
  ideal cell characteristics, 104–105
  immunosuppression, 117–118

Exogenous cell-based remyelination (cont.)
  inflammatory disease
    glial-committed progenitors, 112–113
    human cells survival, 112
    inflammation, 111–112
  iPSCs, 108
  lesion site selection, 116–117
  monitoring, 111
  myelin mutants, 95
  NSCs
    growth factors removal, 108–109
    neurospheres, 108
    schematic diagram, four-step maturation, 104, 105
  oligodendrocytes, 109–110
  OPCs, neuroblastoma cells (B104), 110
  tacrolimus (FK506), 118
  transplantation
    cografting growth factor secreting cells effects, 101, 102
    electron micrographs, 97, 99
    MS lesions, 103–104
    myelin mutants, 103
    neonatal mixed glial cell, 97, 100
    neuronal replacement, 96
    oligosphere cells, 97, 98
    shaking(sh)pup, 97
    shiverer (shi) mouse, 97
  X-irradiation protocol, 95
Exogenous cell therapy, 94, 96, 115
Expanded Disability Status scale (EDSS), 257, 258
Experimental autoimmune encephalomyelitis (EAE)
  beneficial effects, statins, 215
  chronic-relapsing form, 182
  endogenous remyelination, 78
  exogenous cell-based remyelination, 95–96
  IFNβ therapy, 210–212
  NPCs transplantation, 157
  sodium channel blocking agents
    axonal protection, 183–185
    clinical status, 181–183
    functional protection, of axons, 185–187
    inflammatory infiltrates, 186, 188–191
    MOG, 182
    withdrawal, 190–194
Extracellular adenosine triphosphate (ATP), 53
Extracellular matrix, 13, 16, 64, 101, 131, 133, 134, 163

F

Fampridine, 220
Fast fluid-attenuated inverse recovery (FLAIR), 261, 262
Fate mapping, 76, 80, 134, 138, 142
Fc receptors, 29
Fetal dopaminergic cells, 117
Fibroblast growth factor (FGF) signaling pathway, 7, 8
Fibronectin, 36, 131, 134
Fingolimod. See FTY720
Flecainide, 180–187, 189–191, 194, 196
Fluorescence-activating cell sorting, 143
fMRI. See Functional magnetic resonance imaging (fMRI)
Fractional anisotropy (FA), 241, 264
FTY720
  description, 216
  endothelial barrier properties, 218
  G protein-coupled receptors, 217
  neurodegeneration rate reduction, 216
  relapsing treatment, 217
  rodents, audoradiographical analysis, 217
  S1P receptors
    OLs and OPCs, 218–219
    signaling, 218
Fumarate acid esters (FAE), 219–220
Functional magnetic resonance imaging (fMRI), 265–266

G

GABAergic phenotype, 57
Gadolinium enhancement, 263
Gadolinium enhancing lesion, 239, 245, 269
General subpial demyelination (GSD), 25
Gilenya. See FTY720
Glatiramer acetate (GA), 212–213
Glial fibrillary acidic protein (GFAP), 3, 29, 76, 106, 134, 135
Glial-restricted precursors, 114, 142
Glial scar, 133, 162, 163, 218
Glutamate, 11, 213, 215
g ratio, 72–74
Gray matter (GM) demyelination
  extensive demyelination, 32, 33
  meningeal inflammation, 34
Growth arrest-specific gene 6 (Gas6), 56, 60

H

Histone deacetylases (HDACs), 81
HLA-DR+ macrophages, 50
Hospital Anxiety and Depression Scale (HADS), 261

HSVTK. *See* Thymidine kinase of herpes simplex virus (HSVTK)
Human brain endothelial cells (HBECs), 212
Hyaluronan, 16, 37, 38
Hypertrophy, 53, 135
Hypomyelination, 62, 78

I
Ibudilast, 206
IFNβ. *See* Interferon beta (IFNβ)
IgG
  capping, 50
  deposition, 48
Imaging demyelination and remyelination
  clinical applications
    brain, 244–250
    optic nerve, 242–244
  conventional MRI techniques, 234
  non-conventional MRI techniques
    DTI, 240–241
    MT, 236–239
    multi-component T2 relaxometry, 234–236
    MWF, 236
  positron emission tomography, 241–242
Immunoablative therapy protocol, 214
Immunoglobulin, 28, 37, 209
Immunomodulation
  BMP, 165
  broad immune regulatory capacity, 166
  direct cell-replacement therapy, 164
  IFNAR, 59–60
  IL-6, 60
  immune suppressive effect, 165
  MMP-9, 64
  MOG35-55, 164
  RA, 60
  STAT, 60
  stem cell therapies, 164
  TGFβ1, 60
  TLR-2, 61
  TLR-4, 62
  transplantation effect, 164–165
  zymosan, 60
Immunomodulatory agents, 205, 210, 220, 221
Immunomodulatory therapies
  neural cells, indirect and direct effects, 206–210
    α4β1-integrin, 206
    teriflunomide, 206
    therapeutic vaccination, 209
    TNF antibodies, 209
  reparative and neuroprotective functions, 204–205
Immunopanning, 97, 109
Immunosuppression, 108, 117
Induced pluripotent stem cells (iPSCs)
  central nervous system myelin repair, 144
  ESC, 108, 109, 143
Inducible nitric oxide synthase (iNOS), 59
Inflammatory disease, exogenous cell-based remyelination
  glial-committed progenitors, 112–113
  human cells survival, 112
  inflammation, 111–112
  MHV, 112, 113
Insulin-like growth factor-1 (IGF-1), 63–64
Integrated stress response (ISR), 209
Interferon beta (IFNβ)
  cytokine, 210
  EAE, 211–212
  HBECs, 212
  NGF, 212
  P53, 211
  Th17, 211
  TNF, 210
  TRAIL, 210
Interferon-gamma (IFNγ), 58
  DAMPs, 58
  neural cells, therapeutic agents, 209
Interferon type 1 receptor (IFNAR), 59–60
Interleukin-6 (IL-6)
  family, 60
  superfamily receptor complexes, 55
Interleukin-12 (IL-12), 62
Interleukin-1β (IL-1β), 63
Interleukin (IL)-2 receptor, 165, 206
International Cooperative Ataxia Rating Scale (ICARS), 260
Isoprenoid synthesis, 215

J
JAK/STAT signalling pathway, 57

K
Kainate receptors, 11
Krox-20, 135
Kv1.3 channel activity, 55

L
Laminin, 131, 134
Lamotrigine, 181, 183, 185, 186, 190, 196, 220, 272
Laquinimod (ABR-215062), 220–221

Laser-induced disruption, 53
LDL-receptor related protein, 56
Lebers hereditary optic neuropathy (LHON), 243
Lesion staging system, 24
Leukaemia inhibitory factor (LIF), 57
Lidocaine/lignocaine, 180
Lingo-1
　endogenous remyelination, 78
　OPCs, 15
　remyelination promoting therapies, 39
Lipid rafts, 216
Lipolytic enzymes, 59
Lipopolysaccharide (LPS), 60, 62, 186, 188
Liposomes, 61
Low-affinity NGF receptor (p75ntr), 134
Luciferase gene, 111
Lymphocytes, 30–31
Lymphocytic infiltration, 48–49
Lysolecithin, 54, 61, 63, 64, 73, 74, 77, 95, 109
Lysosomal lipids, 55

M
Macrophages, 26–27, 29–30, 56–57
Magnetic bead sorting, 109
Magnetic resonance imaging (MRI)
　conventional techniques, 234
　neuroprotective therapies
　　atrophy measures, 263–264
　　description, 261
　　DTI, 264–265
　　fMRI, 265–266
　　gadolinium enhancement, 263
　　mechanistic evaluation tool, 261, 262
　　MRS, 265
　　MTR, 264
　　T1-weighted sequences, 263
　　T2-weighted sequences, 261–262
　non-conventional techniques
　　DTI, 240–241
　　MT, 236–239
　　multi-component T2 relaxometry, 234–236
　　MWF, 236
　optic neuritis, 242
Magnetic resonance spectroscopy (MRS), 265
Magnetisation transfer ratio (MTR)
　brain
　　baseline lesion enhancement characteristics, 246
　　enhancement duration and contrast dosage, 246
　　lesion enhancement, 246–247

NAWM, 244–245
　patterns, 245–246
　recovery, 247
　spectroscopy, 246–247
　voxel-based analysis, 248–249
MTI, 239–240
optic neuritis, 243–244
Magnetization transfer (MT) imaging
　advantages and disadvantages, 239–240
　brain white matter, 238
　description, 236–237
　MTR, 239
　QMTI, 238
　T1 shortening, 238
Major histocompatibility complex (MHC) class II expression, 30
Matrix metalloproteinase-9 (MMP-9), 64
MBP. *See* Myelin basic protein (MBP)
Medial ganglion eminence, 5
Meningeal inflammation, 34
Mesenchymal stem cells (MSC), 141
Microarray analyses, microglial activation, 58
Microglial function
　CNS neurodegenerative disease, 52
　complement component receptor CR3 (Mac1), 57
　disease heterogeneity
　　distribution, 50
　　oligodendroglial apoptosis, 48, 49
　　pathology, 48–49
　　ramified microglia, 49
　　T-lymphocyte infiltration, 48
　in health and disease
　　demyelination experimental models, 53–54
　　purinergic P2 receptor, 53
　　superior cerebellar peduncles, 54
　　surveillance, 52–53
　　transcranial two-photon microscopy, 52
　heterogeneity
　　developmental origins, 51
　　monocytic cells, 51
　　regional variation, 51–52
　　sources, 52
　immunomodulation, 59–60
　inflammatory activity
　　molecular determinants, 58
　　nitric oxide (NO) synthase, 59
　　NK cells, 58
　　pathogenic effect, 59
　　pro-inflammatory activity consequences, 58
　　reactive oxygen species, 58
　　tolerance, 58

microgliosis, 51
phagocytic activity
  description, 55
  NMDA receptor, 57
  remodelling, 57–58
  targets, 55–57
pro-inflammatory cytokines
  chemokine receptor, 64
  IGF-1, 63–64
  IL-12, 62
  IL-1β, 63
  neurotensin receptor-3, 54
  neurotrophin receptor TrkA, 55
  osteopontin, 63
  TLR-4, 62
  TLR-mediated inflammation, 61–62
  TNFα receptor, 62–63
purinergic receptors P2Y12/13, 54
RAG-/-mice, 63
repair, 47–48
Microgliosis, 51
microRNA (miRNA), 9
Minocycline, 214
Mitoxantrone, 214
MRI. See Magnetic resonance imaging (MRI)
MRS. See Magnetic resonance spectroscopy (MRS)
MS. See Multiple sclerosis (MS)
MT imaging. See Magnetization transfer (MT) imaging
MTR. See Magnetisation transfer ratio (MTR)
Multi-component T2 relaxometry, 234–236
Multiple sclerosis (MS)
  axonal
    injury, 180–181
    mitochondrial metabolism, 180
    pathology, 179–180
  immunomodulatory therapies
    neural cells, 206–221
    reparative and neuroprotective functions, 204–205
  MS plaques, 103–104
  neuroinflammatory lesions, 181
  pathology of, 203–204
  sodium channel blocking agents (see Sodium channel blocking agents)
Multiple sclerosis impact scale (MSIS-29), 259
MWF. See Myelin water fraction (MWF)
Myelin
  clinical applications
    brain, 244–250
    optic nerve, 242–244
  conventional MRI techniques, 234
  debris, 56–57
  microglia/macrophages, 29–30
  non-conventional MRI techniques
    DTI, 240–241
    MT, 236–239
    multi-component T2 relaxometry, 234–236
    MWF, 236
    PET, 241
  phagocytic activity targets, 55
Myelin basic protein (MBP), 13, 14
Myelin oligodendrocyte glycoprotein (MOG)
  induced EAE, 164
  sodium channel blocking agents, 182
Myelin water fraction (MWF)
  advantages and disadvantages, 236
  description, 234

N

Natalizumab, 206
NC-derived stem cells (NCSC), 140
Nerve growth factor (NGF), 212
Neural cell adhesion molecule (NCAM), 15–16
Neural cells, therapeutic agents
  active indirect mechanisms, 206
  development, 206
  direct effects
    chemotherapeutic agents, 214
    FTY720, 216–219
    fumarate esters, 219–220
    lamotrigine, 220
    laquinimod, 220–221
    minocycline, 214
    riluzole, 220
    statins, 215–216
  IFNγ, 209
  immune system, physiological effects, 209
  indirect effects
    GA, 212–213
    IFNβ, 210–212
  ISR, 209
  monoclonal antibodies (mabs), 206, 209
  monoclonal antibodies, CNS, 209–210
  passive indirect mechanisms, 209
  protective autoimmunity, 209
  relapse rate, 219
  TNFγ, 209
Neural stem cells (NSCs), exogenous
  cell-based remyelination
    growth factors removal, 108–109
    neurospheres, 108
    schematic diagram, four-step maturation, 104, 105

Neural stem/precursor cells (NPCs) transplantation
  cell populations candidate comparison, 167
  cell-replacement strategies, 166
  DG, 154–155
  features of, 155
  neuroblasts, 154
  in neurological disorders
    acquired demyelinating diseases, 156–157
    cell delivery route, 158–160
    clinical translation considerations, 157, 158
    direct myelin repair potential, 157
    EAE, 157
    gene defect-free myelin-forming cells, 156
    hereditary defects, 156
    remyelinating properties, 157
    transplanted cells tracking, 160–161
  origin, 153–154
  SGZ, 154–155
  stem cell therapeutic plasticity, 167, 168
  therapeutic plasticity
    broad tissue trophic support, 163–164
    cell replacement, 161–162
    immune modulation, 164–166
    neurogenesis, 164
    therapy, 162–163
    trophic effects, 163
  transit amplifying cell, 155
Neuroinflammatory demyelinating disease, 180–181
Neurological disorders, NPCs
  acquired demyelinating diseases, 156–157
  cell delivery route
    advantage, 160
    direct injection, 160
    inflammation mediators, 159
    MRI, 158
    multipotents, 158–159
    subarachnoid space, 158
    TNF-β, 159
  clinical translation considerations, 157, 158
  direct myelin repair potential, 157
  EAE, 157
  gene defect-free myelin-forming cells, 156
  hereditary defects, 156
  remyelinating properties, 157
  transplanted cells tracking, 160–161
Neuroprotective therapies
  clinical outcome measures
    classification, 257–258
    multidimensional measures, 258–259
    multiple sclerosis functional composite (MSFC), 257, 258
    quality of life, 258
    therapeutic effects, 257
    unidimensional measures, 259–261
  description, 255
  interrelated pathological processes reflection, 256
  MRI measures
    atrophy measures, 263–264
    description, 261
    DTI, 264–265
    fMRI, 265–266
    gadolinium enhancement, 263
    mechanistic evaluation tool, 261, 262
    MRS, 265
    MTR, 264
    T1-weighted sequences, 263
    T2-weighted sequences, 261–262
  neurophysiological measures
    CSF/blood measures, 267
    VEP, 266–267
  neuroprotective trial protocols comparision, 272–274
  paraclinical outcome measures, 261
  patient identification
    age, 270
    disability thresholds, 271
    disease activity, 271
    disease course, 270
    disease duration, 270
    paraclinical parameters, 271
    sentinel lesion, 272
  RNFL quantitative assessment, 266
  trial design
    clinical efficacy endpoints, 268–269
    clinical trials process, 267
    paraclinical endpoints, 269
    safety endpoints, 268
NGF. *See* Nerve growth factor (NGF)
Nodes of Ranvier, 113–114
Normal appearing white matter (NAWM)
  demyelination, 34
  MTR change, brain, 244–245
Notch signaling pathway, 16
NPCs transplantation. *See* Neural stem/precursor cells (NPCs) transplantation

O

OCT. *See* Optic coherence tomography (OCT)
Ofatumumab, 206
Olfactory ensheathing cells (OEC). *See also* Schwann cells (SC)
  advantages, 136–137
  in vitro similarities, 134

Index

migration potential, 135
origin of, 134
transplantation in trauma models, 135
Olfactory epithelium progenitors (OEp), 140
Oligodendrocyte precursor cells (OPCs)
   central nervous system myelin repair, 142
   chemokine CXCL1 signaling, 13
   description, 1
   differentiation control
      adult, 10
      neonatal regulation, 8–10
   endogenous remyelination, CNS
      Cre-lox labelling strategy, 76, 77
      differentiation phase, 78
      EAE, 78
      Lingo-1, 78
      RXRg, 78–79
      SVZ, 76
      transgenic mice, 78
   environmental factors
      neural tube, 6
      Shh, 5–7
      signaling pathways, 7–8
      type 2 astrocytes, 7
   lineage relationships, in vitro
      cell cultures, 2–3
      development stages, 2
      PDGF, 3–4
   locations, CNS
      cell types, 4
      markers and growth factor receptors, 4–5
      notochord, 5, 6
      rostral regions, 5
   migration, 36–37
   migration control
      guidance, 11–12
      presumptive myelinating regions, 12–13
      signals regulation, 11
   molecular markers development, 1–2
   myelination control
      components, 14–15
      Lingo-1, 15
      MBP, 13, 14
      NCAM, 15–16
      Notch signaling pathway, 16
      timing of, 15–16
   neurotransmitter receptors, 11
   noggin, 142
   remyelination, 36–37
   semaphorins, 12, 36, 77
   SV40 transduction, 110
   tenascin, 13
Oligodendrocyte progenitor cells
   exogenous cell-based remyelination, 109–110

   microglial function, 61–63
   neural cells, 218–219
Optic coherence tomography (OCT), 242
Optic neuritis, imaging
   MRI, 243
   MTR, 243–244
   OCT, 242
   VEP, 242
Osteopontin, 63

P
Pelizaeus–Merzbacher disease (PMD), 75
Peripheral nervous system (PNS), myelin repair
   progenitor cells
      adult PNS stem/precursor cells, 140
      boundary cap cells, 138–139
      endoscopic biopsy, 140
      olfactory epithelium progenitors (OEp), 140
   Schwann cells(see Schwann cells (SC))
PET. See Positron emission tomography (PET)
Phagocytic activity
   microglial function
      description, 55
      remodelling, 57–58
   targets
      apoptotic cells, 56
      axons, 55
      myelin, 55
      myelin debris, 56–57
      myeloid cells, 56
      receptor-mediated phagocytosis, 55
      scavenger receptor-AI/II, 57
      synapse removal, 58
      synaptic stripping, 57
      TAM (Tyro 3, Axl and Mer), 56
      Tim4, 56
      TREM2, 56
      tyrosine kinase receptors, 56
Platelet-derived growth factor (PDGF), 3–4
PMD. See Pelizaeus–Merzbacher disease (PMD)
Positron emission tomography (PET)
   markers of myelin, 241
   tracking transplanted cells, 160
Pro-inflammatory cytokines
   chemokine receptor, 64
   IGF-1, 63–64
   IL-12, 62
   IL-1β, 63
   osteopontin, 63
   TLR-mediated inflammation, 61–62
   TNFα receptor, 62–63
Proteolipid protein (PLP), 11, 75

Q

Quantitative magnetization transfer imaging (QMTI), 238

R

Rebif. *See* Interferon beta (IFNβ)
Remyelination
 axons, 35
 description, 72–73
 imaging (*see* Imaging demyelination and remyelination)
 and inflammation, 38–39
 inhibitory pathways
  canonical Notch–Jagged pathway, 37–38
  β-catenin, 38
  glycosaminoglycan hyaluronan, 38
  PSA-NCAM, 37
  TCF7L2 transcription factor, 38
  Wnt pathway, 38
 light microscopy remyelination detection, 35
 oligodendroglial lineage cells, 35
 OPCs
  description, 36
  differentiation, 37
  migration, 36–37
  proliferation, 36
 periventricular and cerebellar lesions, 35
 promotion therapies, 39
 saltatory conduction, 75
Retinal nerve fibre layer (RNFL), 266
Retinoic acid (RA), 60
Retinoid X receptor-g (RXRg), 78–79
Rituximab, 206

S

Scale for the Assessment and Rating of Ataxia (SARA), 260
Schwann cells (SC)
 axonal regeneration, 131
 environment modification, 133–134
 intrinsic property modification, 132–133
 limitation, 131
 meninges, 131
 NCAM polysialylation, 133
 neuregulins discovery, 130
 OEC
  advantages, 136–137
  in vitro similarities, 134
  migration potential, 135
  origin of, 134
  TACE, 134
  transplantation in trauma models, 135
 remyelination
  description, 79–80
  implications, CNS axons, 80
  neural crest lineage, 80
  technological achievements, 130
  viral tracing methods, 130
Schwann cells precursors (SCp), 138
Semaphorins, 12
Signaling pathways, OPCs, 7–8
Skin derived precursors (SKP), 141–142
Sodium channel blocking agents
 axonal protection, 183–185
 clinical status, 181–183
 direct effect on
  axons, 194–195
  immune cells, 195–196
 functional protection, of axons, 185–187
 inflammatory infiltrates
  iNOS expression, 189, 190
  microglial migration, 188, 189
  microglial phagocytosis activity, 186, 188
  Nav1.5 sodium channels, 186
  Nav1.6 sodium channels, 188
  phenytoin, 188, 189
 MOG, 182
 neurological status, 182
 spinal cord axons function
  electrophysiological assessment, 186, 187
  loss, 183, 184
 tetrodotoxin (TTX)-sensitive sodium channels, 180
 withdrawal
  inflammatory rebound, 192–194
  neurological status, 190–192
Sonic hedgehog (Shh)
 BMPs, 6–7
 description, 5–6
 FGF, 7, 8
 inductive effects, 6–7
Sphingosine-1-phosphate (S1P) receptors
 OLs and OPCs, 218–219
 signaling, 218
Spinal cord
 axons function
  electrophysiological assessment, 186, 187
  loss, 183, 184
 injuries, 260
 OPCs, 12, 13
S1P receptors. *See* Sphingosine-1-phosphate (S1P) receptors
Statins, 215–216
Stem cell therapeutic plasticity, 167, 168

Subgranular zone (SGZ), NPCs, 154–155
Subventricular zone (SVZ), endogenous remyelination, 76
Super paramagnetic iron oxides (SPIOs), 161

T
Thymidine kinase of herpes simplex virus (HSVTK), 58
TNF-related apoptosis induced ligand (TRAIL), 210
Toll-like receptor-2 (TLR-2), 61–62
TRAIL. *See* TNF-related apoptosis induced ligand (TRAIL)
Transcription factors, 8
Transplantation
  exogenous cell-based remyelination (*see* Exogenous cell-based remyelination)
  NPCs (*see* Neural stem/precursor cells (NPCs) transplantation)
Tumor necrosis factor alpha (TNFα)
  neural cells, therapeutic agents, 209
  pro-inflammatory cytokines, 62–63
T1-weighted sequences, 263
T2-weighted sequences, 261–262

V
Very late antigen-4 (VLA-4), 206
Visual evoked potential (VEP)
  neuroprotective therapies, 266–267
  optic neuritis, 242
Visual impairment, 259

W
Wnt pathway, 38, 84
Wnt signaling, 7–8